Underwater Technology

Offshore Petroleum

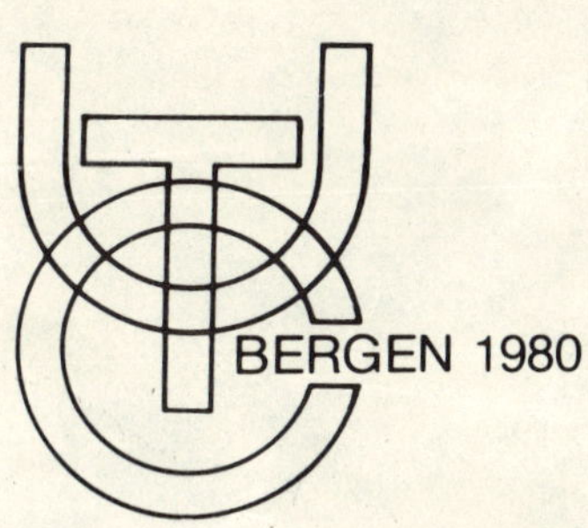

Other Titles of Interest

ANGEL	A Voyage of Discovery
MCCORMICK	Anchoring Systems
MCCORMICK	Port and Ocean Engineering under Arctic Conditions
POND & PICKARD	Introductory Dynamic Oceanography
ROSS	Energy From the Waves
SIMEONS	Hydro-power

Related Pergamon Journals*

Deep-Sea Research

Marine Pollution Bulletin

Ocean Engineering

Progress in Oceanography

**Free Specimen Copies sent on Request*

UNDERWATER TECHNOLOGY

Offshore Petroleum

Proceedings of the International Conference,
Bergen, Norway, April 14-16, 1980

Edited by

L. ATTERAAS
F. FRYDENBØ
B. HATLESTAD
T. HOPEN
Bergen, Norway

PERGAMON PRESS

OXFORD · NEW YORK · TORONTO · SYDNEY · PARIS · FRANKFURT

1980

U.K.	Pergamon Press Ltd., Headington Hill Hall, Oxford OX3 0BW, England
U.S.A.	Pergamon Press Inc., Maxwell House, Fairview Park, Elmsford, New York 10523, U.S.A.
CANADA	Pergamon of Canada, Suite 104, 150 Consumers Road, Willowdale, Ontario M2J 1P9, Canada
AUSTRALIA	Pergamon Press (Aust.) Pty. Ltd., P.O. Box 544, Potts Point, N.S.W. 2011, Australia
FRANCE	Pergamon Press SARL, 24 rue des Ecoles, 75240 Paris, Cedex 05, France
FEDERAL REPUBLIC OF GERMANY	Pergamon Press GmbH, 6242 Kronberg-Taunus, Hammerweg 6, Federal Republic of Germany

---

First edition 1980

British Library Cataloguing in Publication Data

Underwater technology.
1. Ocean engineering - Congresses
I. Atteraas, L
627.7 TC1505 80-40414

ISBN 0-08-026141-8

In order to make this volume available as economically and as rapidly as possible the authors' typescripts have been reproduced in their original forms. This method has its typographical limitations but it is hoped that they in no way distract the reader.

Printed in Great Britain by A. Wheaton & Co. Ltd., Exeter

FOREWORD

These proceedings stem from the UTC-80 (Underwater Technology Conference - 1980) which was held in Bergen, Norway April 14 - 16, 1980.

Although underwater petroleum production and underwater operations have experienced notable advances during the past decade, we know far from enough. This will - in all probability - always be the case. The UTC-80 provided a forum for presenting new developments focusing on the near future, with an overall purpose to contribute towards development of safe and economic underwater operations and systems.

One reason for arranging the conference in Bergen was the growing emphasis on underwater technology there. This has come about partly due to the combination of very deep, sheltered fjords where full scale underwater testing, trials and simulation may take place, the presence of organizations and institutions with specialized, related knowledge; and proximity to deep water offshore fields.

The papers presented are reproduced directly from the authors manuscript. We know of a number of typing errors in these. However, it was felt that rather than publish a perfect volume of proceedings, at a time when some of the topics have lost their actuality, we want the contents to be known as soon as possible. The message - after all - should in proceedings of this kind be of greater importance than the absence of typing errors and linguistic finesse. Whoever disagrees are invited to present their view to the editors before the next volume, which will be the result of the UTC-82, also to be arranged in Bergen.

Lyder Atteraas, Ph.D.
Freddy Frydenbø, B.Sc.
Brigt Hatlestad, Dr.ing.
Terje Hopen, M.Sc.

Address: UTC, Box 4252, N-5013 NYGÅRDSTANGEN/BERGEN, NORWAY

COMMITTEES

The Underwater Technology Conference - 1980 was organized by:

DnV - DET NORSKE VERITAS

NUI - NORWEGIAN UNDERWATER INSTITUTE

CMI - CHRISTIAN MICHELSENS INSTITUTE

BMV - BERGENS MEKANISKE VERKSTEDER - THE AKER GROUP

BB - BERGEN BANK

and the CITY OF BERGEN

CONFERENCE BOARD

Dr. Lyder Atteraas, principal engineer, (Chairman), DnV
Mr. Asbjørn Brandtun, industrial planner, City of Bergen
Mr. Odd Johan Heldal, construction manager, BMV
Mr. Øystein Martinsen, head, NUI
Mr. Robert Nergaard, consultant, BB
Dr. Frode Galtung, director, CMI

PROGRAM COMMITTEE

Dr. Brigt Hatlestad
Mr. Freddy Frydenbø
Mr. Terje Hopen

An ADVISORY PROGRAM COMMITTEE consisted of representatives from Norsk Hydro, Statoil, Amoco, Elf, Esso, Mobil and Phillips; the Norwegian Petroleum Directorate, IMENCO, Kvaerner Engineering, Konsberg Våpenfabrikk, the Continental Shelf Institute and NUF - the Norwegian Underwater Technology Society.

CONFERENCE CHAIRMEN

Mr. Leif Øien, Esso Exploration and Production, Norway
Mr. Francois Sagné, Elf Aquitaine Norge A/S
Mr. Carl M. Rønnevig, Imenco
Mr. Dave Dixon, Amoco Norway Oil Co.
Mr. Kjell E. Haugsten, Kvaerner Engineering

The conference was opened by Mr. Henrik Ager-Hanssen, executive vice president of STATOIL

CONTENTS

Subsea Production Systems

NORTH EAST FRIGG FIELD PROJECT

C. Duvet

Elf Aquitaine Norge A/S, Norway

ABSTRACT

The ELF AQUITAINE NORGE (EAN) NORTH EAST FRIGG FIELD (NEF) STATION is planned to be installed in 100 meters of water, 18 km from FRIGG FIELD FACILITIES. Fig.1

The NEF project is comprised of four major elements. Fig.2.

- A subsea template fig.3, protecting a six wellhead cluster and a manifold. Each x-mas tree is connected to a manifold and is remotely controlled from the Field Control Station by a 2"-20 x 1/2" umbilical.

- A 16" - 20 km subsea line links the manifold to TCP 2 treatment modules. A 1½" subsea line supplies hydrate inhibitor to the Field Control Station.

- A Field Control Station (FCS) fig.4, FCS is an articulated column installed 150 meters from the template and has the following main functions: i) To control x-mas trees through hydraulic direct control lines, ii) To inject/kill wells through 2" Kill & Service lines and iii) To inject hydrate inhibitor in x-mas tree through 1/2" injection lines.

- Treatment, Metering and Boosting Modules are to be installed in TCP 2 platform. They will allow NEF gas (7 to 5 million NM^3 per day) to be injected into the Frigg gas flow.

NEF Production concept is based upon the assumption that reliable equipment & techniques authorize to automatically control subsea gas production and therefore to have NEF installations normally unmanned.

Design and detailed engineering is presently carried out and first offshore works: Template installation and drilling are planned in

1981, followed by sea line and FCS installation in 1982. Well completion should take place in 1983 and start up in early 1984. NEF project is the first major underwater gas production system ever attempted in North Sea. This project will confirm EAN capability to develop sub sea fields in North Sea and will enlarge its expertise for deeper water and higher latitudes.

KEYWORDS

Sub Sea gas production-Sub Sea gas production tree-remote control of sub sea installations - articulated column for control of sub sea production tree.

INTRODUCTION

ELF AQUITAINE NORGE (EAN) has initiated the NEF underwater development project to produce a marginal satellite field and also to develop the capability to produce hydrocarbons in adverse environment.

Preliminary studies carried out in 1978 have been finalized early in 1980 and to date NEF project is in the detailed design engineering phase for major components.
When completed, NEF installations include six producing wells clustered into a sub sea template, connected to a subsea manifold and remotely controlled from a Field Control Station. Above NEF installations are to be located on a Frigg field extension in 100 m of water and 18 km from TCP 2 Frigg platform where NEF gas is processed.

EAN has overall responsibility for the project. EAN expertise obtained from Frigg field development has been combined to Societe Nationale Elf Aquitaine (production) experience acquired in other sub sea developments. FRIGG NORWEGIAN ASSOCIATION participants: NORSK HYDRO, STATOIL, TOTAL MARINE NORSK are directly involved in the project management.

NEF DEVELOPMENT AND PRODUCTION CONCEPT

NEF fields is a marginal field with a recoverable reserve limited to 9 billions NM^3 allowing a daily production starting from 5 to 7 million NM^3 over a 5 year period. In addition NEF gas supply is not mandatory for fulfilling FRIGG gas sale contract. Above particular conditions have induced EAN to favour a non-conventional development concept.

Concept selection: From different possible development schemes: i) Surface wells and steel structure ii) Sub sea wells and permanent production semi submersible, iii) Sub sea wells remotely controlled from FRIGG Facility, iv) The subsea wells clustered in a template and remotely controlled by a local unmanned articulated column has been selected. This selected scheme represents the best optimization in terms of cost (investment & operation), safety/reliability and acquisition of deep water production know-how.

Production philosophy is based upon the assumption that reliable equipment & techniques authorize application of a quasiautomatic operating principle to subsea gas production.

Reliability is planned to be achieved by preferring simple technology and using tested or/and field proven equipment with redundancy when applicable.

Following some typical choices illustrating NEF production philosophy:

- Tubing retrievable safety valves, two in tandem, have been preferred to conventional wire line retrievable ones
- No flow control choke
- Wells will not be individually tested
- FCS is normally unmanned and safety is obtained by automatic loops and back up equipment
- Shut down of well cluster is one of the few actions controlled from FRIGG

Further details are provided when describing here after the components of the NEF project.

GENERAL DESCRIPTION OF THE NEF PROJECT

NEF installations are comprised of four major elements:

- Field Control Station
- Sub sea wells, template and manifold
- Sea line and electrical cables
- Treatment & metering modules

FIELD CONTROL STATION (FCS)

FCS Concept

FCS is designed to control subsea equipment in unmanned normal production conditions and to be in manned condition for specific operations only.

Above concept emphasizes utilization of automatically controlled and highly reliable equipment together with extensive redundancy for critical systems such as power supply and inhibitor injection.

Likewise safety is maintained by automatic safety loops.

Manned operations are limited to:

- maintenance/repair
- resupply on a routine basis
- operations on wells (killing, bleed off and massive methanol injection, SSSV testing)
- experimental operations

These operations are specific and will be performed occasionally.

FCS general lay-out

FCS is an articulated column, Fig.4, from bottom to top it comprises

- A gravity base supporting the universal joint
- An universal joint connecting the base to the lower part of the column
- A column consisting of i) ballast for heavy material,ii) float tanks for buoyancy & stability,iii)decks in shaft with necessary machinery and equipment.
  The column supports risers for umbilicals and fendering at sea level
- A head with work deck, classified areas, control rooms, living quarters and helideck

The articulated column oscillations depend on water conditions; It has been evaluated that a maximum inclination angle of 2° will occur less than 10% of a normal year when sea conditions reach 7 and more on Beaufort scale.

FCS Main Function, System, Equipment.

Field Control Station (FCS) is the check point in the NEF production control system. All FCS functions, systems and equipment are finally aiming at ensuring that control will be safe and reliable; here below are described the most representative components of FCS.

Remote control system

Remote control system is comprised of:

- A direct hydraulic control system of operation subsea x-mas tree & manifold valves. In addition this system include monitoring of well head pressure and sand erosion.
- Control is normally operated from FCS control panel, and each subsea valve is connected to above panel through one control line enclosed into an umbilical.
- An automatic monitoring of the critical equipment and systems that are self operated on FCS ie: hydraulic control system, inhibitor injection system, power supply etc...
  This local monitoring is achieved through a logical network of detections and actions
- A communication system by radio or cable link from FCS

to QP FRIGG platform transmits necessary signals related to above monitoring and receives from QP to FCS shut down commands related to major events on FRIGG.

Inhibitor Injection System

Methanol is stored on TCP 2 and pumped to FCS through a 1½" line to a local tank and then is injected into each x-mas tree through ½" line contained in umbilicals.

It has been evaluated that 8 m^3 of methanol will be daily injected in the whole system to prevent hydrate formation in x-mas trees, manifold and sea line.

Killing/Service System

A killing and service system is designed for injecting either special fluids (killing of well) or methanol (massive hydrate destruction) in the wells and to test Subsea safety valves. This injection is made through a 2" killing & service line connecting the x-mas tree to FCS auxiliary production equipment located in the shaft: mud pump and tank.

In addition this 2" line can be used to bleed off well and manifold through a flare on the work deck.

Utilities

FCS is equipped with all necessary utilities for manned and unmanned working conditions, it includes in particular:

- Power supply
  Power will be supplied either by an electrical subsea cable or by diesel generators installed in the column.

  In manned condition power consumption is about 200 KW, reduced to 50 KW in unmanned condition and to 25 KW in emergency condition.
- Living quarters
  Living quarters are designed for 6 persons in manned conditions and could accomodate 12 in emergency conditions.
- Helicopter deck
  Helicopter deck is designed for SIKORSKY S61 N. From FCS motions, wind or fog conditions it has been evaluated, that helicopter landings will not be possible from 5 to 15 days a year.

Safety and reliability

NEF installation safety and reliability are particularly dependent of FCS systems. Safety and reliability of FCS is organized as follows:

<u>Automatic safety</u> of FCS equipment and systems is based

upon self actuated loops linking detection to corrective action through a logic network and to automatic shut down when no correction is possible to critical systems(fire detection,gas detection, low pressure detection)in unmanned condition .

Redundancy is also part of the safety organization and back -up solutions have been provided to critical systems ie: methanol injection, power supply, telecommunication, lighting etc.. In addition, emergency safety systems complete the organization.

Conventional safety & protection systems have been provided: vent & flaring, emergency evacuation means, fendering...

Preventive Maintenance and Survey have been taken into consideration, in particular with regards to structural stability. All floats and capacities can and will be periodically surveyed.

SUB SEA WELLS, TEMPLATE AND MANIFOLD

Subsea Production Concept

NEF Field is developed with six production wells clustered into a Template installed on the sea bottom. One well is vertical and 5 are directional. Each x-mas tree is connected to a manifold through a flowline installed on the Template and is controlled from Field Control Station through a control/service umbilical.

Subsea Drilling Completion and Production Equipment are designed for being reliable and easy to be controlled and maintained.

Drilling and completion equipment and techniques are conventional and field proven; utilization of tubing retrievable safety valves is one of few deviations from common practice.

Wellhead, x-mas tree and connector are as simple as possible and instrumentation is limited to three hydraulic transducers and one sand probe per well, there are neither choke nor any electrical sensors. Direct hydraulic control is used for remote controlling valves.

Tree to manifold connection is designed to be performed by divers with conventional flanged joint for avoiding sophisticated and less reliable automatic connector. Some innovation is necessary with regards to umbilicals and umbilical connections which are not conventional pieces of equipment and require particular quality assurance prior to utilization.

Subsea Wells, Drilling and Completion

NEF wells are drilled from a semisubmersible and drilling operations are similar to exploration drilling operations down to the NEF gas bearing formation top.

NEF well completion, fig.5, is very specific with regard to safety, in particular i) the sand screen hanger-packer assembly is provided with an extra landing nipple/plug for increasing safety during running of the production string ii) Two tubing retrievable safety valves (TRSSSV) controlled from FCS. TRSSSV are preferred since more reliable though work over operations are required for their replacement.

Subsea Well Head and X-Mas Tree

NEF drilling well-head includes a conventional 13 5/8 5000 PSI WP wellhead housing and a 9 5/8 casing hanger with vertical accesses to tubing and annular spacing.

NEF x-Mas Tree, fig.6, is a solid block tree including wing valve following main requirements apply:

- Compliance with API 6A & 14D specifications
- 5000 PSI WP & 1000 PSI test pressure
- Valves are 4 1/8 and 2 1/16 5000 PSI Cameron gate valves with hydraulic fail safe close operator and manual override. The production lower master valve and the annulus wing valve are not connected with hydraulic lines but locally operated by divers.
- Vertical access to 4 1/8 production bore and 2 1/16 annulus bore for wire line operations
- Soft landing jacks
- Protective shield

Tree Connector is a 13 5/8 - 5000 PSI Cameron collet connector hydraulically actuated and equipped with diver operated mechanical locks.

Manifold

The manifold includes all piping from x-mas tree to pipeline anchor flange. Its main characteristics are:

- Straight 4 1/2 flowlines with solid tees and arranged with high flexibility for taking care of expansion and subsea installation problems.
- 4 1/2 - 5000 PSI remote controlled isolation valve connects each flowline to header
- A 16" header receives all flowlines and is connected to the pipeline by a 16" 5000 API gate valve.

No pig trap/pig launcher has been provided.

Template

The template Fig.3, accommodates 6 wells plus two spare slots and necessary supports for manifold piping. Its main characteristics and functions are:

- To support conductor pipes & guide bases at an early stage
- To protect x-mas tree and manifold at a later stage
- Template is designed for easy access and large well spacing. Overall dimensions are 28 x 16 7 and weight is about 250 T
- Sub systems are included, such as: a piling/levelling system, a pipe line anchoring device, and necessary installation equipment.

SEALINES AND CABLES

NEF installations are connected to FRIGG TCP 2 platform through a 16" SEALINE; A 1½" supply line and, optionally, one electrical & communication cable.

16" Gas Line

This line transports NEF gas to TCP 2 for treatment, it has the main following characteristics:

- Pipe: 16" OD, 15.88 wall thickness in API 5Lx60 and concrete coated
- Length: 18 km
- Working pressure: 170 bars

1 1/2" Supply Line

This line supplies FCS with a daily rate of 8 T of methanol

- Pipe: 1 1/2" OD, 6.35 wall thickness in API grade B
- Length: 18 km
- Working pressure: 2 bars

Electrical Cable

An electrical cable is an optional solution to supply power to FCS and to back up radio teletransmission between FCS and TCP 2/QP. It would avoid active power generation on board FCS but is considered as easily damaged and would required an 100% back up diesel power supply.

TREATMENT & METERING MODULES

Process Description

NEF gas is processed (compressing and dehydratation) with FRIGG gas after being passed through a slug catcher, a scrubber and a gas meter. Liquid effluents from slug catcher/scrubber is separated and condensate is pumped into the FRIGG system after being metered. Water & methanol mixture is flashed and stored before being disposed of.

Location of Equipment

NEF gas treatment module is planned to be integrated with the ODIN one and located on TCP 2 deck where sufficient provisions have been made.

Quantities to be Processed

NEF treatment equipment are designed for processing 6 to 7 millions NM^3 of gas, 35 to 40 m^3 of condensate and about 10 m^3 of water plus 8 m^3 of methanol on daily basis.

EXPERIMENTATION

It has been contemplated to have one NEF well remotely controlled from QP platform by two experimental control systems:

A long distance multiplex control system (LDCS), comprising:

- A control/display panel on QP
- A power supply and transmission cable from control panel to a pool receptacle
- A pod locked into a pool receptacle located on the FCS base and containing multiplex & hydraulic equipment
- A lifting/guiding device between FCS base and deck for pod handling
- Hydraulic & electrical link to x-mas tree through the LDCS-SDCS automatic switching device

A short distance sequential hydraulic control system (SDCS) comprising:

- A control/display panel on FCS
- An electrical supply, a sequential hydraulic umbilical and an electro-acoustic transmission between FCS panel and pod receptacle
- A pod locked into a pod receptacle located on the FCS base and containing sequential hydraulic and electro-acoustic equipment

- Same lifting/guiding device as for the multiplex pod
- Same hydraulic & electrical link to well head as for LDCS.

The direct hydraulic control system will be used for controlling the "experimental" well when not controlled by the experimental control systems.

PROJECT STATUS AND PROGRAM

After several months of preliminary studies carried out in 1978-1979, NEF preliminary concept has been approved in 1979 and a special task force has been mobilized in Stavanger by the end of 1979.

To date the development concept is finalized and detailed design engineering of all components has commenced. Most of the offshore works are planned for the 1981-1983 period and, with commonly applied reservations, NEF project should be commissioned by the end of 1983 and start up is forecasted early in 1984.

CONCLUSION

When completed NEF Project is to be considered as a step forward in developing subsea equipment and techniques for production of oil & gas in deep water.

Though always preferring known technology we still face some problems for which new solutions have to be applied. This challenge applies potential risks but will reward all parties concerned by building up an advanced technological solution for diversifying production concepts in North Sea environment. In addition, partners and involved contractors will acquire experienced personnel, an essential factor for future technology of subsea production.

ACKNOWLEDGEMENT

Author and EAN express their appreciation to their partners NORSK HYDRO, STATOIL, TOTAL, for their permanent support which has resulted in the present status of the project.

We also wish to thank Norwegian authorities, contractors and Consultants whose understanding and constant effort greatly contribute to quality and progress of the work.

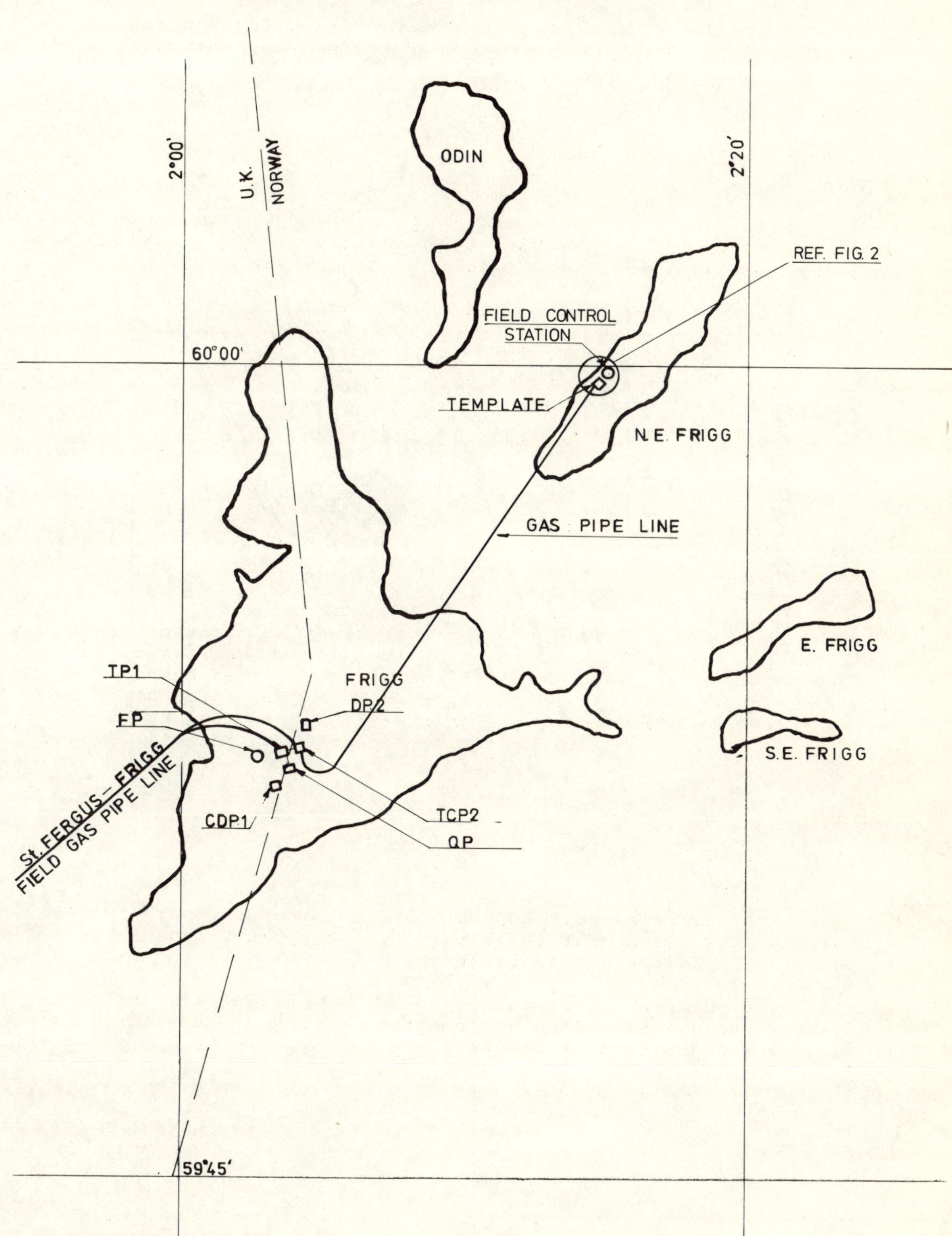
NORTH EAST FRIGG
LOCATION MAP
FIG. 1
2°00'
U.K.
NORWAY
ODIN
2°20'
REF. FIG. 2
FIELD CONTROL
STATION
60°00'
TEMPLATE
N.E. FRIGG
GAS PIPE LINE
E. FRIGG
TP1
FRIGG
DP2
FP
S.E. FRIGG
St. FERGUS - FRIGG
FIELD GAS PIPE LINE
CDP1
TCP2
QP
59°45'

NORTH EAST FRIGG

INSTALLATION ARRANGEMENT

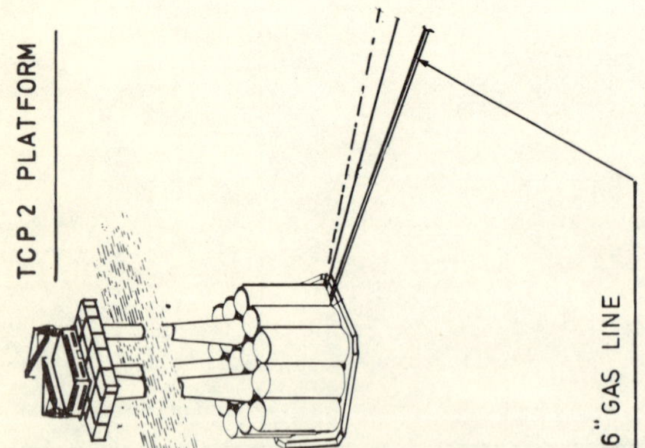

FIG. 2

NORTH EAST FRIGG
TEMPLATE & MANIFOLD FIG. 3

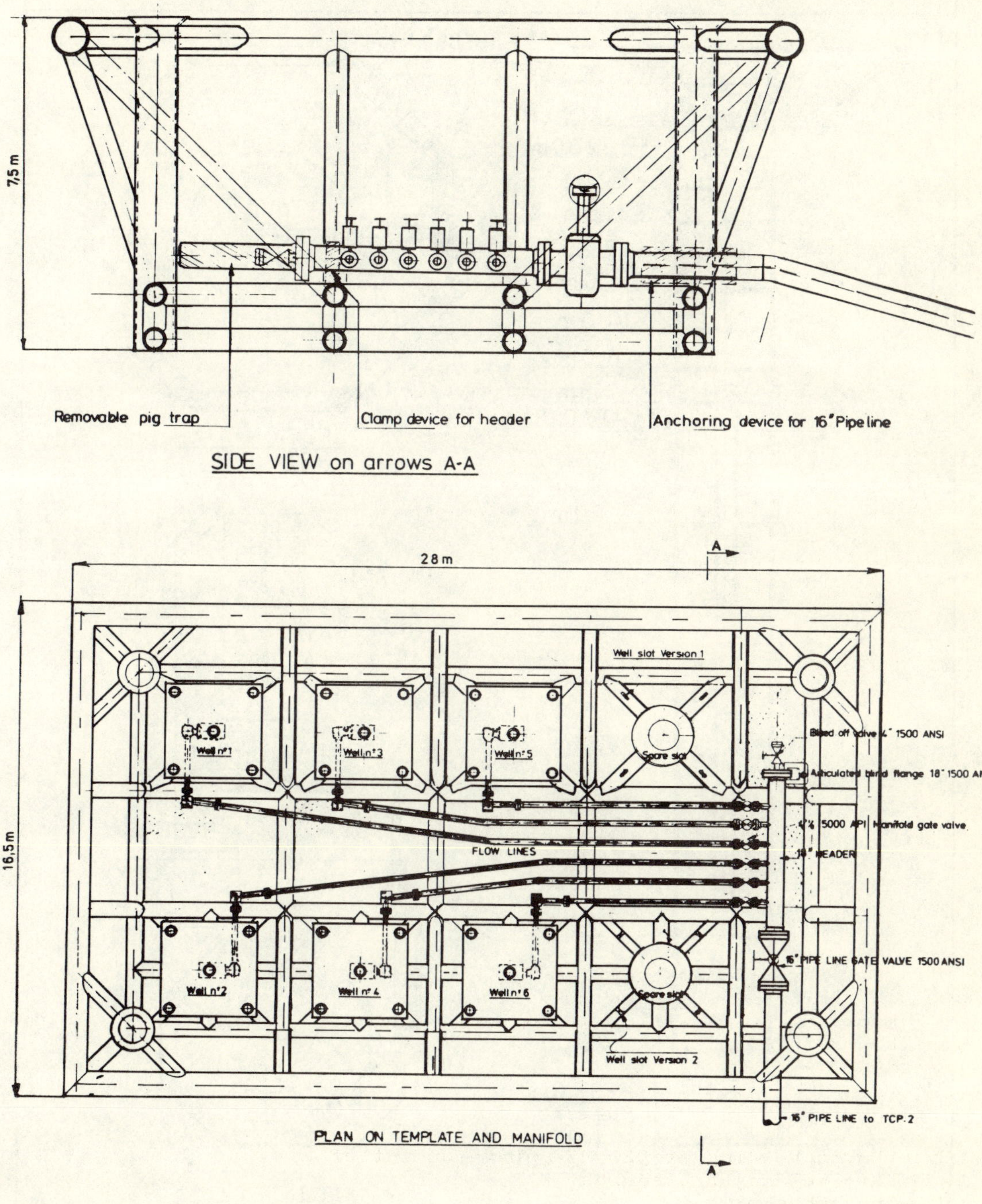

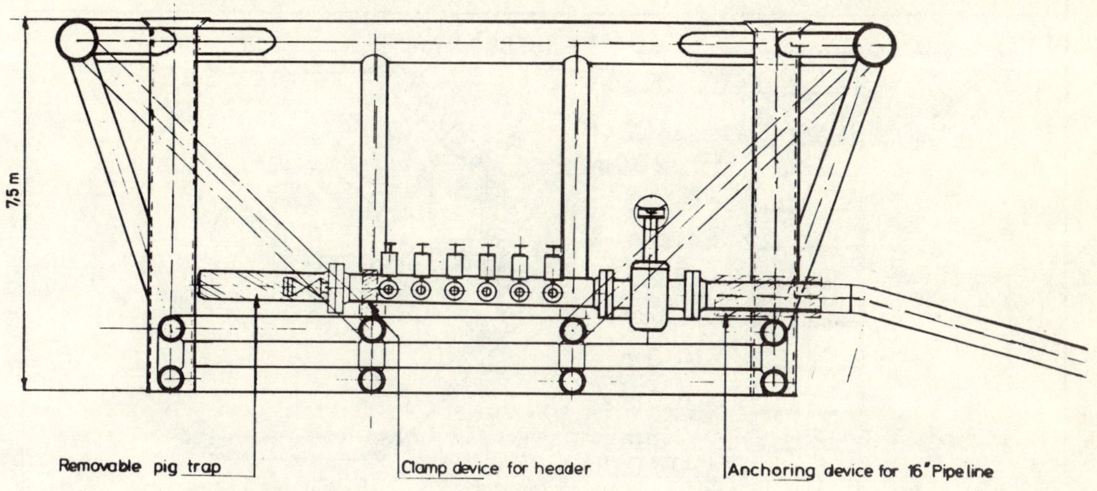

SIDE VIEW on arrows A-A

28 m
A
16,5 m
Well slot Version 1
Well n°1
Well n°3
Well n°5
Spare slot
Bleed off valve 4" 1500 ANSI
Articulated blind flange 18" 1500 ANSI
4" 1/16 5000 API Manifold gate valve
18" HEADER
FLOW LINES
16" PIPE LINE GATE VALVE 1500 ANSI
Well n°2
Well n°4
Well n°6
Spare slot
Well slot Version 2
16" PIPE LINE to TCP.2
A

PLAN ON TEMPLATE AND MANIFOLD

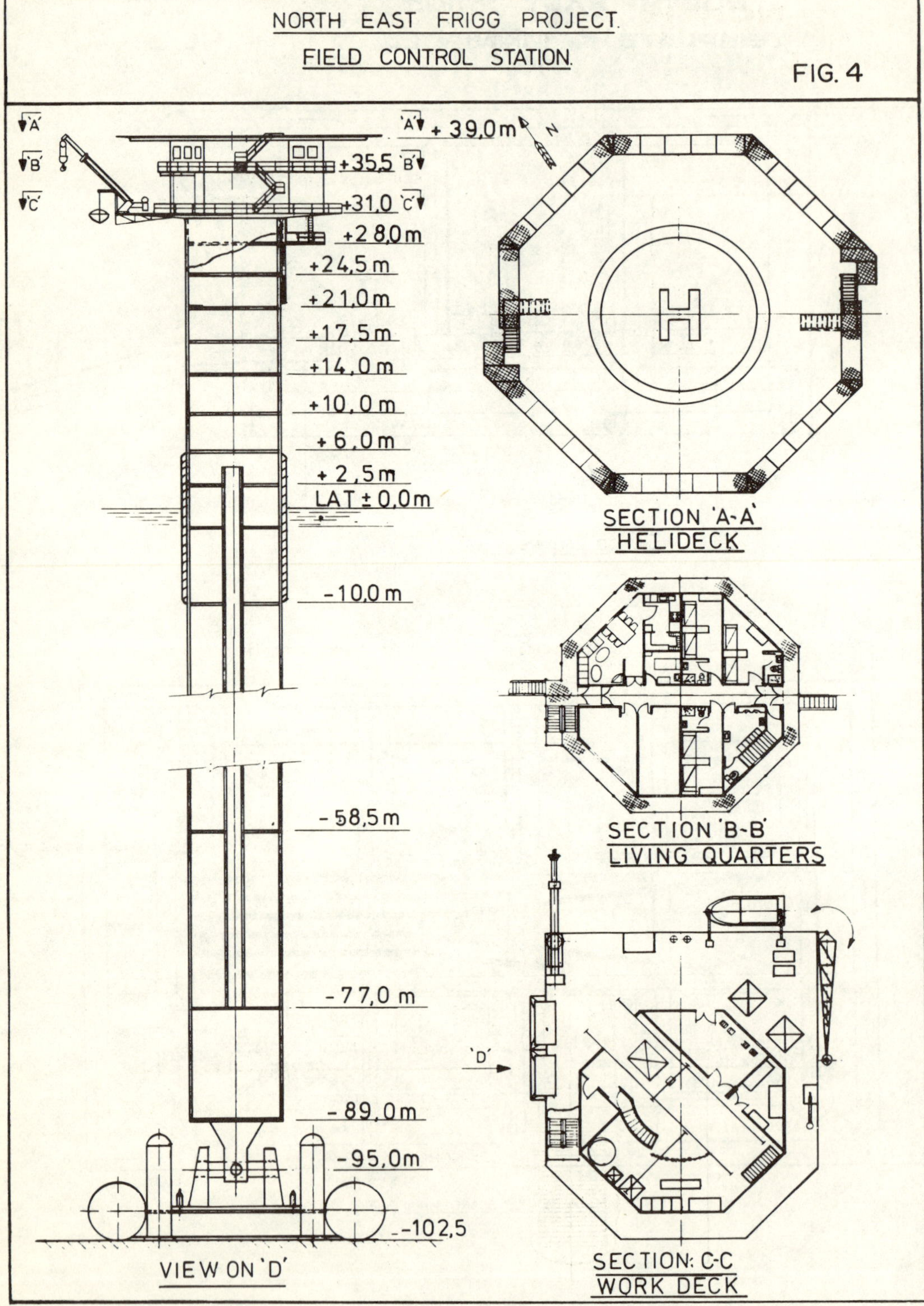
NORTH EAST FRIGG PROJECT.
FIELD CONTROL STATION.
FIG. 4
'A'
'B'
'C'
+ 39,0m
+35,5
+31,0
+28,0m
+24,5m
+21,0m
+17,5m
+14,0m
+10,0m
+6,0m
+2,5m
LAT ± 0,0m
-10,0m
-58,5m
-77,0m
-89,0m
-95,0m
-102,5
VIEW ON 'D'
N
SECTION 'A-A'
HELIDECK
H
SECTION 'B-B'
LIVING QUARTERS
'D'
SECTION: C-C
WORK DECK

NORTH EAST FRIGG

DRILLING COMPLETION SCHEME FIG.5

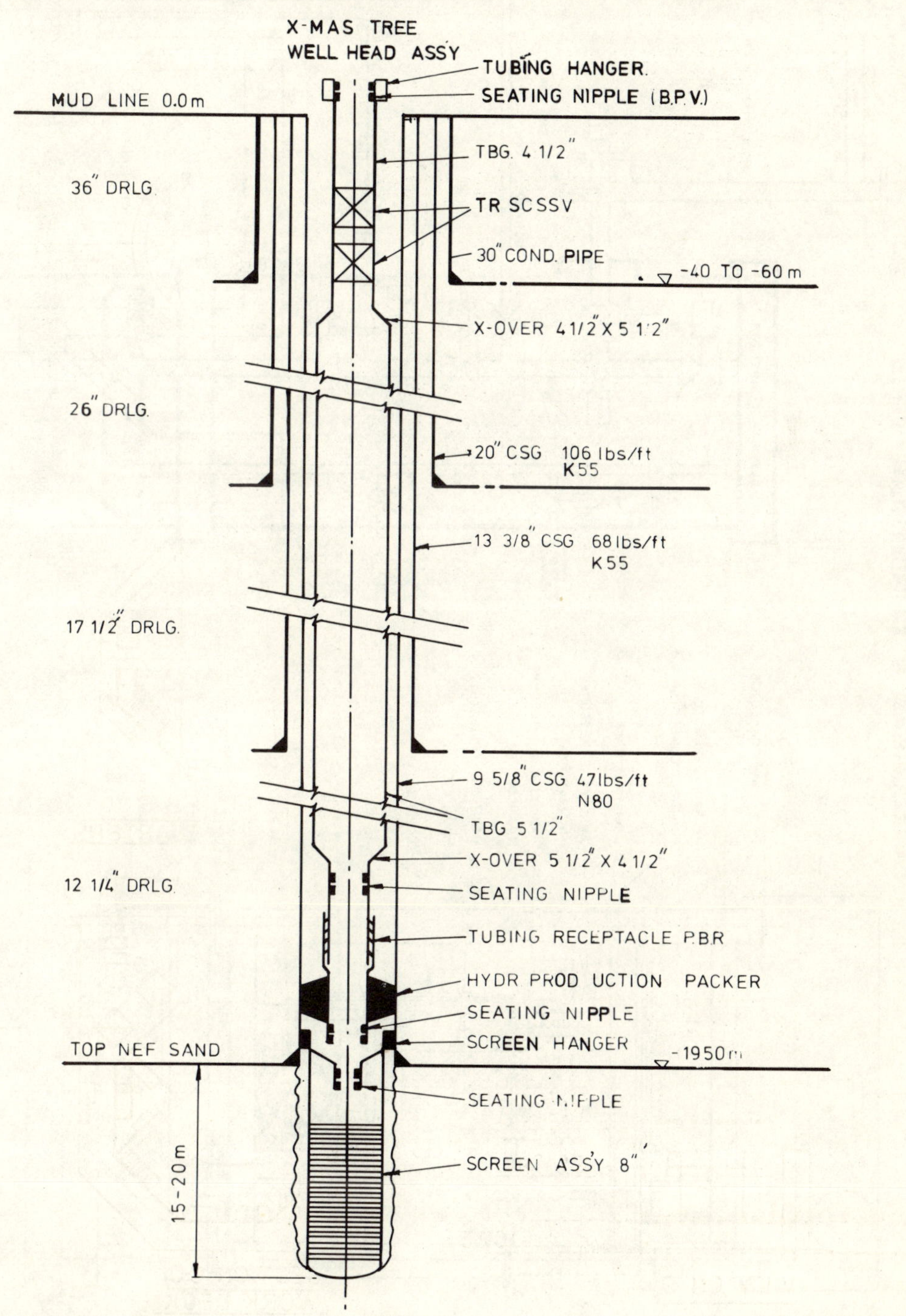

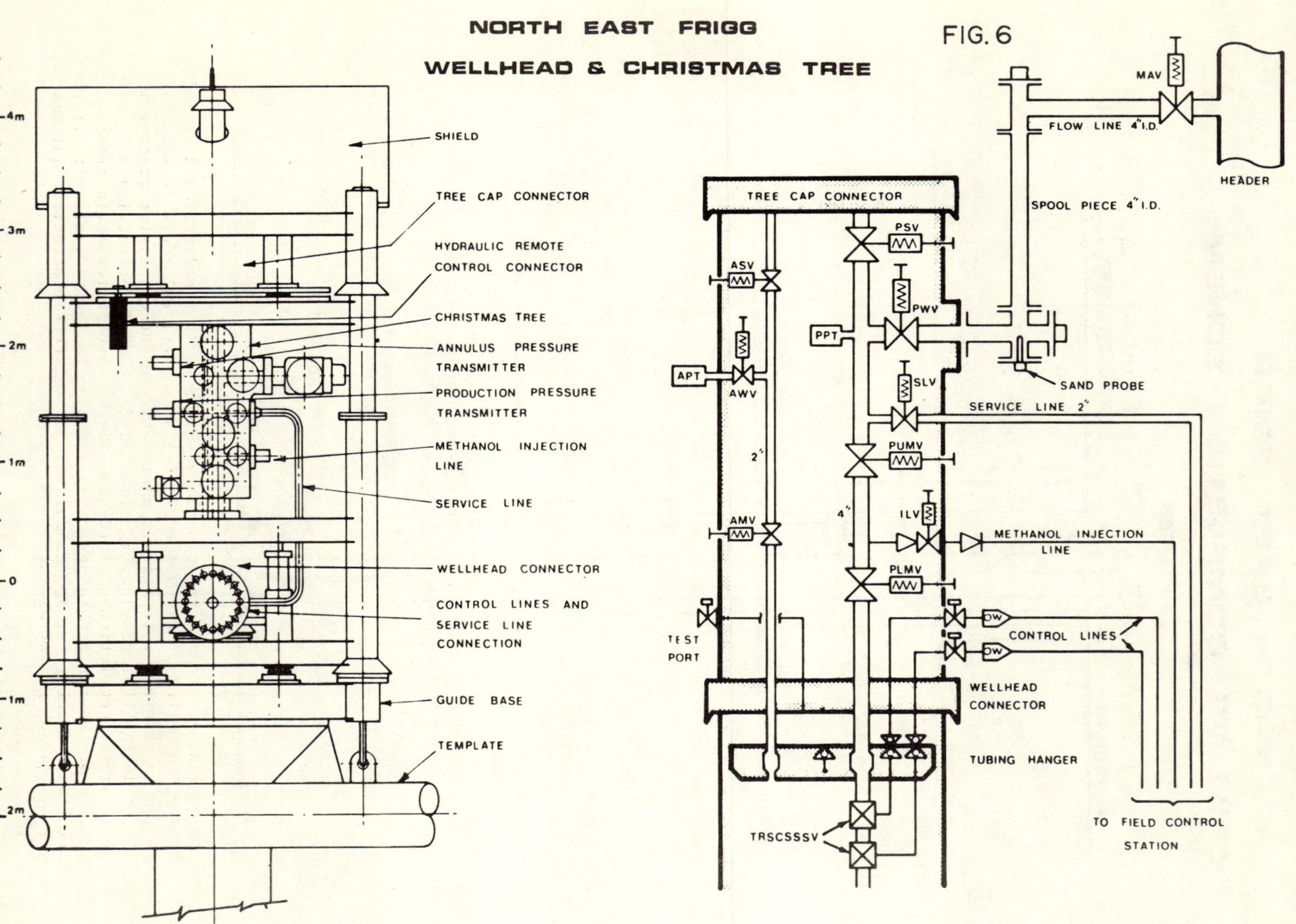
NORTH EAST FRIGG
WELLHEAD & CHRISTMAS TREE
FIG.6
4m
3m
2m
1m
0
1m
2m
SHIELD
TREE CAP CONNECTOR
HYDRAULIC REMOTE CONTROL CONNECTOR
CHRISTMAS TREE
ANNULUS PRESSURE TRANSMITTER
PRODUCTION PRESSURE TRANSMITTER
METHANOL INJECTION LINE
SERVICE LINE
WELLHEAD CONNECTOR
CONTROL LINES AND SERVICE LINE CONNECTION
GUIDE BASE
TEMPLATE
TREE CAP CONNECTOR
ASV
APT
AWV
2"
AMV
TEST PORT
PSV
PWV
PPT
SLV
PUMV
4"
ILV
PLMV
TRSCSSSV
MAV
FLOW LINE 4" I.D.
HEADER
SPOOL PIECE 4" I.D.
SAND PROBE
SERVICE LINE 2"
METHANOL INJECTION LINE
CONTROL LINES
WELLHEAD CONNECTOR
TUBING HANGER
TO FIELD CONTROL STATION

DESIGN AND APPLICATION OF A DRY CHAMBER FOR SUBSEA PRODUCTION

M. C. Tate and D. L. Miller

Cameron Iron Works, Inc., Houston, Texas, USA

ABSTRACT

Deepsea petroleum reserves has motivated the development of advanced subsea production systems which must provide safe and economical operations despite the hostile underwater environments. This paper presents such a system.

Extending the state-of-the-art of dry chambers for subsea production, an advanced vertical access chamber has been designed for dry containment of wellhead production equipment and controls to a depth of 1200 feet (365 meters). Det Norske Veritas Rules for the Construction and Classification of Diving Systems, 1975, supplemented by American Society of Mechanical Engineers Boiler and Pressure Vessel Code, Section VIII, Division 2, was utilized in the chamber design. Inherent advantages, such as standardization and modular designs, accommodations for all marine wellheads, trees and flowlines as well as transport or service vehicles, diver access, and direct vertical access for surface wireline and downhole operations are addressed.

In addition to the presentation of design parameters, chamber installation procedures and wellbore fluid control operations are also described.

KEYWORDS

Subsea production; dry system; vertical access chamber; production control system; design loads; stress analysis techniques.

INTRODUCTION

From the beginning, the dry system precept has been to complement and extend wet tree completion methods. Using the field-proven practical ideas of wet completion and incorporating techniques particular to the dry system creates a synergistic design that provides a more powerful tool for achieving safe and economical subsea completions despite the hostile underwater environment. Utilizing state-of-the-art technology, such a design has been developed: an advanced Vertical Access Chamber for dry containment of wellhead production equipment and controls to a depth of 365 meters.

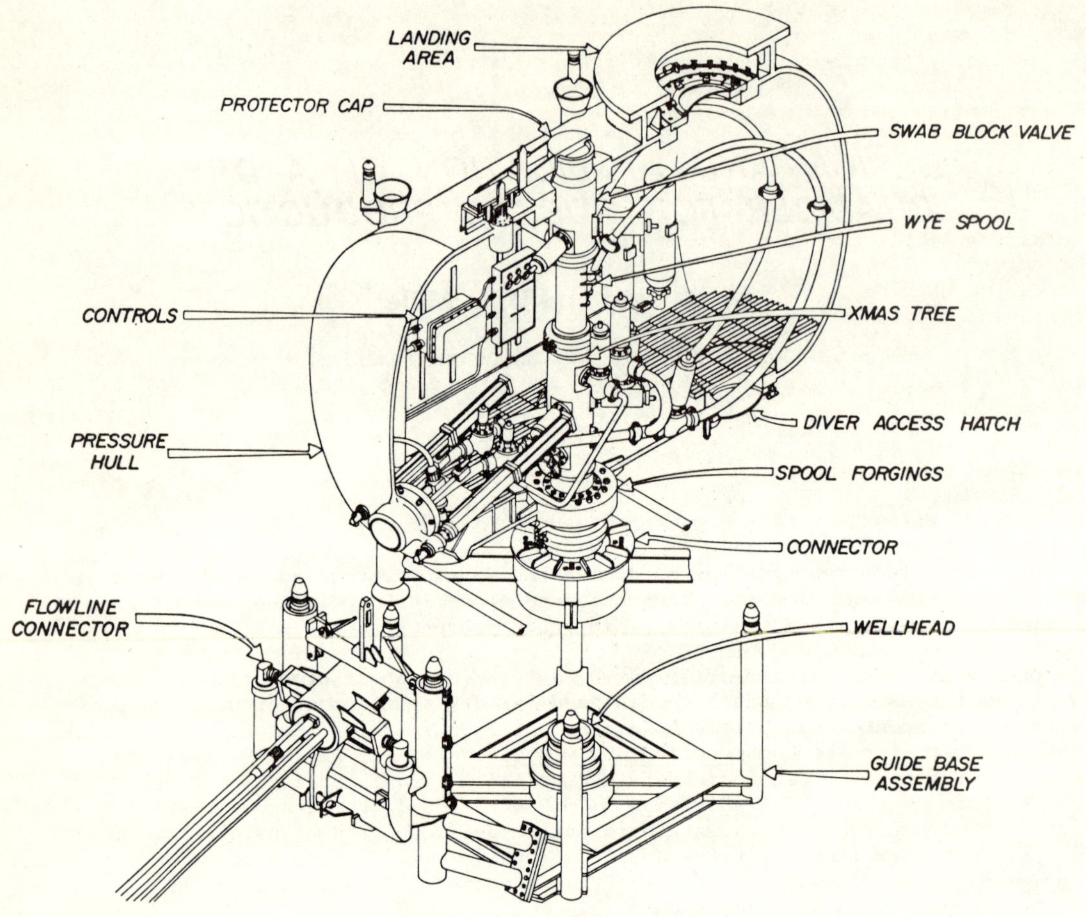

Fig. 1. Major components of Vertical Access Chamber.

APPLICATION

The application of the dry system characterizes that of the wet tree but becomes more advantageous when its inherent characteristics can or must be utilized.

The ability to provide "hands-on" repair and maintenance of components in a controlled one atmosphere environment is one such characteristic. Although appearing superficial, this ability does have its ramifications. The work required can be handled by personnel specifically trained and familiarized with system operation and components. A reduction in the time required to complete the task is therefore possible with the degree of success and safety greatly enhanced over typical intervention methods. The economics become readily apparent when such capabilities exist.

Another aspect related to the utilization of the dry environment is the ability to provide dry electrical hookup, thus avoiding the problems associated with wet electrical connections. The added factor that the sequencing and electrical control pod are also enclosed, promotes an increase in the general reliability of the overall control system.

With the ever present concern to minimize ecological impact, the chamber can also be used in an auxiliary capacity for early detection of oil or gas leaks. By enclosing the tree and immediate flowlines, the chamber provides for the initial containment of such leaks allowing early detection and automatic shut-in. The occurring isolation would define the problem well and direct the available energies needed for repair.

These are but a few of the more obvious and useful characteristics that can be exacted from a dry system. In addition, although not without limits, the system permits a large degree of latitude to pursue the problems associated with subsea production. It is this flexibility that has been instilled in the vertical access chamber design and that which will be discussed in more detail in the remaining sections of this paper.

SYSTEM COMPONENTS

The major components of the system are best illustrated in Fig. 1. Located at the lower section are the guidebase and wellhead with standard six foot center for the posts. The flowline connector posts are also integral with the base assembly. The collet connector used to attach the chamber to the wellhead is mated to the chamber through the spool forging. This forging consists of two parts, one permanently attached to the hull and the other that adapts the tree to the connector and wellhead. The chamber's integral guidepipes are directly connected to the collet connector, providing the principal load path for moments applied to the chamber. More than half of all moment reactions are carried through this path, thus avoiding the spool forging. A number of tree options are available with the use of this adapter spool; shown is a tree with TFL and vertical access capabilities. The hydraulic and electrical controls are also enclosed by the pressure hull. A landing area at the top permits the transfer of service personnel by submersible or bell. This particular model also has the feature of a diver access hatch for intervention by a free swimming diver - applicable in shallower waters. The last major components of this system is the flowline connector carriage which can be run before or after the chamber is set.

These components combine to produce a working system suitable for a wide variety of operating condition and installation locations. They provide the means for continual monitoring of well and tree with the ability to conduct repairs in a one atmosphere environment. With the addition of the workover riser, Fig. 2, downhole problems can be accommodated from the surface.

INSTALLATION

With guidelines established, the chamber is run with the use of the workover riser. The chamber is set, and the collet connector engaged on the wellhead. Standard testing of the various control valve seals is then performed to insure proper equipment operation. Transfer of service personnel can now be initiated for preparation of well production equipment. This is of course assuming that flowlines have been run. If not however, the chamber preparation can be postponed until the flowline connection and bundle have been established. At that time the docking and transfer can take place for the necessary adjustments of the production components and pull-in of the flowlines. To insure safety, within the entry hatch area beneath the docking plate exists penetration taps which enable evaluation of the chamber interior pressure, gas content, liquid level and temperature before the transfer hatch is opened. Having performed

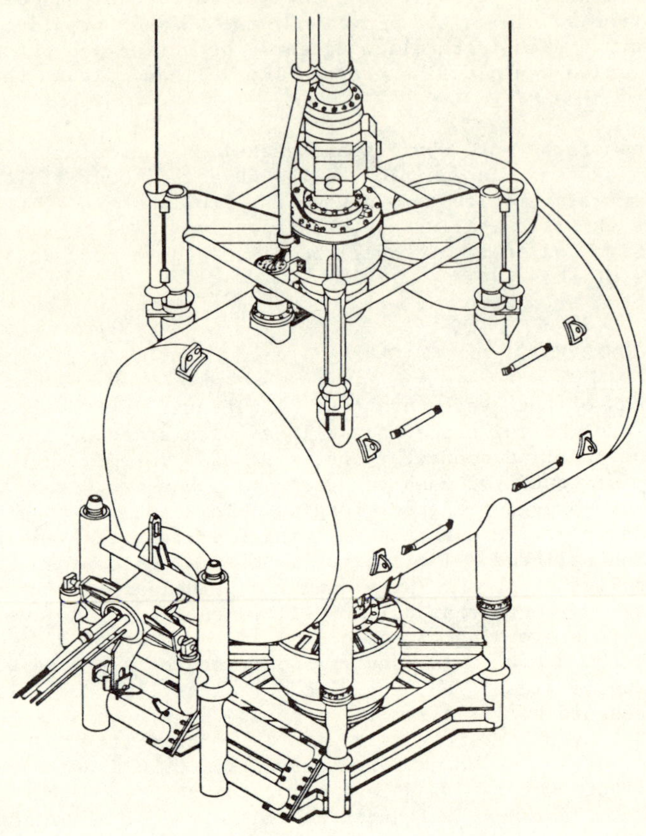

Fig. 2. Workover riser configuration.

flowline pull-in and final checkout, the system is ready for production.

WELLBORE FLUID CONTROL

Table 1, with the schematic diagram, illustrates a typical method for hydraulic sequencing control. However, the system can be modified to accommodate specific operating criteria from the customer.

DESIGN AND ANALYSIS CRITERIA

The Vertical Access Chamber was designed to meet the Det Norske Veritas Rules for the Construction and Classification of Diving Systems, 1975, Oslo, Norway. These rules provide formulas for determining external pressure ratings and requirements for reinforcement of openings as well as allowable primary general membrane stresses. A second code, the American Society of Mechanical Engineers, Boiler and Pressure Vessel Code, Section VIII, Division 2, Rules for Construction

TABLE 1 Hydraulic Sequencing Control System

MODES OF OPERATION									
MODE	DESCRIPTION	CROSSOVER	ANNULUS MASTER	PROD. MASTER 1	PROD. MASTER 2	WING NO. 1	WING NO. 2	DHSV NO. 1	DHSV NO. 2
1	Shut-In or Pig Flowlines	O	C	C	C	C	C	C	C
2	Monitor Annulus Pressure	O	O	C	C	C	C	C	C
3	Preparation for Production	C	C	O	O	O	O	C	C
4	Production from String 1	C	C	O	C	O	C	O	C
5	Production from String 2	C	C	C	O	C	O	C	O
6	Production from Strings 1 & 2	C	C	O	O	O	O	O	O

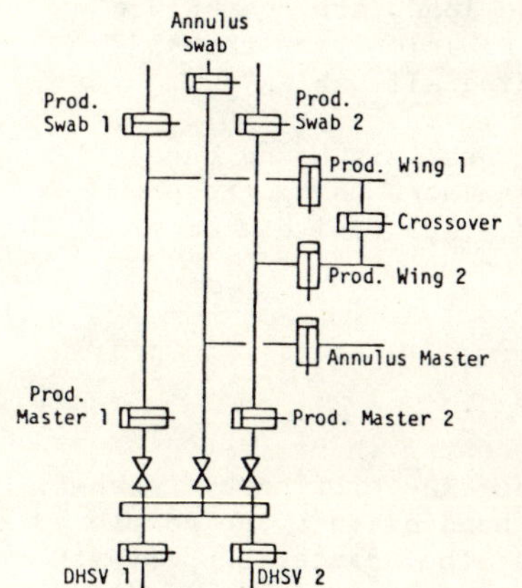

MODE 1 - Well is shut in, crossover valve is open for pigging operations.

MODE 2 - Crossover valve remains open. Annulus master valve is opened to bleed annulus pressure.

MODE 3 - The crossover, annulus master and downhole safety valves are closed. All other valves are opened. The production manifold to the tree is opened reducing the differential pressure across the downhole safety valves in preparation for production.

MODES 4, 5, and 6 - These are production modes and any one of these modes may be selected from Mode 3. The downhole safety valves are opened to obtain production from both strings. If production is required from only one string, the production master, production wing, and downhole safety valves are closed on the opposite string.

NOTE: If the system is operating through any Mode 2 through 6, the tree may be shut in by selecting Mode 1. A restrictor (time delay) is incorporated in closing circuit of the downhole safety valves to insure that production wing and master valves close first equalizing pressure across the downhole safety valves before they close.

of Pressure Vessels, 1977, through Summer 1979 Addenda, New York, N.Y., was added to provide procedures for calculating stresses and to provide limits for primary local membrane, primary general membrane plus bending, and primary plus secondary membrane plus bending stresses. Reviewing these two codes for the lowest allowable stresses gave the following allowables:

Primary General Membrane = Lesser of 1/3 Ultimate or 1/1.8 Yield = σ_t

Primary Local Membrane = 1.5 σ_t

Primary General Membrane + Bending = 1.5 σ_t

Primary Plus Secondary (Total Membrane Plus Bending) = 3 σ_t

As required by the ASME Code, the Maximum Shear Stress Theory of Failure was used to combine the various stress components into a single value that was compared to the allowable stress. Twice the maximum shear stress is termed by the Code as the "stress intensity", and is defined as the algebraically largest principal stress minus the algebraically smallest principal stress.

Stress Intensity = $S_1 - S_3$, where the principal stresses are S_1, S_2, and S_3, and $S_1 > S_2 > S_3$

ANALYSIS TECHNIQUES

The structural analysis of the Vertical Access Chamber was performed for loads applied to the various components of the system. These loads are summarized and presented in Table 2. The rules of Section VIII, Division 2 of the ASME Code were used for classifying stresses and corresponding allowables.

Most of the stresses were determined by finite element analysis using three dimensional models, while some shell stresses were calculated using the procedures given in Welding Research Council Bulletin 107. The finite element techniques used for the structural analysis are detailed below.

Analysis of Upper Hatch

The top hatch consists of a hemispherical head at the center (hydrostatic pressure acts on the convex side) that is secured to an adapter ring by latches. The design is such that pressure tends to tighten the head against the adapter and, therefore, the O-ring seal is pressure energized. The adapter supports the head, and this assembly is then bolted to the reinforcement ring. The 10.5 in. (267 mm) X 9 in. (229 mm) ring provides reinforcement to the opening in the shell. The outer edge of the ring welds to the chamber. The cylindrical neck of the opening is welded to the chamber 9.25 in (235 mm) outside the reinforcement ring. The top landing plate is welded to the cylindrical neck with twelve equally spaced gussets located between the top plate and neck. The purpose of these gussets is to provide support for the inner portion of the landing plate when a submarine is mated to the chamber.

The top hatch is loaded by pressure which produces an axial load acting along the opening centerline and bending when a submarine is attached. There can also be an impact load in the axial direction due to initial contact by the submarine.

TABLE 2 Summary of Design Load Cases

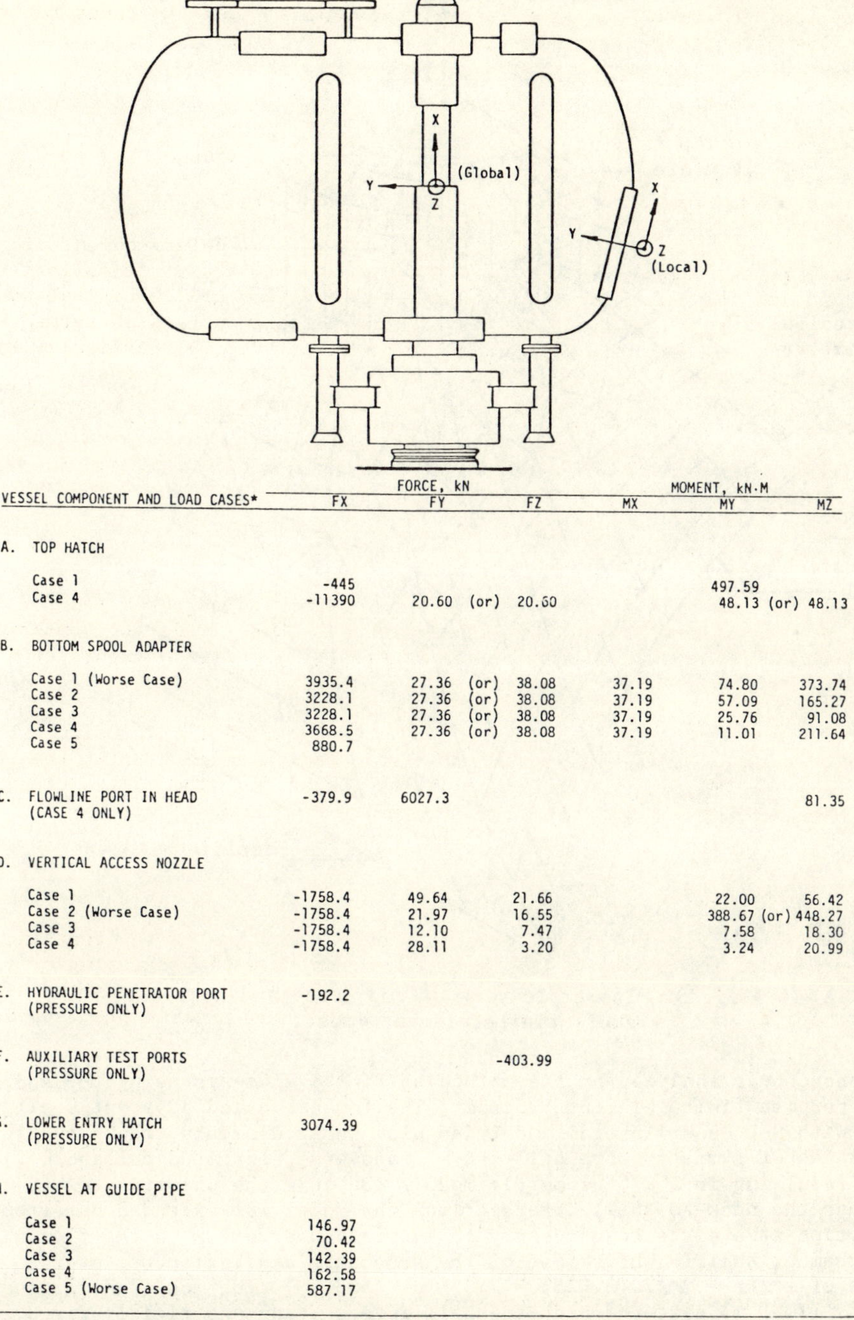

VESSEL COMPONENT AND LOAD CASES*	FORCE, kN FX	FY	FZ	MOMENT, kN·M MX	MY	MZ
A. TOP HATCH						
Case 1	-445				497.59	
Case 4	-11390	20.60 (or)	20.60		48.13 (or)	48.13
B. BOTTOM SPOOL ADAPTER						
Case 1 (Worse Case)	3935.4	27.36 (or)	38.08	37.19	74.80	373.74
Case 2	3228.1	27.36 (or)	38.08	37.19	57.09	165.27
Case 3	3228.1	27.36 (or)	38.08	37.19	25.76	91.08
Case 4	3668.5	27.36 (or)	38.08	37.19	11.01	211.64
Case 5	880.7					
C. FLOWLINE PORT IN HEAD (CASE 4 ONLY)	-379.9	6027.3				81.35
D. VERTICAL ACCESS NOZZLE						
Case 1	-1758.4	49.64	21.66		22.00	56.42
Case 2 (Worse Case)	-1758.4	21.97	16.55		388.67 (or)	448.27
Case 3	-1758.4	12.10	7.47		7.58	18.30
Case 4	-1758.4	28.11	3.20		3.24	20.99
E. HYDRAULIC PENETRATOR PORT (PRESSURE ONLY)	-192.2					
F. AUXILIARY TEST PORTS (PRESSURE ONLY)			-403.99			
G. LOWER ENTRY HATCH (PRESSURE ONLY)	3074.39					
H. VESSEL AT GUIDE PIPE						
Case 1	146.97					
Case 2	70.42					
Case 3	142.39					
Case 4	162.58					
Case 5 (Worse Case)	587.17					

*Case 1 = Pressure + Current + Weight + Service vehicle impact + Flowline bundle and piston
Case 2 = Pressure + Current + Riser + Flowline bundle and piston
Case 3 = Pressure + Current + Guideline shear + Flowline bundle and piston
Case 4 = Pressure + Current + Weight + Flowline connection + Flowline bundle and piston
Case 5 = Pressure + Running load

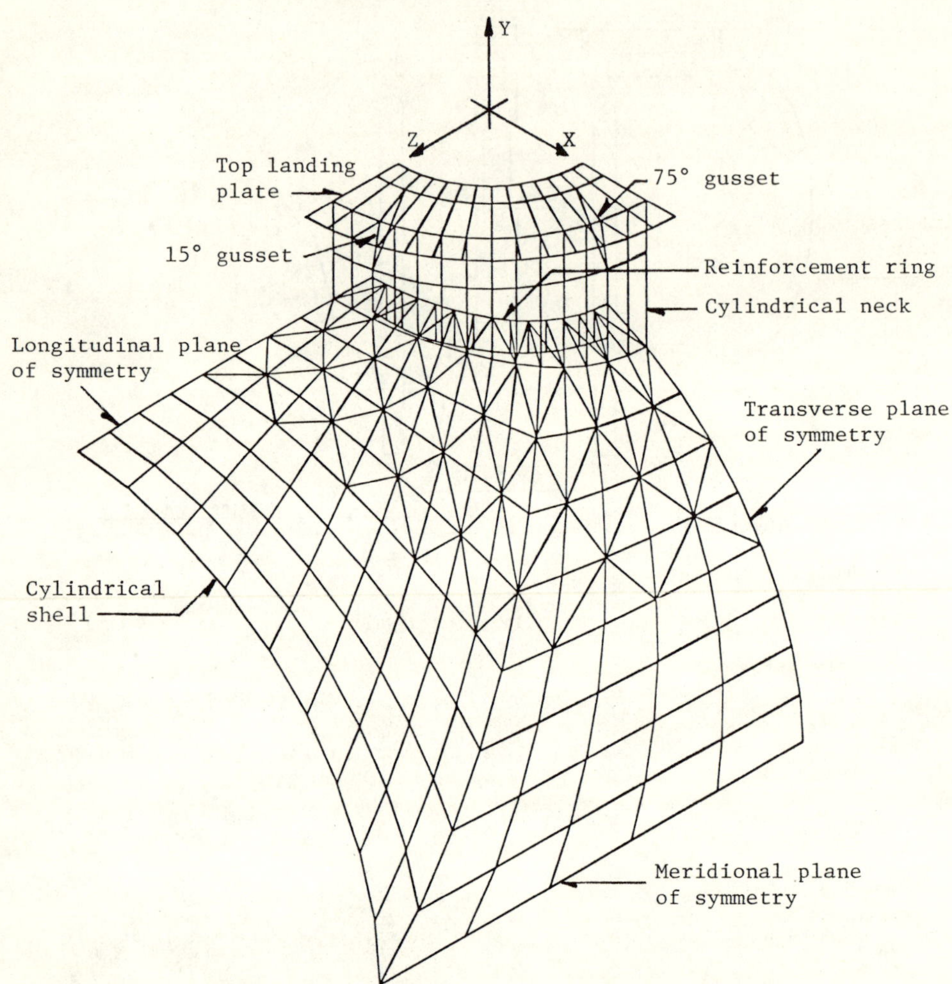

Fig. 3. Isometric view of top hatch and adjacent shell finite element model.

The top hatch was analyzed by the method of finite elements using the ANSYS general purpose finite element program. The top hatch and adjacent shell were modeled with 3-D quadrilateral and triangular shell elements, shown in Fig. 3. Three planes of symmetry were utilized - transverse, longitudinal and meridional - resulting in a 1/8 geometric model. Because the primary area of concern was around the neck-to-shell intersection, the model was extended out from the intersection at a distance necessary to insure conservatism in the flexibility of the chamber shell. While most of the secondary bending occurs over a distance of $2\sqrt{Rt} = 2\sqrt{72(2.625)} = 27.5$ in. (698 mm), the model was extended 62.75 in. (1.59 m) past the neck-to-shell intersection and rigidly fixed. This representation is conservative in view of the facts that an elliptical head weldment is 6.25 in. (159 mm) from the neck on one side, and an internal stiffening ring is 5.75 in. (146 mm) from the neck on the other side.

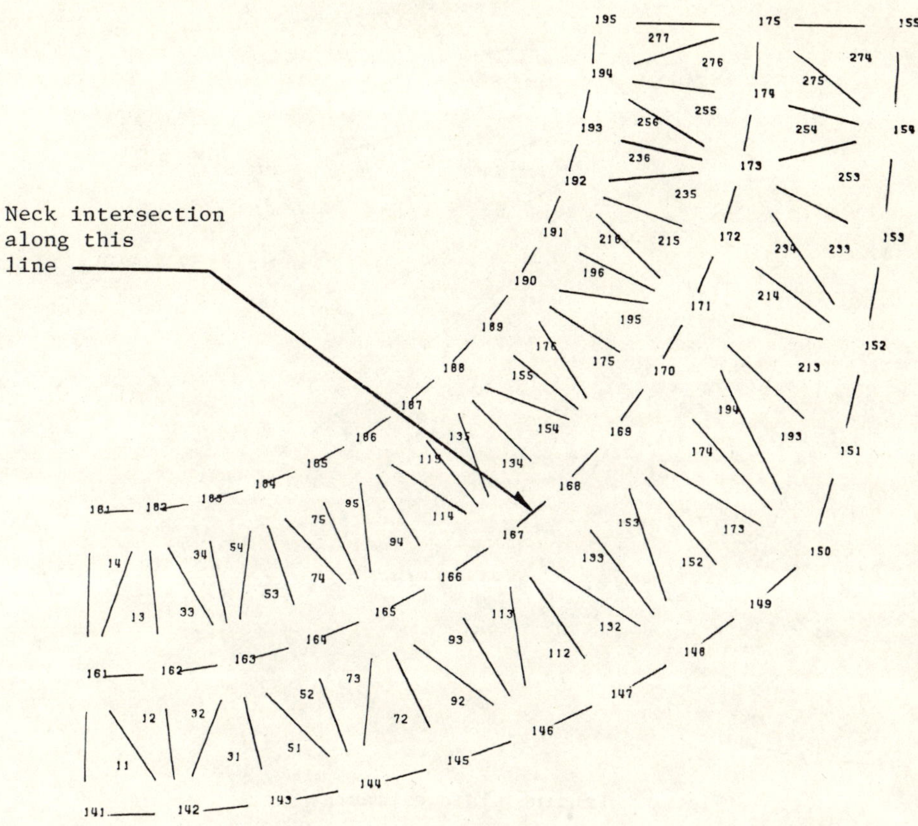

Fig. 4. Cylindrical shell elements bordering hatch neck.

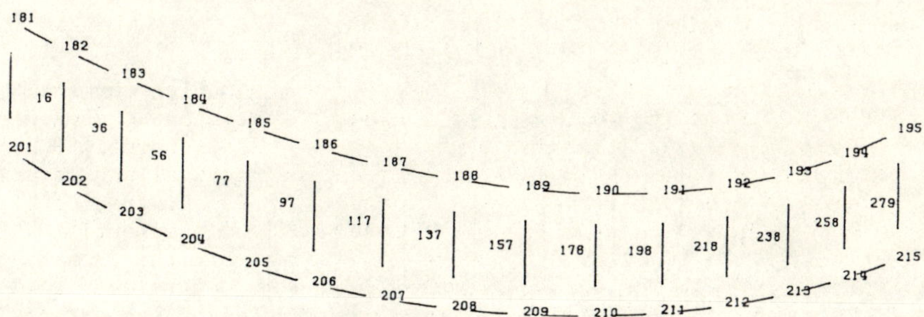

Fig. 5. Reinforcement ring elements.

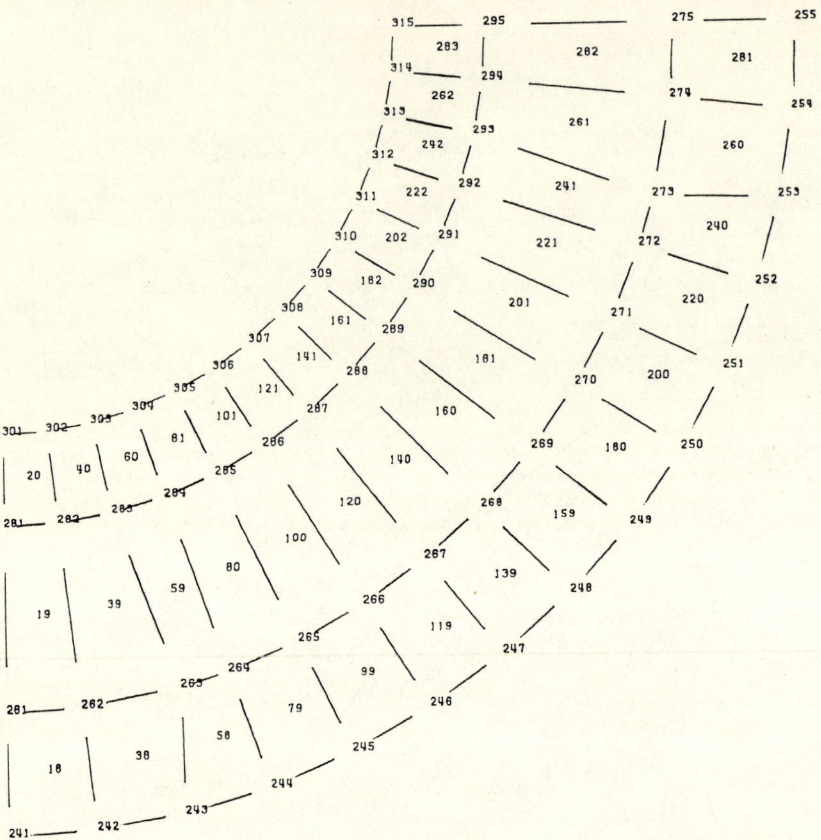

Fig. 6. Landing plate elements.

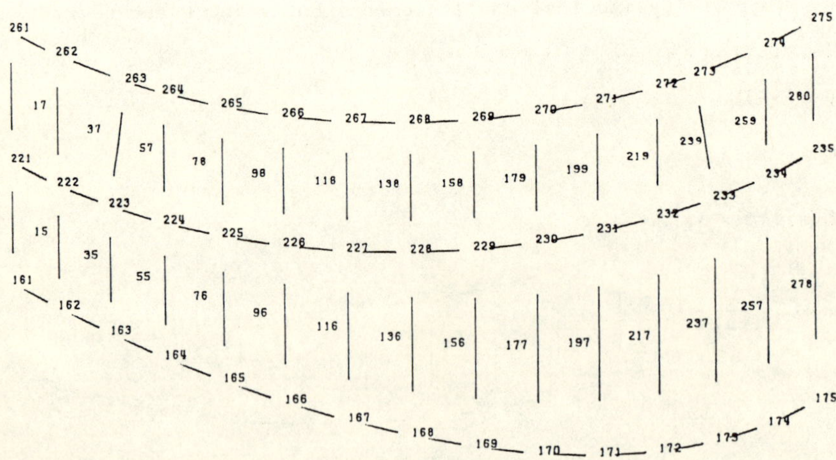

Fig. 7. Cylindrical neck elements.

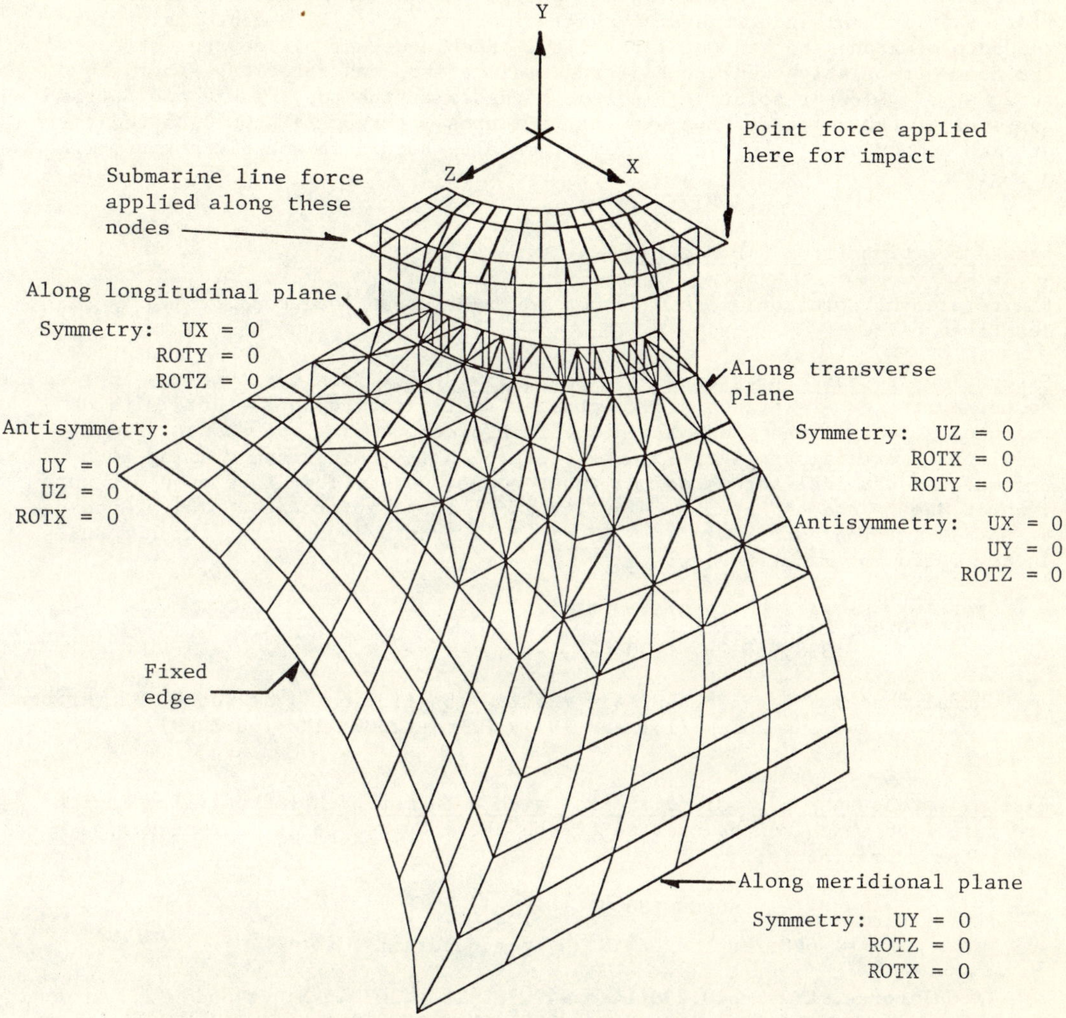

Y
Z
X
Point force applied here for impact
Submarine line force applied along these nodes
Along longitudinal plane
Symmetry: UX = 0
ROTY = 0
ROTZ = 0
Antisymmetry:
UY = 0
UZ = 0
ROTX = 0
Along transverse plane
Symmetry: UZ = 0
ROTX = 0
ROTY = 0
Antisymmetry: UX = 0
UY = 0
ROTZ = 0
Fixed edge
Along meridional plane
Symmetry: UY = 0
ROTZ = 0
ROTX = 0

Fig. 8. Top hatch and adjacent shell finite element model boundary conditions.

F
F
F
F
2F
Symmetry
Antisymmetry
Point force

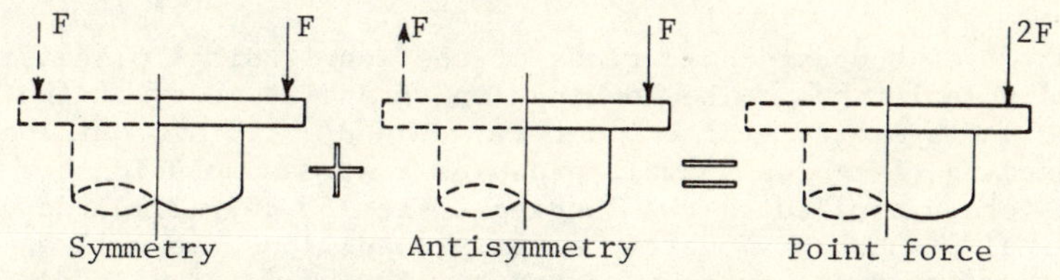

Fig. 9. Illustration of symmetry and antisymmetry load solutions.

The geometry consisted of thin shell and plates and was modeled with elastic flat quadrilateral and triangular shell elements (ANSYS STIF 63). This element has both membrane and bending capability. Each node has six degrees of freedom - translations in the X, Y and Z directions and rotations about the three axes. Element solutions include stresses at the top, middle and bottom surfaces at the center of the element. Figures 4 through 7 show isolated views of each component of the finite element model. Both node and element numbers are given.

Stress Solutions from 3-D Models

Finite element solutions were made for four different load conditions, as described below.

External hydrostatic pressure = 550 psi (3.79 MPa). This pressure was applied to the entire outer surface of the chamber shell, including the area within the neck. Because the area within the reinforcement ring was absent from the model, the vertical pressure loading effects on the reinforcement ring were included as vertical forces on the inner nodes. Other loading components were ignored due to the high relative stiffness of the reinforcement ring. Spacing of the 15 internal nodes was at 6.43 deg (.112 rad.) increments, which resulted in the following input forces:

$$F_{pressure} = PA/4 = -(550/4)\,\pi\,(56)^2/4$$
$$= 338,660 \text{ lbs } (1506 \text{ kN})$$

Nodal force, FY = -338,660/14 = -24,190 lbs (107.6 kN) at nodes 182 through 194. At two end nodes (181 and 195), FY = -12,095 lbs (53.8 kN).

External pressure = 550 psi (3.79 MPa) with a submarine attached. The maximum pressure load which can be exerted on the shell by dewatering the neck area below the submarine is:

$$F_{pressure} = PA/4 = -550\,\pi\,(38.5)^2/4$$
$$= -640,290 \text{ lbs } (2848 \text{ kN}) \text{ per quadrant}$$

Nodal force, FY = -640,290/14 = -47,735 lbs (203.4 kN) at nodes 262 through 274. At two end nodes (261 and 275), FY = 22,868 lbs (101.7 kN).

The pressure only and pressure plus submarine load conditions both had uniform loadings and therefore symmetrical boundary conditions were used at the three planes of symmetry. These conditions are shown in Fig. 8. The location of the applied submarine loads are also shown.

Figure 8 also gives boundary conditions at the longitudinal plane for anti-symmetry loading solutions. The combination of antisymmetry at the longitudinal plane and symmetry at the other two planes, with a point load applied at the top landing plate (as shown), produces a stress solution for two equal and opposite forces applied to the landing plate 180 deg (3.14 rad.) apart. If the longitudinal plane has symmetry boundary conditions (as well as the other two planes) the same point force produces a solution for two equal forces in the same direction and 180 deg (3.14 rad.) apart. By adding these two solutions algebraically, the result is a point load on the model. This is illustrated in Fig. 9 on a flange. The sum of the two load solutions, each having a point load, F, results in a solution for one point load on the 3-D model equal to 2F. This procedure was used to determine the solutions for impact loading, as described below.

Impact load. The results of the symmetry and antisymmetry solutions were saved on tape. These two solutions were then combined using a postprocessor (Post 27) which gave stresses for an equivalent single force of FY = -1×10^5 lbs (-445 kN) applied to the landing plate. This solution was used for analyzing impact loads due to a mating submarine.

The point load was applied on the transverse plane (as opposed to the longitudinal plane) from considerations of known behavior. A circumferential bending moment on a cylinder produces higher stresses than an equal moment in the longitudinal direction, thus favoring the force on the transverse plane.

Current against the submarine produces a moment and shear on the hatch. As explained above, the circumferential moment was chosen as the worse case loading direction and was therefore used in the analysis. The procedure used to determine the solution for bending is described below.

Submarine induced bending. Antisymmetry boundary conditions, as shown in Fig. 8, were imposed on the model. Vertical loads were applied downward on the other nodes of the landing plate to simulate the bending moment, as illustrated in Fig. 10. The moment produced by this loading is 123,200 in-lbs (13,920 N·M). The stresses from this solution were then multiplied by the ratio 48.13/13.92 = 3.458 and added absolutely to the pressure plus submarine load induced stresses.

Several of the quadrilateral elements in the model had a slight amount of warpage. Warpage causes an additional moment on the element, higher than actual stresses and lower than actual stiffnesses. However, these element stresses were consistent with the stresses in the adjacent elements and therefore this effect was considered negligible.

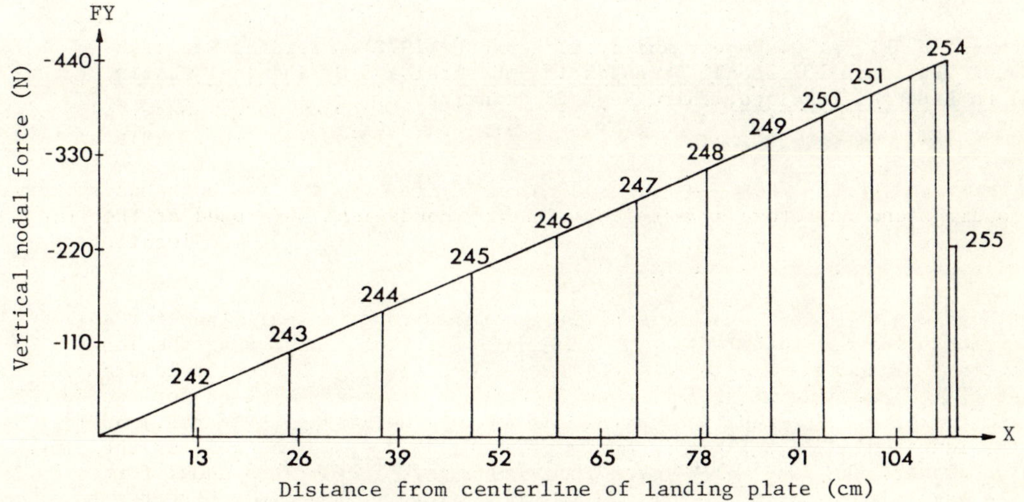

Fig. 10. Nodal forces simulating submarine induced bending load on landing plate.

All of the generate stress intensities for all load cases proved to be lower than the allowable values and acceptable.

CONCLUSION

The dry system is a unique tool that can expand the options available to an operator when subsea development is considered. The flexibility of the system provides a high degree of latitude for functional variations. It is capable of adapting to a wide range of operating conditions while maintaining the necessary safety and reliability required for subsea installation. The dry system can be considered a viable alternative for subsea completions.

ACKNOWLEDGEMENTS

The authors wish to recognize the members of the Oil Tool Research and Development group of Cameron Iron Works, Inc. The efforts of all these people were responsible for the development of this design. We are also very grateful to Mr. R. D. Jolly for his help in creating the illustrations, and to Ms. Rhonda Anderson for her help in preparing this paper for publication.

REFERENCES

ASME Boiler and Pressure Vessel Code, Section VIII, Division 2, Rules for Construction of Pressure Vessels, 1977 Edition through Summer 1979 Addenda, New York, N.Y.

Det Norske Veritas, Rules for the Construction and Classification of Diving Systems (1975), Oslo, Norway.

Swanson Analysis Systems, Inc., ANSYS Engineering Analysis System User's Manual, Revision 3, August 1, 1978.

Wichman, K. R., A. G. Hopper and J. L. Mershon (1972). Welding Research Council Bulletin 107, Local Stresses in Spherical and Cylindrical Shells due to External Loadings, Third Revised Printing.

INSERT TREE COMPLETION SYSTEM

K. W. Brands

Shell Internationale Petroleum Maatschappij B.V., Offshore Research and Development, The Hague, The Netherlands

SUMMARY

This paper briefly describes the concept philosophy and the present state-of-the-art of the Insert (or Caisson) Tree U.W. Well Completion System, designed and manufactured by Cameron Iron Works for Shell Internationale Petroleum Maatschappij (S.I.P.M.).

For operational reasons a site offshore Brunei has been selected for the installation of a prototype tree in 1980.

KEYWORDS

Underwater X-mas tree; underwater completion; through the flowline (TFL); insert tree; offshore oil production; tubing hanger; slimline connector; flowline connector; retractable hydraulic stab.

INTRODUCTION

It is SIPM's conviction that in areas with much fishing activity the profile of the underwater trees has to be reduced from the present height above seabed of approximately 30 feet to a height which makes protection of the structure by for instance a dome technically feasible. The ultimate goal is to eliminate the exposed structure above the seafloor completely. This is presently seen as a follow-up to the existing designs for an insert (or caisson) tree.

In brief the advantages of an Insert Tree (Fig. 3) may be summarised as follows:

1. Lower profile above seabed.

2. Increased safety as tubing hanger and the tree valves are sub-mudline inside the conductor.

3. Easier installation and handling of the streamlined equipment from a drilling rig.

4. Tree configuration is simplified, e.g. swabvalves and wye spool are eliminated. Vertical re-entry is still possible in case of emergency, without having to pull the complete X-mas tree.

   However, experience with 3" TFL tools has shown that all routine well-servicing operations can be carried out with TFL techniques. Non-routine operations, such as fishing jobs and sand-washing work, have successfully been completed by the TFL techniques in a test well.

5. Reduction of the 270 deg. well-head loops to 90 deg. loops.

6. Simplification of the required control system.

As principal disadvantage can be mentioned:

1. The X-mas tree block is inaccessible for divers which rules out minor repair jobs like rectifying possible leaks in controlline piping; however access could be obtained by using special techniques.

The design is based on a conventional two stack drilling system (20 3/4" and 13 5/8" BOP stacks); however, it could be adapted for a single BOP stack system in another application.

The dual 3½" x 3½" completion strings will be installed simultaneously (dual running) which is dictated by the use of a solid tubing hanger.

The remote controlled flowline connector is, for the Brunei application, designed for a vertically initiated first end connection.

X-MAS TREE LAYOUT

A typical underwater tree configuration is shown in Fig. 1. It comprises 12 valves and two wye spools with diverters.

In order to install the X-mas tree block inside the conductor it has to be investigated if the number of valves can be reduced without compromising on safety.

Each valve or component in Fig. 1 will be taken in turn and analysed as to its requirements. However, prior to this analysis it is necessary to evaluate the reasons for operating the valves on an underwater tree.

In addition to the valves on the satellite U.W. tree there will be valves and chokes mounted on the platform (or in an U.W. manifold) to which the well is connected. Latter valves will always be used for all normal routine well operations and shut-ins and will close automatically, in event of a platform emergency or flowline breakage. The U.W. tree valves will consequently never be closed for routine operations under flowing conditions.

Various operations or conditions necessitating stroking of the X-mas tree valves can be listed as:

1. Testing that valves close satisfactorily.

2. Flowline break or leakage.

3. TFL operations or pigging of flowlines.

4. Platform emergency shut-down. Primary shut-down would be the valves on the platform and a secondary operation might be closure of the U.W. tree.

5. Closure of the well for a long period.

6. Preparation for workover or well repair.

7. Annulus pressure blowdown.

Hereafter, where analysing each valve requirement, the above operational philosophy should be borne in mind where applicable.

Swabvalves

The swabvalves are operated during all vertical entries into the well during initial completion and subsequent major workovers.

The valves are never operated while the well is producing; they assure that the well fluid cannot leak into the sea. Nevertheless, in the event that they might leak, wireline plugs have generally been fitted above the closed swabvalves in U.W.C.'s.

Prior to doing a major workover, or remedial work to the X-mas tree, it is necessary to kill the well before the X-mas tree can be removed. The swabvalves permit safe reconnection of a riser to the well so that plugging or killing of the well can be carried out from a workover vessel. Alternatively, where the well has TFL capability, killing or setting of plugs can be done from the platform prior to arrival of the workover vessel, provided the flowlines are in good condition. Moreover in the insert tree configuration the 90 deg. flowloads may be removed, leaving the master valve block installed, which allows a riser reconnection with valves in place.

Failure of any of the swabvalves does not directly affect well productivity and may well go unnoticed until future re-entry.

Wye Spool

The Wye spool is at the junction of the vertical access and flowline access to the tree. A diverter is fitted in the Wye spool to provide TFL access via the flowline.

As a result of favourable experience with TFL techniques and increased capabilities, e.g. sand-washing, confidence in the system has grown. Therefore routine wireline work in full TFL completions will not be required.

The Wye spool and its diverter can be eliminated by connecting the flowlines to the top of the tree. However, the flowloops must be removable independently from the master valve block in case the flowlines are damaged.

Annulus Cross-over Valve

The normal U.W. tree is usually provided with an access to the annulus of the well through at least one of the flowlines. The annulus cross-over valve is fitted to prevent pressure in the TFL circulating lines from passing to the annulus via a unidirectional annulus master valve. A bi-directional annulus master valve will eliminate the requirement for an annulus cross-over valve in this case, i.e. without a dedicated annulus line to the platform.

Master Valves

On surface completed wells, dual master valves are mandatory in some areas. The U.S.G.S. O.C.S. order No. 6, August 1969 states that two master valves shall be installed on the tubing in wells with a surface pressure in excess of 5,000 psi. In the U.K. D.T.I. proposals for offshore installations call for two master valves to be installed on tubing in wells. No regulations exist yet for underwater wells.

Based on the operating philosophy stated at the beginning and taking into consideration that completions can be equipped with a tubing retrievable surface controlled sub-surface safety valve (TRSCSSV), it may be argued that the TRSCSSV could be considered as a back-up master valve and that, therefore, a single master valve would be adequate.

Wing Valves

On a conventional surface X-mas tree the wing valves are the primary working valves used for all routine operations and initial shut-ins of the well. On an underwater tree all these routine operations are done at the platform and the X-mas tree valves are only operated infrequently thus being subjected to little wear. Due to the above, wing valves are not necessary on an underwater satellite tree from which the vertical access through swabvalves, etc. is eliminated.

Cross-over Valve

This valve provides three functions:

1. TFL circulation through the flowlines with the master valves closed until the tools reach the neighbourhood of the X-mas tree. The down-hole circulating path is used as soon as the TFL tools have entered the tubing string.

2. Production of one tubing string through two flowlines or commingling production of two tubing strings through both flowlines.

3. Pigging or paraffin cutting in the flowlines.

The cross-over valve is a necessary part of the X-mas tree valving.

Gaslift and Chemical Injection Valves

These are not shown in Fig. 1 as dependent on the completion these may or may not be required, but if they are, it would necessitate additional valves to the basic ones shown in Fig. 2.

Sub-surface Safety Valve

It is Shell policy that this item should be installed in all offshore wells including U.W.C.'s for reasons of good safe operating practice.

Conclusion

The valve configuration evolved from this analysis and shown in Fig. 2 is the minimum requirement for an underwater X-mas tree, which in no way jeopardizes the overall safety of the system and in fact should give improved reliability due to the smaller number of components.

The valve configuration shown in Fig. 2 is equally applicable to an above mudline X-mas tree as it is to a below mudline X-mas tree, although technical implications can make this configuration unattractive for an above mudline tree.

This philosophy was adopted for the insert tree, as it was considered that the master valve package and possibly also the cross-over valve could be installed inside a standard well conductor.

Based on favourable experiences with TFL work in Shell's operations, it appears that vertical re-entry for wireline purposes is not required. However it must be possible to establish vertical communication with all three bores (3" x 3" x 2") after the flowloops are disconnected from the master valve and pulled to surface.

The basic configuration of the SIPM proposed insert tree is given in Fig. 3.

Cameron Insert Tree (Fig. 4)

The Cameron Insert Tree has a stack-up height from the 30" housing to the top of the guideposts, of only 10.81 ft., whereas of previous underwater trees of comparable capability, the stack-up height has been as much as 35 ft.

This offers the possibility to cover the entire exposed assembly with a low profile protective dome structure. It would thus afford protection of the equipment from trawler boards, ships' anchor lines, and falling objects. Additionally the master valve block and well-head are located in excess of twenty feet below the mudline. However should a catastrophic occurrence take place and the entire upper tree structure be destroyed, the well will still be failsafe closed and prevent any possibility of a blow-out.

The setting depth below the mudline can be specified by the operating company as part of the design considerations, at any practical depth required.

The insert tree consists of the following major components from bottom to top:

1. Tubing hanger

2. Tree connector

3. Master valve section

4. Intermediate connector

5. Lower guide structure and upper tree structure

6. Outboard hub and running frame

7. Cross-over test cap

These components and their operations are described in further detail below.

Tubing Hanger (Fig. 5)

The solid block tubing hanger and tubing hanger running tool have been designed to reduce the installation roundtrip to one.

It requires only hydraulic pressure to lock. There is only one seal in the entire tubing hanger (ref. Fig. 5). This seals between the body of the tubing hanger and the 9 5/8" casing seal assembly. It is identical to that used in the weight set seal assembly system and is bi-directional. Thus, a pressure test from above will guarantee the seal's integrity.

Before the tubing hanger is landed, the seal element is in the relaxed or de-energized position. In this position, it is protected from possible damage by the back-up ring from below and the junk ring from above.

To set the tubing hanger, once it is landed, force from the tubing hanger running tool is exerted on top of the actuator ring pushing it down. As pressure builds up, the shear pins in the back-up ring are sheared which then allows all components to move down. At this time the seal element is pushed out over the energizing shoulder squeezing the seal between the body of the 9 5/8" seal assembly and the body of the tubing hanger.

Concurrently the tubing hanger lock ring is forced out into the lock groove in the 9 5/8" seal assembly by the locking taper on the actuator ring.

The tubing hanger is locked in place by the lock ring moving under the snap ring. To release the tubing hanger the snap ring is retracted by the running/retrieving tool which releases the tubing hanger lock ring freeing the tubing hanger from the seal assembly.

The body of the tubing hanger is prepared for a 3" x 3" x 2" completion with four controlline preps.

The orientation key will be automatically guided into the groove in an orienting bushing, installed prior to installation of the completion strings.

The hanger is provided with profiles for Otis well-head plugs.

Tree Connector (Fig. 6)

The tree connector is one of four practically identical connectors, i.e. 13 5/8" tie back connector, tree connector, the intermediate connector and the manifold connector. Additionally a 19¼" connector is provided with a 19¼" tie back spool to the 20¾" BOP stack.

The only difference between the four 13 5/8" slim line connectors is the addition of a specially tailored upper body insert, incorporating the three bores, in the tree- and intermediate connectors.

There is one hydraulic lock mechanism and three unlocking mechanisms on this connector. The first is the primary hydraulic release via the middle hydraulic port in the connector body. When pressure is applied the actuator ring moves up opening the locking fingers. Should this fail, a secondary hydraulic release is available in the form of the floating piston. If hydraulics will not function a third release mechanism is available in the form of a mechanical override. The studs shown on the locking/unlocking sleeve are connected to tie-rods, which extend above well-head at seabed level, where pull from the rig can be exerted on a manual override fishing neck.

The well-head connector has a lock groove cut on the ID of the body for use with an orienting key. The body insert incorporates an adapter flange which bolts to the top of the connector. Included with this flange are tubing hanger stab subs for 3" x 3" x 2" x 1" x 1" x 1" x 1" bores. The stab sub retainer is keyed to the connector body. The retainer also incorporates an orienting slot, for the tubing hanger orienting key.

Master Valve Section (Fig. 7)

The insert tree master valves incorporate a downstream sealing, floating plug ball valve. A 90 deg. rotation will move the ball from the open to the closed position.

The required torque is dependent on the pressure differential across the ball for a given size. This torque is provided by a hydraulically powered rack and pinion operator.

The master valve section containing 3" x 3" x 2" valves is a solid block, with an overall length of 99" and a total weight of 5,500 lbs. The complete assembly including the hydraulic operators will fit inside 30" casing (28" ID). The valve and operator are rated for 5,000 psi working pressure.

A schematic drawing of the operator is given in Fig. 8.

Primary closing force is supplied by belleville springs; however provision has also been made for well pressure to assist in closing by acting on the tailrod of the operator.

At well pressures below hydrostatic pressure the floating tailrod pressure balances the operator and the closing torque is provided by the spring cartridge only.

Intermediate Connector

The intermediate connector is bolted directly to the bottom of the flowline extension spool and is identical to the well-head connector apart from the increased number of stab subs: two 3", one 2" and eleven 1 3/8" stabs.

Lower Guide Structure and Upper Tree Structure (Figs. 9 and 10)

The lower guide structure also referred to as outboard hub guide frame, is run latched to the upper guide structure (= inboard hub guide frame) with hydraulic latching cylinders. The upper tree structure incorporates the flowline loops, intermediate connector and wing valve block assembly.

The lower structure serves several purposes, the most important of which is to place the two flowline connector outboard posts. After locking the intermediate connector to the master valve block, the lower guide structure is spaced out by means of 4 hydraulic jacks and locked to the permanent guide base by remote control.

Should it become necessary to retrieve the upper tree structure at a later date, the lock between the two structures is released, and the intermediate S.L. connector unlocked and the upper guide structure is pulled. This leaves the lower structure with the outboard hub in location for reconnection at a later date.

Outboard Hub and Running Frame (Flowline Connector)

This component is very similar to that used in an earlier U.W. development project offshore Brunei. The outboard hub facilitates a vertical initiation, first end connection of the flowlines to the tree. The two guide funnels will locate on the outboard posts of the lower guide structure and the outboard hub is locked down to the tree by means of two latches, which snap into locking grooves cut into the tops of the posts. As flowline laying progresses, the outboard hub assembly will swivel to the horizontal position. A horizontal retractable hydraulic connector provides access to the flowline connector control pod for final remote connection of the inboard and outboard hub. After this connection is made, communication with the tree control pod is established through the hubs. The flowline connector hydraulic hoses become a back-up for the tree control hydraulic hoses after the flowline connection is made.

Cross-over Test Cap

The cross-over test cap facilitates testing of the flowlines prior to initiation of the flowline connecting sequence from the platform. It also facilitates identification of the lines with TFL tools.

DOWNHOLE COMPLETION (Fig. 11)

The downhole completion will incorporate dual 3½" x 3½" tubing strings with an H-member at the bottom of the well and an emergency side pocket H-member above the top packer.

Incorporated in the downhole safety system are two tubing retrievable safety valves with lock open sleeves and separate TFL safety valve sliding side doors.

If necessary the tubing retrievable safety valve can be replaced without disturbing the packers. This is facilitated by the installation of the dual locator head (or detachable packer head) above the top packer.

COMPONENT TESTING

Experience has shown that prior to land testing and installation offshore it is required and in some cases essential, to test various critical components under simulated deep sea conditions in a pressure vessel.

Also component testing provides an opportunity to compare products of various manufacturers under identical test conditions.

The components tested to date are listed in the Appendix.

ACCEPTANCE TESTING

Both the downhole 3" TFL completion equipment as well as the well-head and tree equipment including running tools, etc. are tested under observation by Shell engineers at the suppliers premises prior to acceptance by Shell.

LAND TESTING

A special well has been drilled onshore Brunei which accommodates the insert tree completion and well-head. All the equipment used offshore is installed in this well. This allows: the equipment to be checked thoroughly under actual operating conditions which already has proved to be very valuable, to verify installation procedures and it permits operating personnel to familiarise themselves with the new equipment in an actual working situation.

OFFSHORE COMPLETION

The well will be drilled and completed in the South China Sea offshore Brunei during August/September 1980.

APPENDIX

Component Testing

The following components have been tested at the Koninklijke Shell Exploration and Production Laboratory in Rijswijk, Holland.

1. Cameron Insert Tree ball valve with hydraulic operator.

2. Cameron retractable hydraulic stab.

3. Various control system components of different manufacturers.

4. Otis tubing retrievable safety valves.

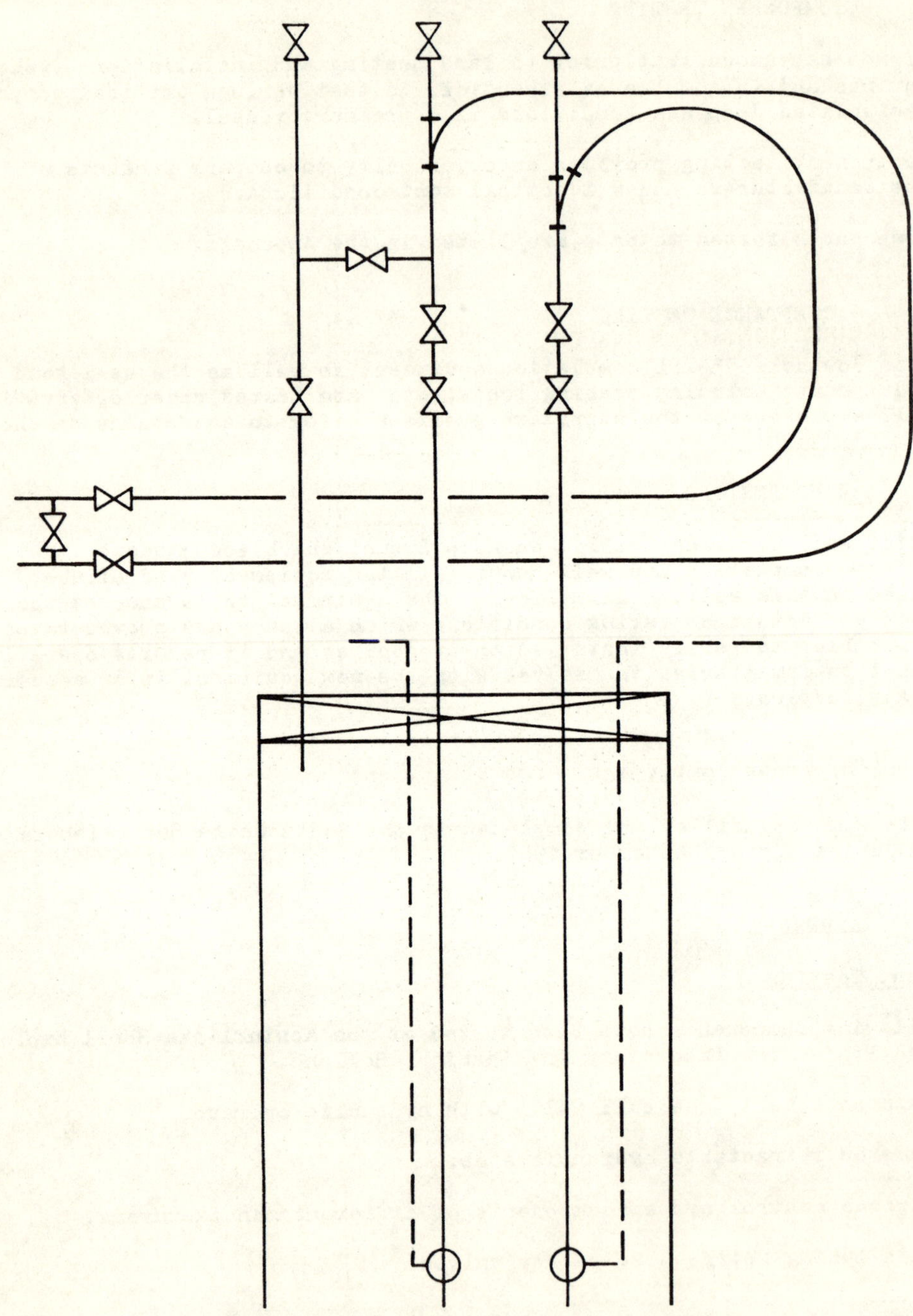

UW TREE: MAXIMUM VALVE REQUIREMENTS

FIG. 1

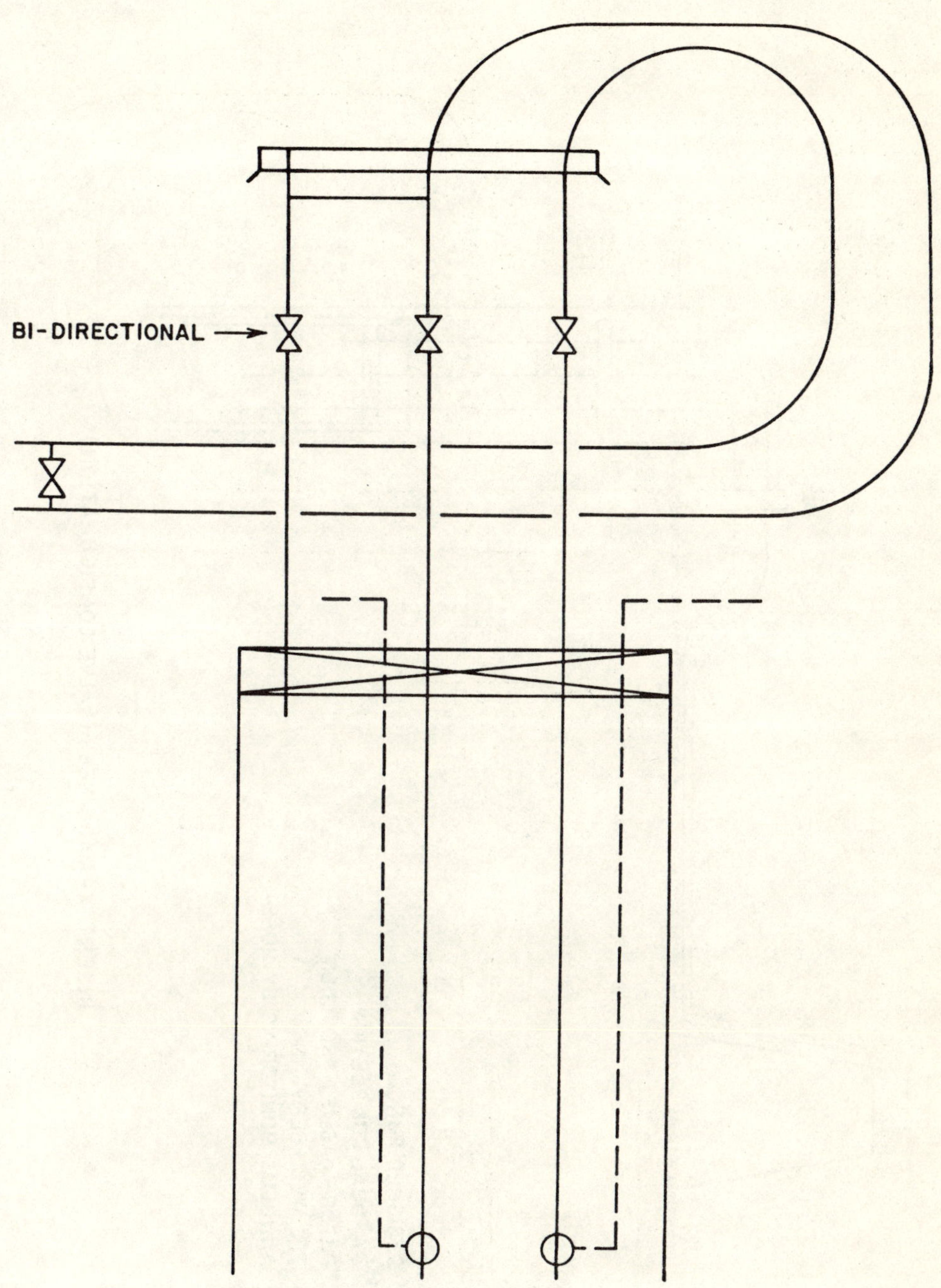

UW TREE: MINIMUM VALVE REQUIREMENTS

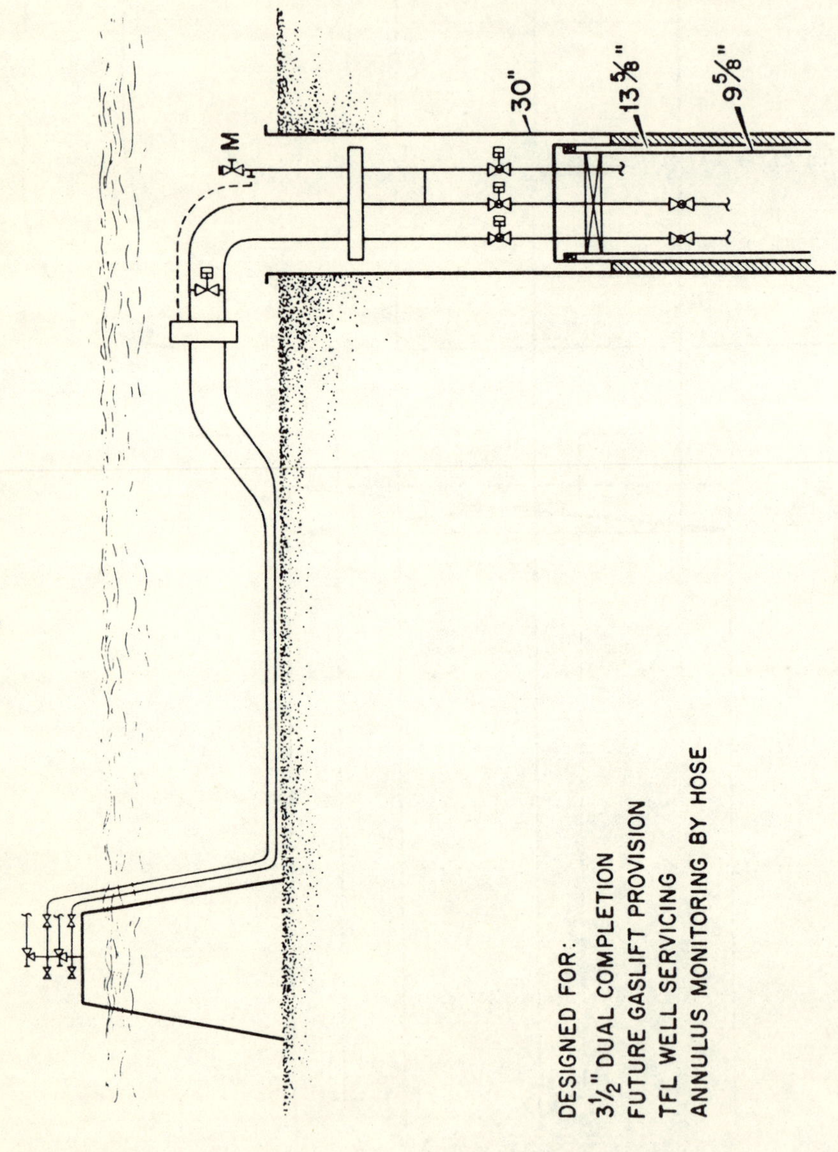

INSERT X-MAS TREE (VALVE CONFIGURATION)

FIG. 3

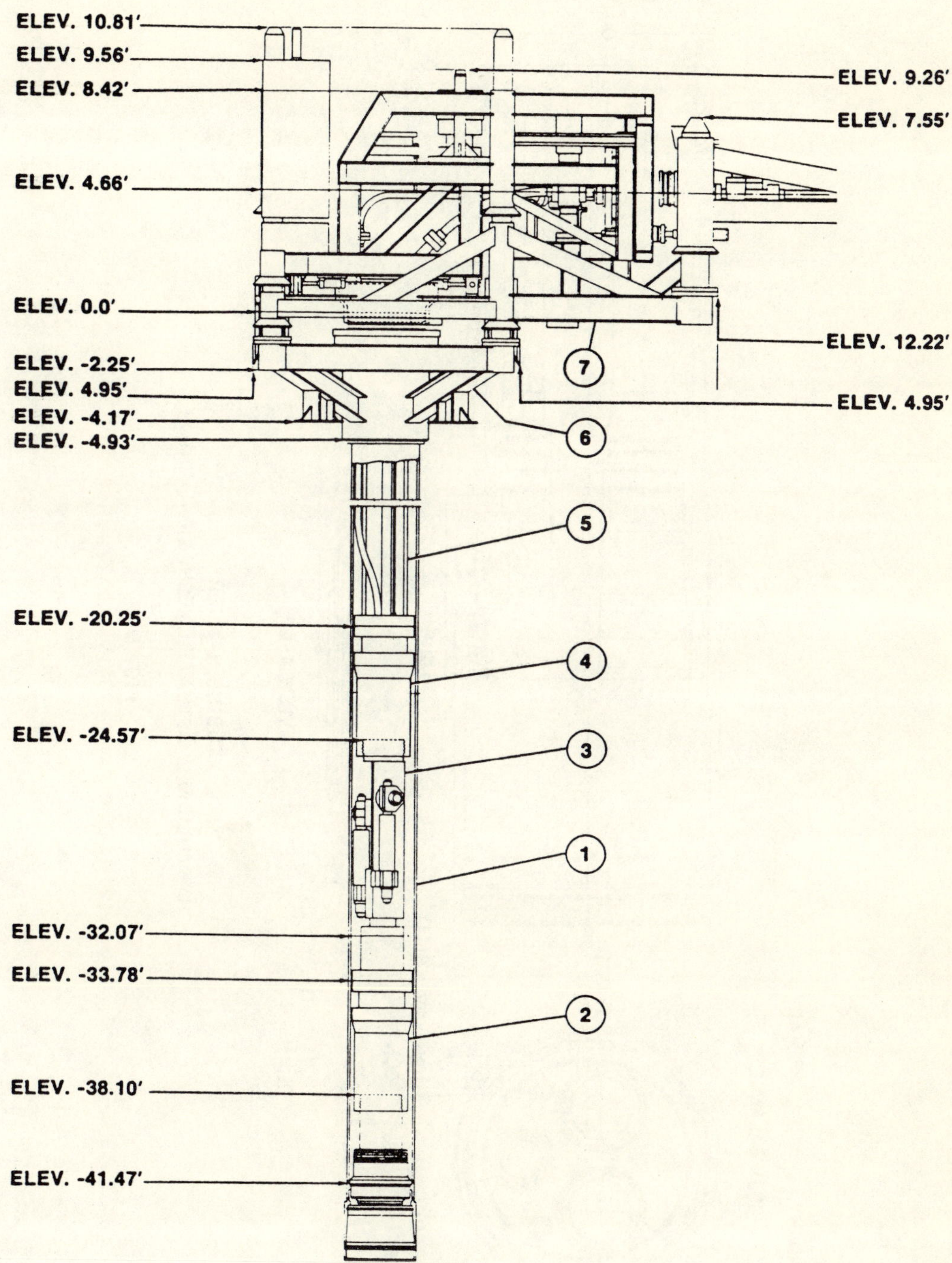

CAISSON STACK UP

FIG. 4

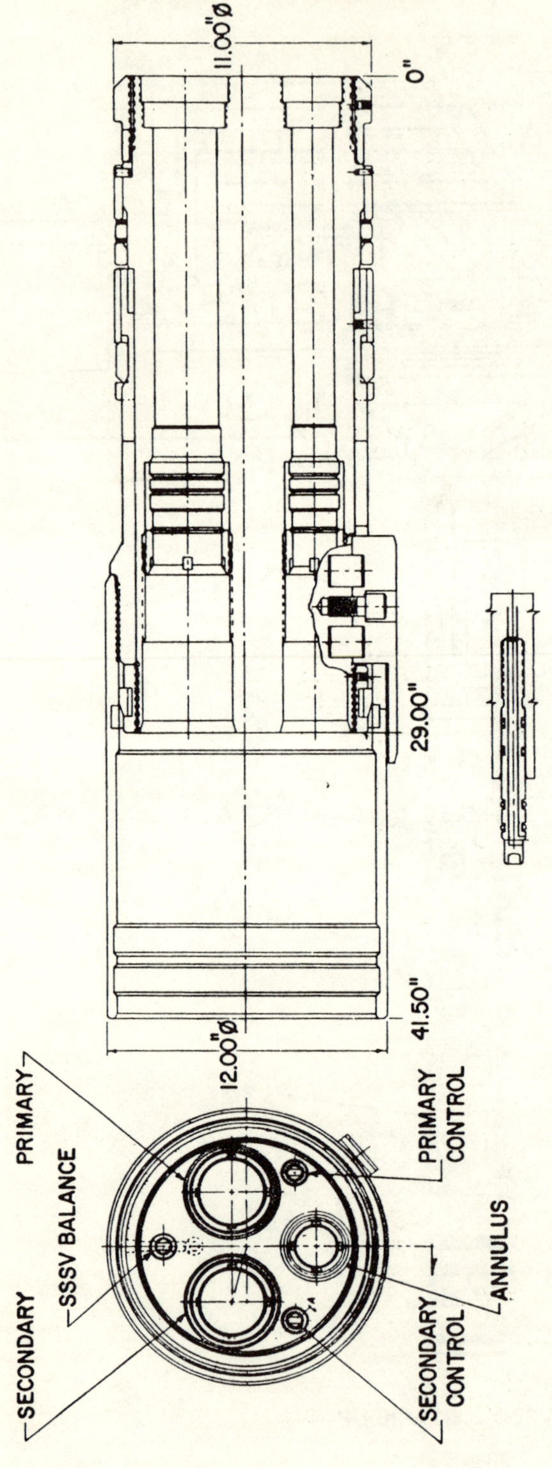

TUBING HANGER
HYRAULICALLY ACTUATED MECHANICAL LATCH

FIG. 5

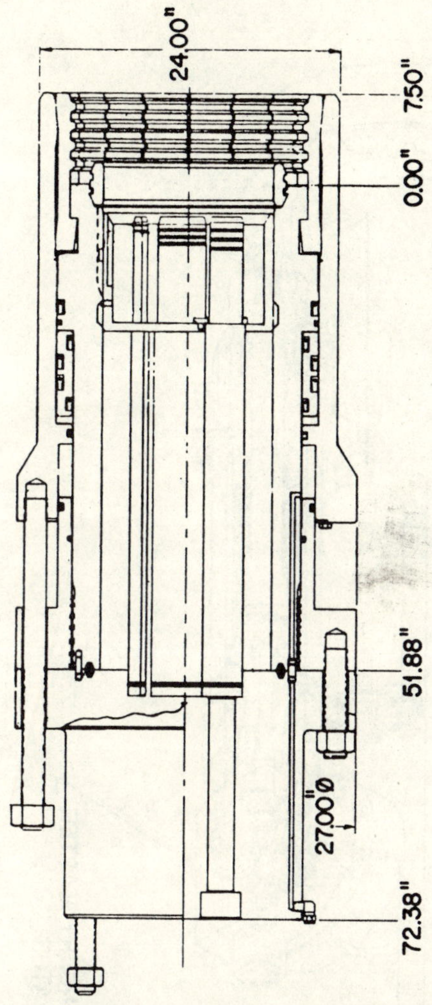

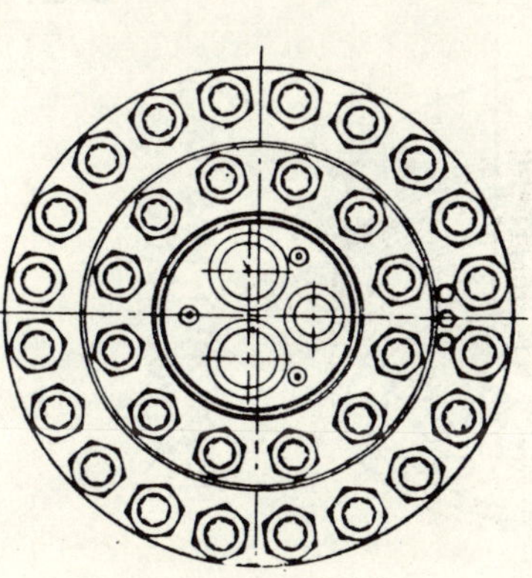

TREE CONNECTOR 13 5/8 10,000 PSI WP

FIG. 6

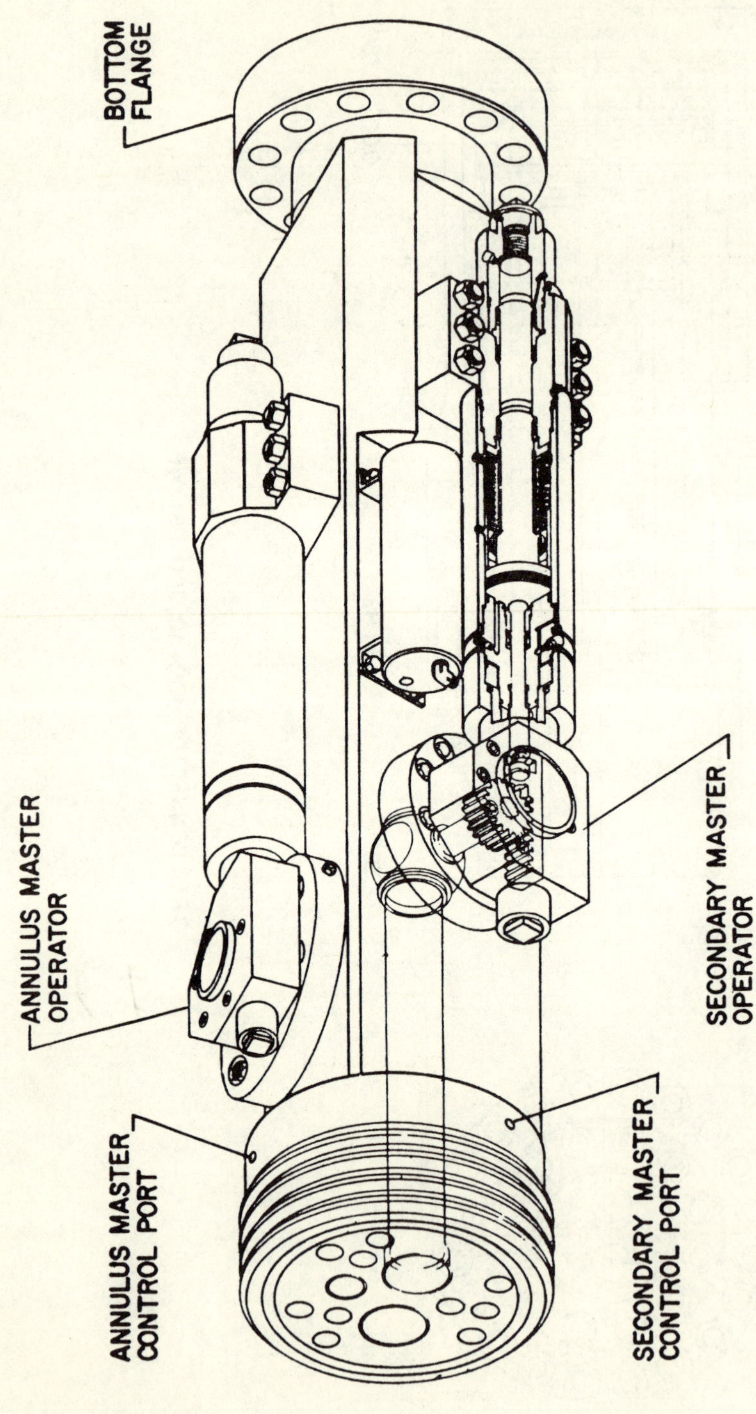

INSERT TREE MASTER SECTION ASSEMBLY

FIG. 7

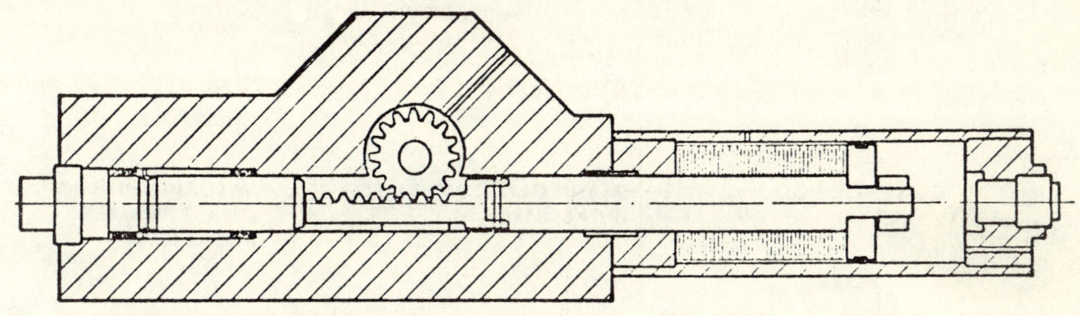

SCHEME OF BALLVALVE OPERATOR

FIG. 8

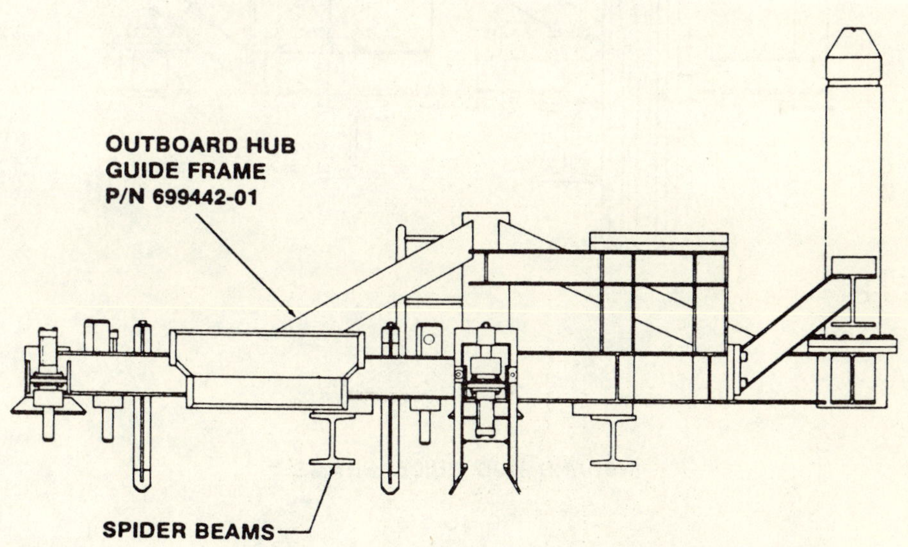

OUTBOARD HUB GUIDE FRAME

FIG. 9

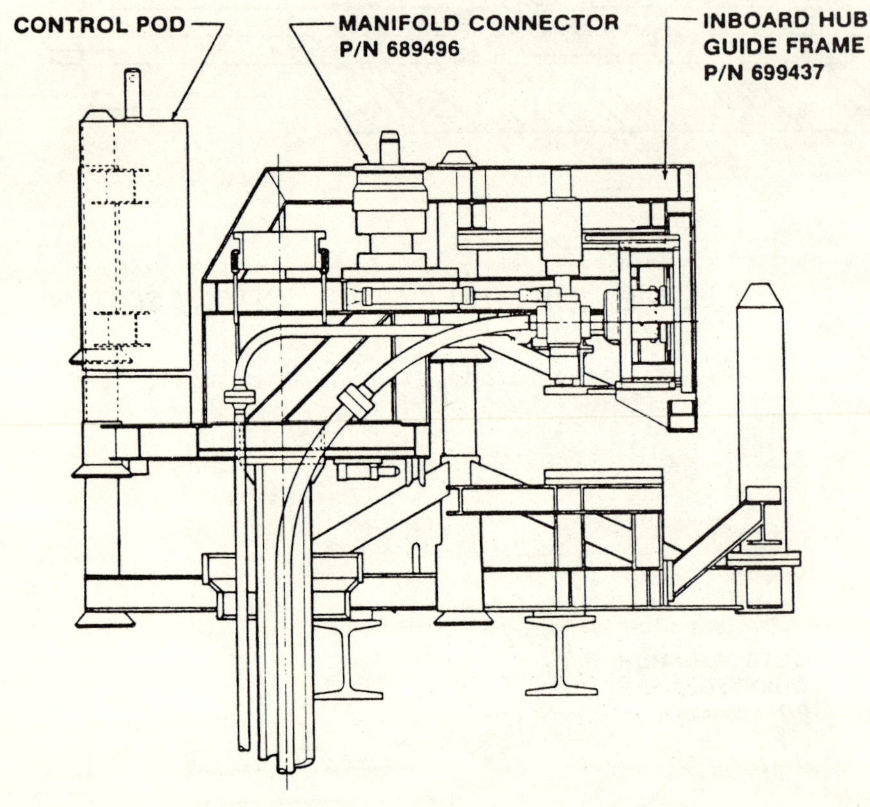

INBOARD HUB GUIDE FRAME

FIG. 10

INSERT TREE COMPLETION DIAGRAM
S.S.
P.S.
DUAL CONTROLLINE
T.R.B.V.
TFL. BALLVALVE NIPPLE SLEEVE
5 GASLIFT MANDRELS/ STRING (see below)
TELESCOPIC JOINTS
ACTIVATOR NIPPLE
S/P. G/L. MANDREL + GLV.
LOCATOR NIPPLE (N)
S.S.D.
SIDE POCKET UPPER 'H' MEMBER LOCATOR NIPPLE
DETACHABLE PACKER-HEAD
RDH. PACKER
LOCATOR NIPPLE
BYPASS S.S.D.
BLAST JOINTS
BYPASS S.S.D.
RDH. PACKER
BLAST JOINTS
BYPASS S.S.D.
RDH. PACKER
BYPASS S.S.D.
LOWER 'H' MEMBER
KN. NIPPLE

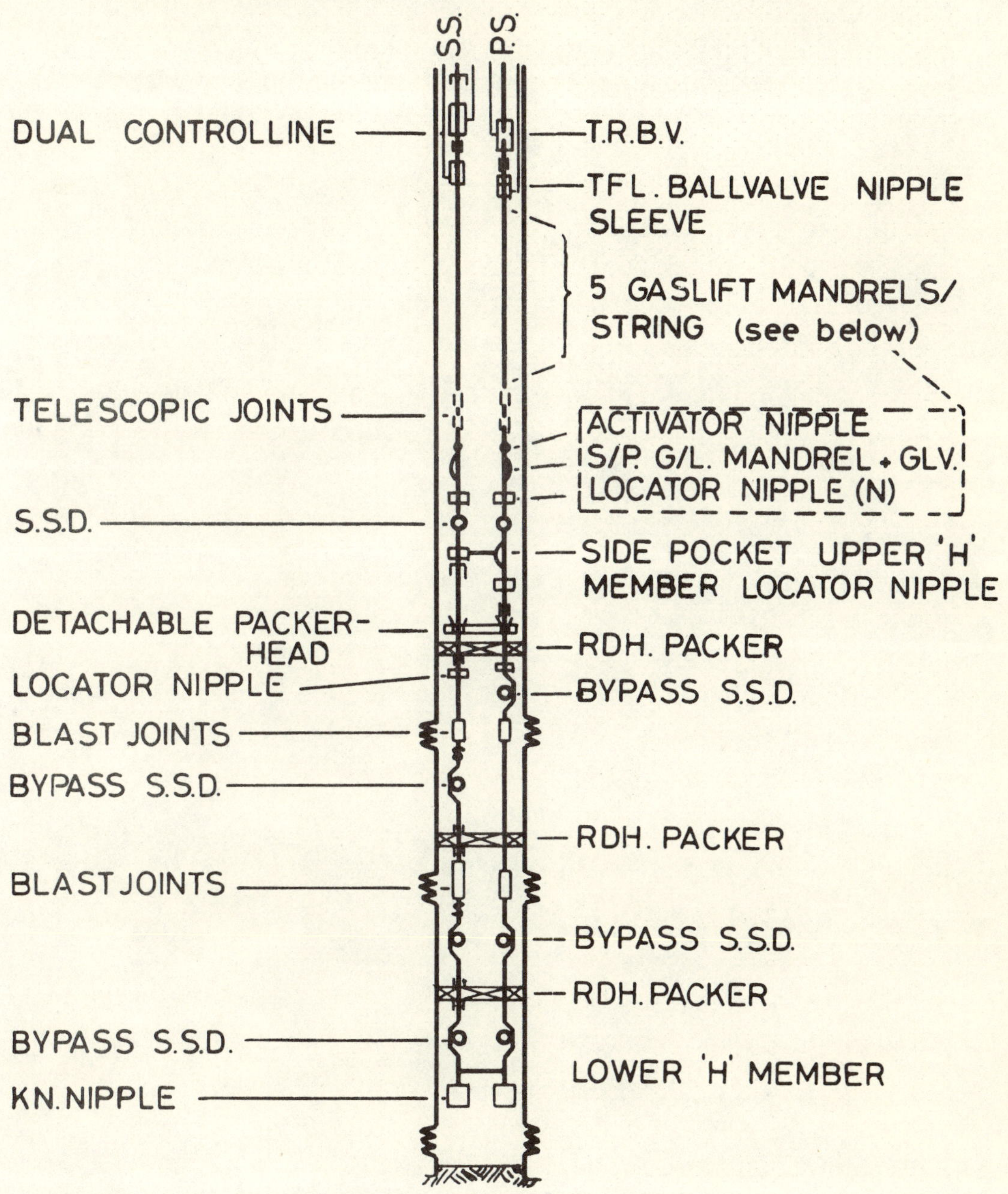

FIG. 11

ABOUT THE ONE-ATMOSPHERE MANIFOLD CENTER

J. G. English

CanOcean Resources Ltd., New Westminster, BC, Canada

ABSTRACT

The first, dry one-atmosphere subsea manifold was installed in the Gulf of Mexico in 1975. A second, larger manifold was installed in the Garoupa field offshore Brazil in 1978.

This paper describes these manifolds, their functions, design considerations, fabrication methods and standards, internal components and sub-systems, and an installation technique. In comparison to other subsea completions, the dry, one-atmosphere manifold offers a proven flow line connection method, wide flexibility in the selection of internal components and sub-systems, and the use of skilled, but non-diver trained, personnel for installation and servicing of internal equipment.

Although the manifolds in use today are part of satellite completion schemes, the concept is readily adaptable to the clustered well and template schemes now in vogue. Typical possibilities are briefly reviewed as they relate to the North Sea area.

This paper summarizes the state-of-the-art of one-atmosphere manifold design, the present and potential uses and the inherent benefits.

KEY WORDS

Garoupa; manifold; template; dry pull-in; one-atmosphere; keel-haul; bullnose.

HISTORY

The one-atmosphere concept was developed by CanOcean Resources Ltd.[1] in order to provide a method of utilizing proven oil patch technology on the ocean floor in an environment where experienced oil patch technicians could be utilized for service work. Basically this meant encapsulating the equipment in a steel chamber and providing a breathable one-atmosphere environment inside for service personnel.

---

[1] (formerly Lockheed Petroleum Services Ltd., and no longer associated with the Lockheed Corporation)

CHAMBERS	INSTALLATION DATE	PRODUCTION LINE	INTERNAL EQUIPMENT	LOCATION/ (OWNER)	WATER DEPTH
WHC - 1 Vertical	JULY 1972	2-in. TFL	Triple Segmented Tree Hydraulic Sequencing Control System	Gulf of Mexico (Shell)	375-ft (114-m)
WHC - 2 Horizontal	SEPT. 1975	2-in. TFL	Triple Block Tree (Dual Master) Hydraulic Sequencing Control System	Gulf of Mexico (Shell)	245-ft (73.5-m)
WHC - 1 Horizontal	OCT. 1975	2-in. TFL	Triple Block Tree (Dual Master) Electro-hydraulic Multiplex Control replaced by Hydraulic Sequencing	Gulf of Mexico (Union)	210-ft (63-m)
MC - 1 (Three Wells) Input	AUG. 1975	2-in. inputs 4-in. outputs TFL	Electro-hydraulic Control System Three-phase Test Separator 2-in. Multi-Orifice Chokes 4-in. Production Header	Gulf of Mexico (Shell)	245-ft (73.5-m)
WHC - 1 Horizontal	SEPT. 1976	2-in. TFL	Triple Block Tree (Single Master) Hydraulic Sequencing Control System	Gulf of Mexico (Tenneco)	110-ft (33-m)
WHC - 1 Horizontal	FEB. 1977	2-in. TFL	Triple Block Tree (Single Master) Hydraulic Sequencing Control System	Gulf of Mexico (Shell)	205-ft (61.5-m)
MC - 1 (Nine Wells) Input	JAN. 1978	4-in. inputs 2 x 10-in. outputs	Electro-hydraulic Multiplex Control System with Mini-Computer Processor Multi-orifice Choke 5-in. Test Header 12-in. Production Header	Brazil (Petrobras)	420-ft (128-m)
WHC - 8 Horizontal	OCT., NOV., DEC. 1977 JAN., FEB., APR., OCT., 1978 JAN. 1979	4-in.	Double Block Tree (Dual Master) Electro-hydraulic Multiplex Control System with Hydraulic Override	Brazil (Petrobras)	400 - 535-ft (122 - 163-m)
PRC - 6 FRC - 8	AUG. 1976	2 @ 12-in. 4 @ 16-in.	Pup Joint - Flow Line to Riser	North Sea Thistle A Platform (BNOC)	530-ft (161-m)

Fig. 1. One-atmosphere chambers installed to March 1980

The first wellhead and manifold chambers were built as part of a joint development program with a consortium of oil companies, with Shell as the operator. These were followed by installations for other oil companies, all in the Gulf of Mexico (Refer to Fig. 1). Petrobras then placed an order for the largest subsea one-atmosphere production system utilizing ten chambers.

This paper discusses the largest of the one-atmosphere chambers - the Shell and Garoupa Manifold Centers.

A manifold is designed to gather and commingle the hydrocarbon flow from several wells and to transfer it to a production riser or pipeline to shore.

The main advantage offered by any subsea completion, whether wet or dry, is the opportunity for early production.

Early production is possible because the drilling can begin while the wellhead and manifold chambers are still being fabricated. Wellhead chambers are installed as required and within 24 months a manifold can be ready for installation. Shortly thereafter, flow lines are connected from wellheads to manifold and to the production riser. Thus production can begin before or at about the same time as a conventional North Sea platform can be installed ready to begin drilling.

The one-atmosphere manifold field development using satellite wells avoids the expense and risk of directional drilling. Satellite wells also allow marginal fields to be produced to existing platforms or Floating Production Facilities (FPF) in adjacent areas without the need for a dedicated platform.

The Garoupa manifold subsea production system was selected for both these reasons: early production and satellite wells. Figure 2 illustrates the Garoupa subsea system 155 miles

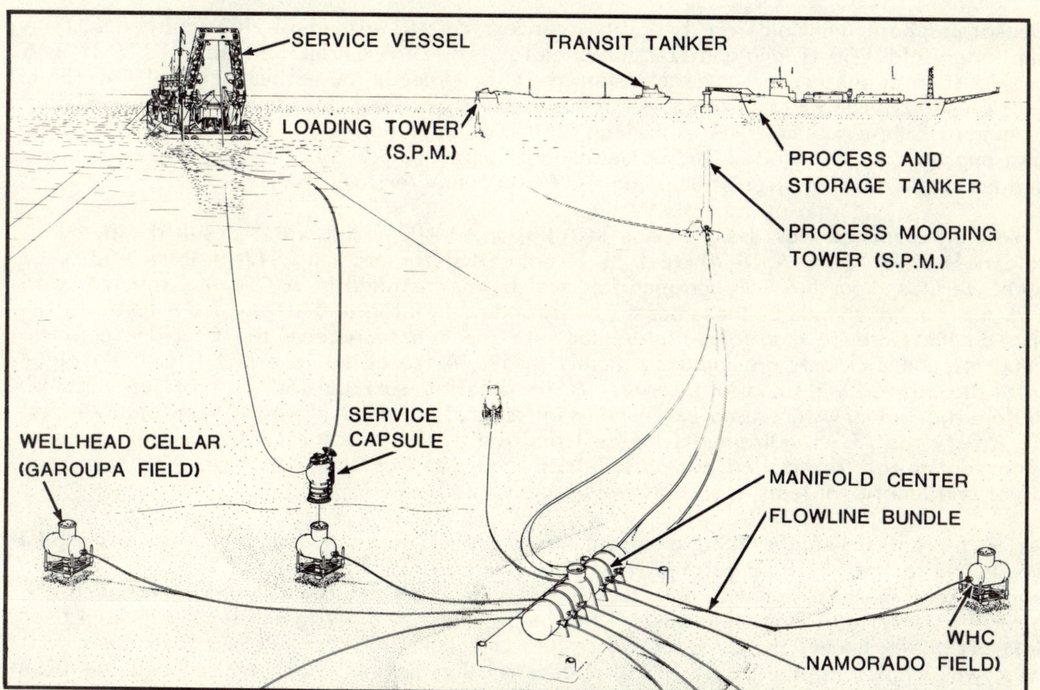

Fig. 2 Subsea Production System

offshore from Rio de Janeiro, Brazil. Production is gathered from wells in the Garoupa and adjacent Namorado fields by a nine-well manifold in the Garoupa field. Two production lines transfer the flow to a single point moor (SPM) production riser with a converted tanker permanently attached. After processing and storage on the converted tanker, production is routed subsea to a second SPM for loading onto shuttle tankers. In the spring of this year, a permanent platform will be installed and the floating process tanker and SPM will be recovered. The manifold production will be re-routed to the platform.

The Shell manifold was the prototype for the Garoupa manifold. Its primary purpose was to test the one-atmosphere manifold concept for future deepwater use. Designed to gather production from three satellite wells in the Eugene Island block offshore Louisiana, it produced to an existing platform. Two wells were completed with one-atmosphere chambers and production from a third existing well on the platform was re-routed from the platform to the manifold. The area was selected to provide easy access by divers to assist in solving any unexpected developmental problems.

Because the one-atmosphere concept is not depth limited, it provides distinct advantages in deepwater applications where diver accessibility is limited or non-existent. The internal components and service personnel are never exposed to water or to pressures above that at sea level regardless of the water depth. This simplifies component selection, allows the use of specialist technicians (electronic, hydraulic, downhole) during start-up debugging and any future maintenance work, and provides great flexibility in subsea operations.

Basically the Shell and Garoupa one-atmosphere manifolds are the same in function, internal components and basic structure. Major differences occur only in size, assembly techniques, and installation methods (Refer to Fig. 3). Where there are major differences, they are identified in the appropriate following sections.

STRUCTURE AND ASSEMBLY

One-atmosphere manifolds are basically cylinders capped with semi-elliptical heads. The Shell manifold is 30-ft long and 12-ft in diameter, while the Garoupa manifold is 80-ft long and 15-ft in diameter. Structural design meets or exceeds the requirements of the ASME Section VIII Division 2 code for Boilers and Pressure Vessels. Both manifolds are certified as manned submersibles by the American Bureau of Shipping (ABS). The Shell manifold was also approved by the United States Geological Survey (U.S.G.S.). These requirements are similar to Lloyd's Registrar of Shipping and Det Norske Veritas (DNV).

The Shell manifold was designed for installation in 400 ft of water resulting in a hull thickness of 1.1/4-in. A516-70 steel. External belt stiffeners and 1.1/2-in. thick A537 CL I steel were used for the Garoupa manifold which was installed in 388 ft of water (refer to Fig. 4). Funnel shaped bullnose ports (BNP) ranging in internal diameter from 10-in. to 18-in. on the Garoupa manifold, are welded into the hull to receive the flow-line bundles. Nine spare 10-in. ports are included in this array. A truncated cone 7 ft 1 in. I.D. welded into the top of the manifold serves as the mating surface for the service capsule. Automatic welders were used extensively in fabrication of the Garoupa manifold cylinder. To satisfy the ABS requirements for certification of the structural integrity for use as a manned habitat, a detailed computer stress analysis was submitted and ABS witnessed proof and functional tests as well as weld procedures.

After fabrication of the Shell manifold chamber, the internal equipment was assembled inside and tested. In order to improve on the Garoupa manifold delivery time, the internal components were assembled and tested on a separate support structure, dubbed the birdcage (refer to Fig. 5) while the hull was being fabricated. With one semi-elliptical head left off the cylinder, the birdcage was rolled on tracks into the hull and then the second head was welded in place. This parallel construction technique greatly reduced the total fabrication and assembly time of the Garoupa manifold.

Purpose	Proto-type) Concept Evaluation		Early Production, Satellite Wells
Manifold Size	30-ft x 12-ft dia.		80-ft x 15-ft dia.
No. of Wells	3		9
Design Production Rate (BOPD)	7,200		80,000
Installation Date	August 1975		January 1978
Installation Method	Free Descent		Keel-haul
Assembly	pre-fabricated		birdcage
Wellhead Production Bundles	2 - 2-in.	Production	1 - 4-in.
	1 - 2-in.	Annulus	1 - 2-in.
	1 - 1.1/4-in.	Hydraulic	1 - 1-in.
	----	Electrical	1
Lines to Platform/ or Process Tanker	1 - 4-in.	Production	2 - 10-in.
	2 - 2-in.	Annulus	1 - 2.1/2-in.
		Test/Scraper	1 - 4-in.
	1 - 2-in.	Gas Lift	----
	2 - 1.1/4-in.	Hydraulic	2 - 1-in.
	1 - 1.1/4-in.	Spares	----
	1 - 2-in.	Spare	----
	1	Electrical	1
Well Test	Platform or Internal Test Separator		Process Tanker
Piping	TFL		Non-TFL

Fig. 3 Manifold Characteristics

Fig. 4 Garoupa Manifold

Fig. 5 Birdcage with MC Internals

INTERNAL SYSTEMS

Production

The primary function of the manifold is to gather the hydrocarbon production from the satellite wells and produce to the surface. Both the Shell and Garoupa manifolds have similar piping arrangements.

The Garoupa manifold receives oil through a single Coflexip 4-in. production line from each well and manifolds the combined flow into a 10-in. header to either of two 10-in. Coflexip production lines to the floating process tanker's SPM riser (refer to Figure 2). The flow from any well can be diverted to a separate test header and then to the SPM for individual well testing. The annulus and production lines can be pigged from the wells through separate headers in the manifold to the SPM. Pigs can be launched from the manifold production headers to the surface process facility.

There are three basic differences between the Shell and Garoupa manifold production systems. The Shell manifold has an internal test separator, and its piping is set up to receive Through Flow Line (TFL) tools. Its wells are dual completions with two production lines and one annulus line from each well while Garoupa wells are single completions.

The internal test separator allows well testing to be done subsea. This separator is a vertical cylinder 36 in. I.D. x 136 in. with remotely operated valves and sonic level detectors for phase level indications (refer to Fig. 6).

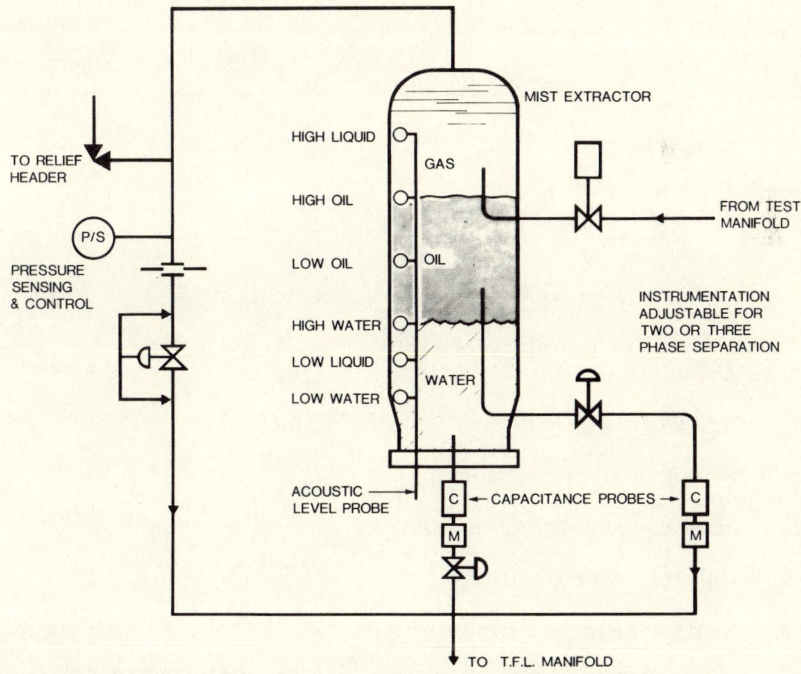

Fig. 6 Manifold Test Separator

Two TFL headers with 60-in. radius pipe bends bypass the chokes and provide access to each Shell well for TFL tools pumped from the platform.

The piping scheme (refer to Fig. 7) for both manifolds is designed to facilitate gas lifting of any individual well with the addition of pre-fabricated spools when needed.

Piping design in both manifolds is in accordance with the Refinery Piping Code B31.3. The Garoupa manifold piping has a design working prssure of 3,600 psig, while the Shell system is designed for 3,000 psig.

In the manifolds, pipe joints are made with Gray-Loc clamp-type connectors. These were

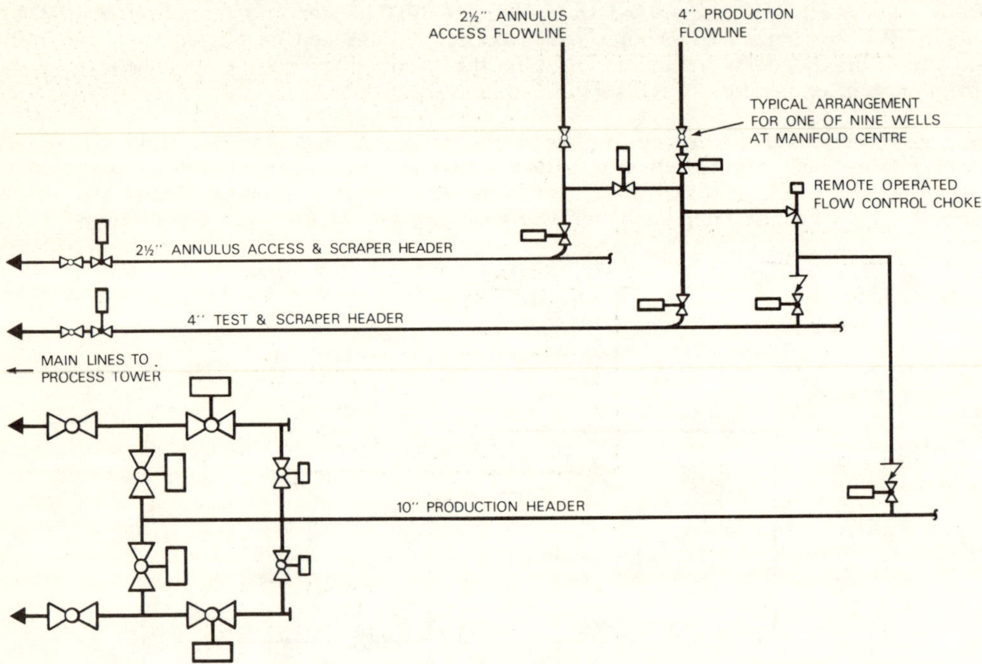

Fig. 7 Garoupa Manifold Production Schematic

selected for reliability, ease of installation and their reduced space requirements, and have proven to be very satisfactory.

The subsea manifold production systems provide:

- collection of satellite well output
- large single (or dual) lines to the surface
- individual well test capability
- pig or scraper operations from the wells
- future well gas lift operations
- TFL capability (Shell)

SUPERVISORY AND HYDRAULIC CONTROL

The Shell and Garoupa manifolds use mini-computer based electro-hydraulic control systems with sequential hydraulic control for back-up (refer to Fig. 8).

All remotely-activated production valves and the chokes in the manifolds have hydraulic actuators. Each manifold has a complete hydraulic package with 7.1/2 hp electric motor,

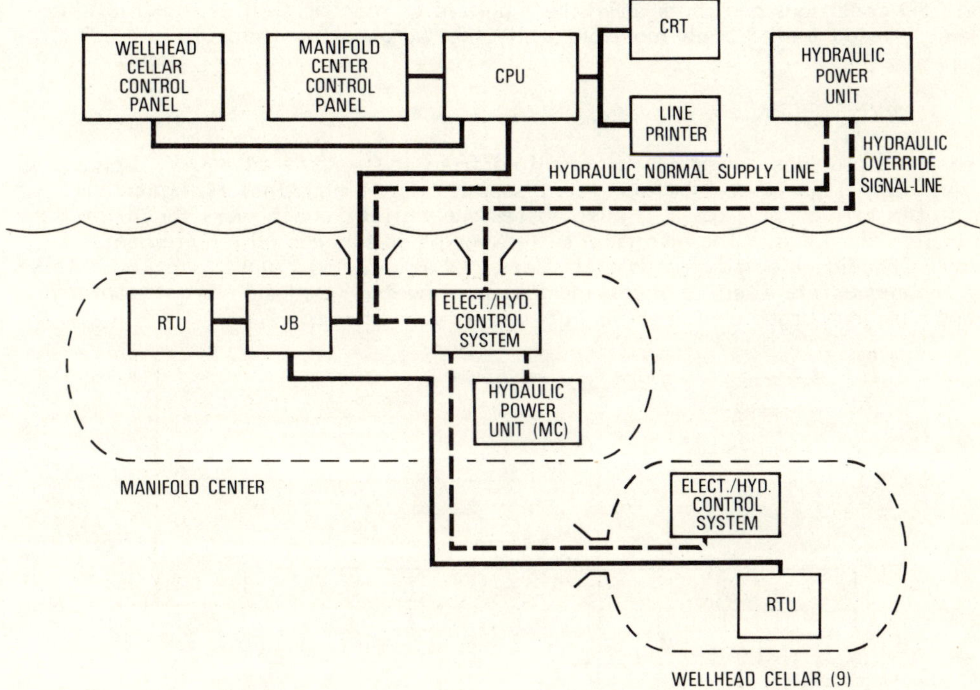

Fig. 8 Control System Layout - Garoupa

pump, bladder-type accumulators and reservoir. Modularized solenoid valve packages are distributed throughout the manifold. The Petrobras manifold hydraulic system also recharges the accumultors in each wellhead chamber. Make-up fluid from the surface facility via a subsea line maintains the hydraulic fluid level in the manifold.

A second hydraulic signal line from the surface to the Petrobras manifold allows production to be maintained in the event of electrical failure by sequential hydraulic control. This sytem has a primary and secondary mode. The primary mode controls only the wells while the secondary mode includes the manifold.

The Shell manifold uses a water based hydraulic fluid while the Garoupa manifold uses mineral oil. Oil offers better lubricity and less deterioration of components but the water based fluid is bio-degradeable and presents no environmental hazard.

The subsea core of the supervisory system is the remote terminal units (RTU). RTUs in the manifold gather valve status indications, pressure and temperature readings, liquid levels, and choke position data. Both digital and analog signals are coded for multiplex transmission via a twisted shielded pair in the subsea cable to the mini-computer in the process control center at the surface. Commands issued from the mini-computer are decoded by the appropriate RTU in the manifold. The Garoupa manifold has over 254 status points, 54 analog points, and 255 command points. At the surface, it incorporates a mimic panel with push button controls for manual control but a CRT and keyboard are available. The mini-computer provides controlled (routine) shutdown (CSD), emergency shutdown (ESD), well test, pigging, choke control, start-up and other programs for automatic operation as well as data logging.

Each manifold has a 28 V.D.C.battery to maintain production in the event of failure in the electrical power or multiplex systems. The battery powers a logic system which detects ESD and CSD conditions and shuts down the manifold or specific well as required. Hard-wired logic is used in the Shell manifold while the Garoupa manifold utilizes a microprocessor.

ELECTRICAL

A subsea cable provides power to the manifold from the surface at 440 V, 3-phase, 60 Hertz. An internal power distribution system supplies the electric motors, lights, fans and DC bus (refer to Fig. 9). The DC bus charges the battery and powers the supervisory control system at 28 VDC. The electrical system meets or exceeds all requirements of the U.S.A. and Canadian electrical codes for Class I Division 2 Hazardous locations. Three primary techniques are used to meet these requirements: explosion proof containers, sealed and nitrogen-purged containers or intrinsically safe devices.

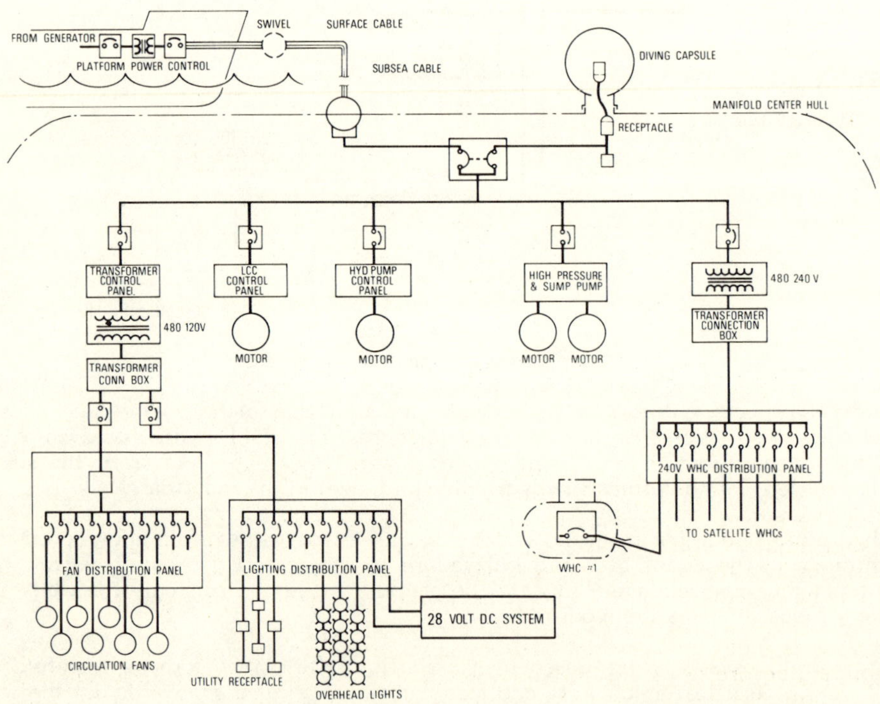

Fig. 9 Garoupa Manifold Electrical System

The electrical system also includes three communication modes for use during manned visits. These systems link the manifold chamber, service capsule, surface support vessel (SSV) and the surface process facility in a communication net. Voice communication is possible through a powered paging system or an independent sound powered telephone system. TV monitors can also be mounted in the manifold to give a video indication of chamber activities and conditions during manned visits.

ENVIRONMENT

The basic characteristic of the one-atmosphere manifold is its ability to continuously provide a clean, dry, inert and low pressure atmosphere for the internal equipment and a similar but breathable atmosphere during manned service re-entries.

Each manifold has a leakage containment system for the removal of liquids and gases (refer to Fig. 10). An 8 g.p.m. 3 hp sump pump charges a 15 hp high pressure pump which

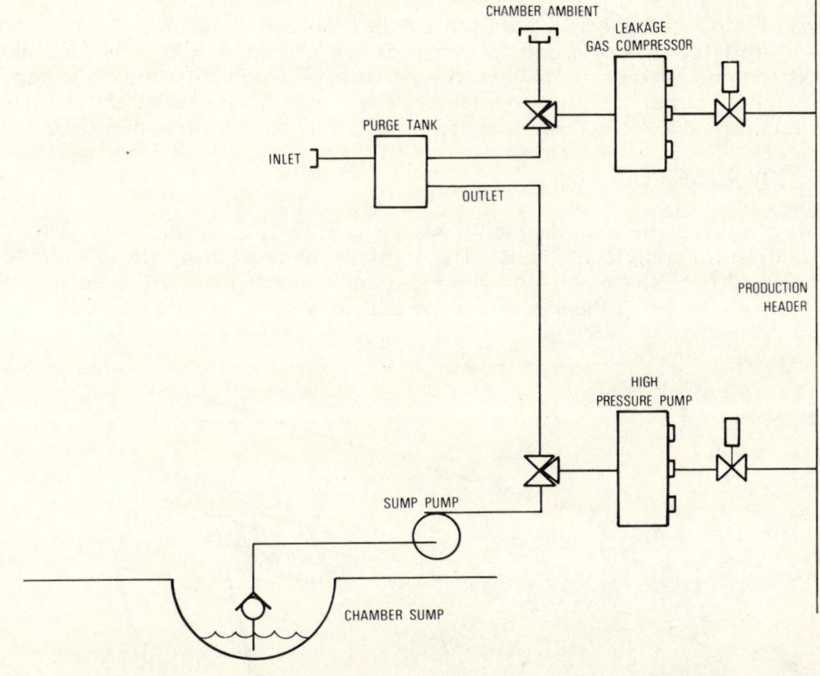

Fig. 10 Leakage Containment System

discharges the excess fluids into the main 10-in. production header at up to 2,000 psi. It the chamber ambient pressure rises above 17 psia, a 15 hp waste gas compressor discharges the excess gas into the same production header. Both these systems function automatically or they may be operated manually from the surface or locally when the chamber is manned.

A purge tank is included in these leakage containment systems to allow service personnel to flush piping before disassembly. This reduces the hydrocarbon contaminants which may get into the atmosphere. Ambient pressure sea water is used to flush the fluids and gases in the pipes into the purge tank which is then emptied by the compressor and high pressure pump.

During manned service visits, technicians breath the chamber's ambient air at 15 psia supplied from the SSV's compressor via an umbilical to the service capsule. The capsule air supply is distributed throughout the manifold and a homogeneous mixture is maintained by fans. The umbilical also provides vacuum exhaust lines from the capsule to the SSV.

The exhaust lines are also used to purge the chamber and to backfill it with nitrogen at the end of each service visit. Thus the chamber is inerted with nitrogen when unmanned.

Service technicians take 45 minute air back packs with full face masks into the chamber on each visit. These are put on if the atmosphere suddenly becomes contaminated with hydrocarbon gases or when the chamber is inerted and only a short duration "screwdriver" task is to be performed.

These masks can be plugged into air stations located along the central aisle of the chamber. These provide a closed loop breathing system - air supply and exhaust -which prevents the unused oxygen from forming an explosive mixture with any hydrocarbons in the ambient chamber atmosphere.

Part of the environmental system is the portable gas analyzers carried into the chamber by the technicians. These are used to monitor oxygen, carbon dioxide and hydrocarbon or lower explosive limit (LEL) levels and to alarm at preset conditions. The LEL detectors are set to first alarm and then to trigger the release of freon into the chamber and to disconnect electrical power to the chamber. The freon will extinguish any fires and prevents a hydrocarbon and oxygen mixture from becoming an explosive mixture.

FLOW LINE PULL-INS

Each of the manifolds is tied to its wellheads by a flow-line bundle and to its surface facility by a number of individual lines. The method of connecting the bundles and the lines to the manifold is called a pull-in. The main components involved in each pull-in are the bullnose, the bullnose port, the seals, and the pull-in winch (refer to Fig. 11).

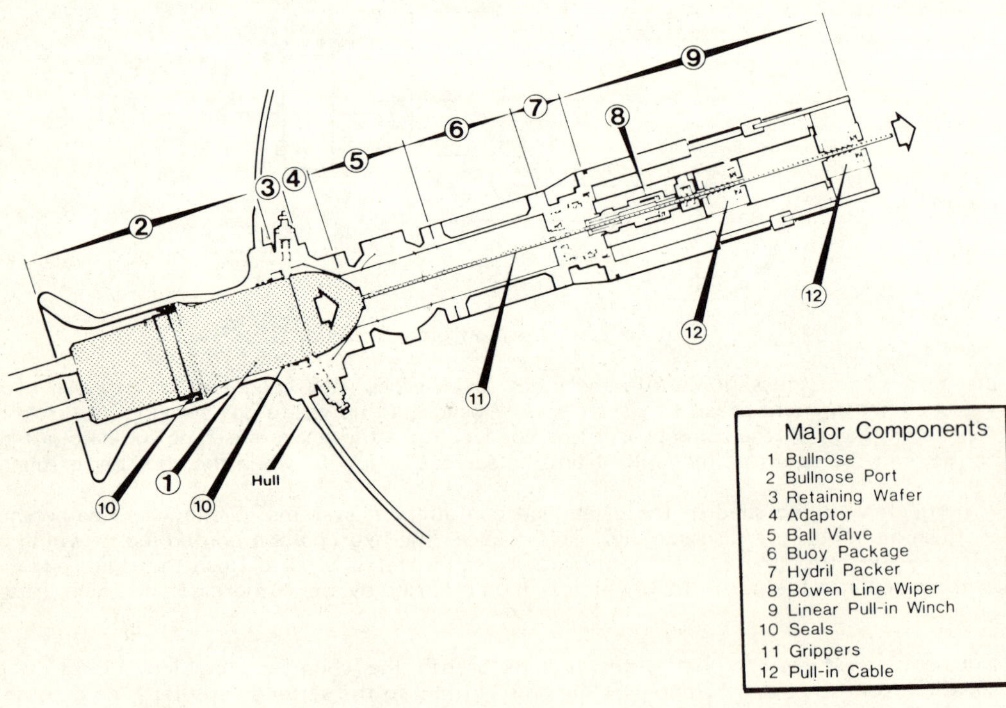

Fig. 11 Pull-in Equipment

Each wellhead flow-line bundle contains the production line or lines, annulus line, hydraulic line, and electrical cable. The Garoupa manifold receives nine bundles, each with a 4-in. production line, 2.1/2-in. annulus line, 1-in. hydraulic line and electrical cable. These bundles all terminate in an 18-in. bullnose. The bullnose is the male counterpart to the mating bullnose port (BNP) on the manifold. This manifold has two separate 18-in. BNPs for the 10-in. production lines to the SPM and separate BNPs for the cable and hydraulic lines and for the test and annulus lines. Each bullnose is pulled in to its predesignated BNP using a patented dry pull-in technique.

The pull-in is initiated by extending a buoy attached to a pilot line out through the chamber BNP. In a lay-away, or first-end connection, this line is used to pull the actual pull-in cable to the surface where it is attached to the bullnose onboard the laybarge. In the manifold, the cable passes through a Bowen hydraulically-actuated wireline stripper and a Hydril annular line wiper. Both of these provide a tight seal against the plastic filled pull-in cable and the Hydril can pack off the bore completely. A linear reciprocating winch ismounted behind the Bowen and Hydril units and pulls in the cable and bullnose (refer to Fig. 12). It provides a steady and constant pulling force and is capable of a 30,000-lb pull. An orientation or king post buoy is attached to the rear of the bullnose to provide gross orientation of the bundle with the manifold piping. Before being permanently latched in place, the service technicians attach a hydraulic orientation tool to the bullnose, which is capable of rotating the bullnose inside the BNP until the bullnose piping is aligned with the internal piping. Once latched in place, and after pressure testing, the bullnose can be hooked up to the manifold piping.

Fig. 12 Linear Pull-in Winch

Lay-to or second-end connections are also executed. Here, any one of three techiques is used to retrieve a cable attached to the bullnose, which is lying on the sea floor. Once retrieved this cable is attached to the cable received from the manifold and the pull-in proceeds.

Depending upon circumstances in the field, either retrieval technique may be used:

1. acoustic buoy on the bullnose brings cable to surface

2. divers directly connect the two cables subsea

3. an RCV (remote controlled vehicle) directly connects the cables subsea.

The first technique creates a lot of slack which takes a long time to pull-in using the linear winch. When divers or an RCV are available, the pull-in time can be significantly reduced. Typical pull-ins encounter loads of 1,200 to 1,800-lb and are completed in four to eight hours. The patented dry pull-in system has been proven successful on both manifolds, on eleven satellite wellheads and on the Thistle "A" platform.

TESTING

Extensive testing of both manifolds occurred before they were installed subsea. All components were tested individually either at the supplier's plant or at the CanOcean Resources Ltd. assembly plant in Vancouver, Canada. Assemblies and sub-assemblies of components were tested as they were assembled. Before shipment an integrated functional and proof test was carried out to ensure that the production, electrical, hydraulic, environmental and supervisory systems performed together.

After the manifolds were mounted onto their bases, a dry land test (DLT) was carried out at the launch site before the tow-out to sea. This involved simulation of actual subsea conditions. A dummy well input to the manifold was used to ensure that the manifold, wellhead and surface equipment interfaced and operated well together. The Garoupa manifold was thoroughly checked during a 12 week DLT in Rio de Janeiro.

After successful completion of their DLTs, the manifolds were towed out for installation.

MANIFOLD BASES AND INSTALLATION

Each of the manifolds was set using a different technique and their bases were designed with the respective technique in mind. The Shell manifold used a free descent method while the Garoupa manifold was keel-hauled and lowered to the sea floor.

In both cases, the manifold base has several functions. It provides: buoyancy and stability during tow-out, a soil-bearing surface, and provides variable buoyancy for the different stages of installation. Typically the base has soft tanks, hard tanks, trim tanks, scour skirts, pile slots and concrete ballast. (Refer to Fig. 13). The soft tanks provide added buoyancy to increase freeboard during tow-out. When these are flooded, the manifold unit remains positively buoyant but is semi-submerged with only the top of the chamber and teacup above water. As the trim tanks are flooded the unit goes negatively buoyant. After settling on the sea floor and verification of the position and orientation, the hard tanks are flooded to increase the bottom bearing pressure.

The Shell manifold was installed using a controlled free descent method. Extensive model testing was used to verify the stability of the unit during tow-out and during the free descent. An electrical umbilical allowed control of flood valves on the base and consequent control of buoyancy. The corner trim and hard tanks contained compressed air to allow recovery by venting water from the other tanks and to provide stability during descent. With the soft and trim tanks flooded by remote control, the manifold sank untethered to the bottom. After a visual check for damage the hard tanks were flooded. This installation was completed in 1975.

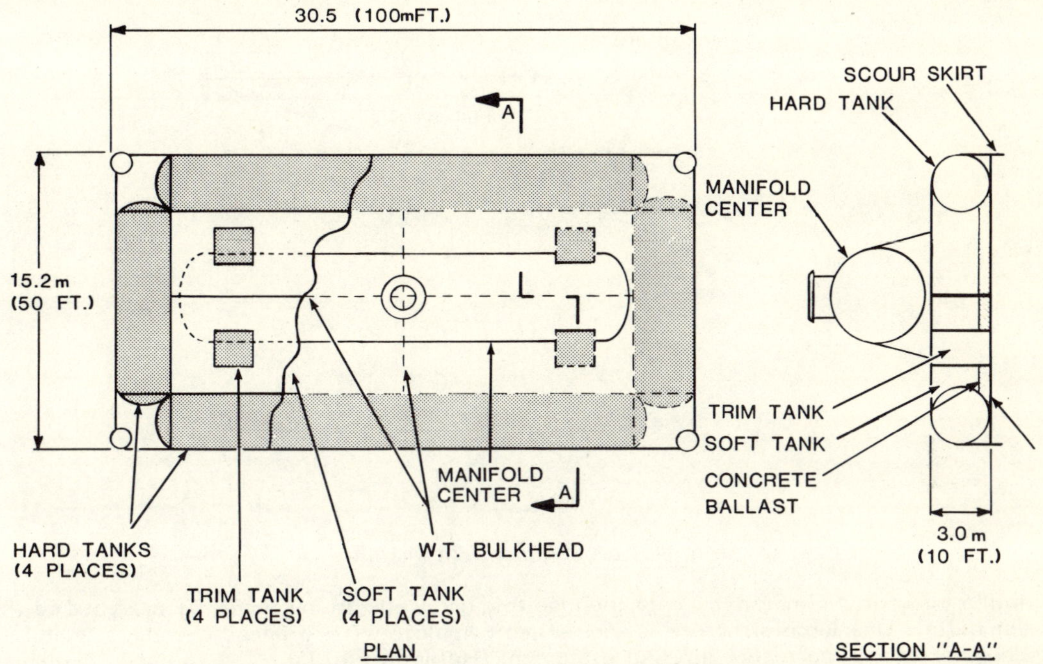

Fig. 13 Garoupa Manifold Base Structure

The Garoupa manifold and base were designed for installation by lowering . When careful dynamic analysis of the loads during sinking indicated that the available crane capacity was marginal, a modified keel-hauling technique was used (Refer to Fig. 14). A travelling block was attached between two mooring winches on the derrick barge and connected to the MC with a 7-in. nylon rope sling. First the soft tanks and then the trim tanks were flooded and the MC swung beneath the barge. The rendering capability of the winches and the resiliency in the nylon sling compensated for the dynamic loads imposed by the swing underneath and vessel heave. In January 1978, the Garoupa MC was lowered to the bottom, lifted and rotated slightly, the hard tanks flooded and the cables released via a hydraulic control umbilical. The complete operation was accomplished in eight hours with the unit only 45 ft off ideal target and within 11° of ideal orientation. The MC was then ready for commissioning.

Both MCs were found in good shape during the subsequent check-out and commissionoing service visits.

Neither manifold was cemented or piled in place. This will facilitate recovery and re-use in other fields.

MANIFOLD TEMPLATES

Templates provide both the guide for drilling operations and a structural base for the subsea completion equipment. Manifolds can be used in three different template schemes employing both template or clustered wells and satellite well completions.

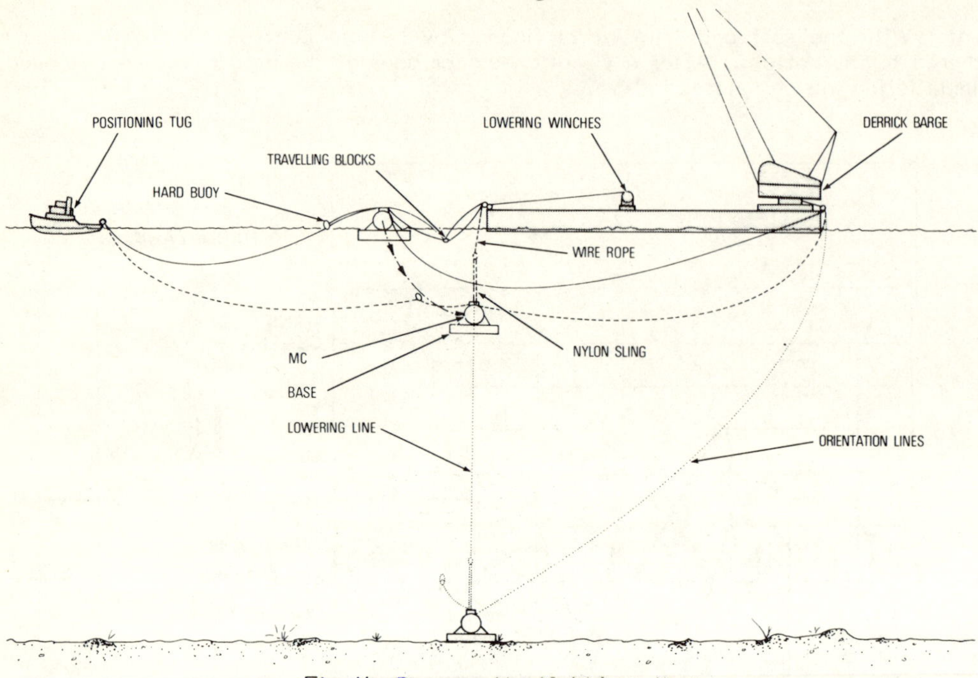

Fig. 14 Garoupa Manifold Installation

In the integrated template concept the manifold and wells are permanently attached to the template. One large structure is submerged complete with wellhead chambers, manifold chamber and all necessary internal equipment (Refer to Fig. 15). The wellhead chambers are initially set up to allow drilling through the chambers with tree and piping hook-ups completed later by subsea technicians as required. This integrated system has been designed to reduce the number of subsea chamber installations and flow-line connections to be done subsea. It is specifically directed at deep water (2,000-ft) applications, to minimize risk at the installation and connection stages.

The second configuration has the template and chambers separated (Refer to Fig. 16). The template is set without the wellhead or manifold chambers in place. As the wells are drilled and completed, the wellhead chambers are run and set. Finally the manifold is run and latched in place. Each well is connected to the manifold by an extendible conduit. This concept allows drilling to begin before the well and manifold chambers have beenfabricated, thus deferring the commitment of a large capital expenditure. Should the first wells indicate a poor field, the wellhead and manifold chambers can be installed elsewhere on a new template.

A further permutation of the manifold chamber is its potential use as a riser base chamber (Refer to Fig. 17). Here the riser terminations and connections to the manifold can be made up and serviced by the technicians working in the manifold chamber. This chamber performs a manifold function as well as serving as the base for the riser.

The one-atmosphere template manifold chambers offer the same advantages as in satellite completions. Satellite wells can also produce to the template manifold.

SUMMARY

The Shell manifold of 1975 (Refer to Fig. 19) and the Garoupa manifold of 1978 (Refer to

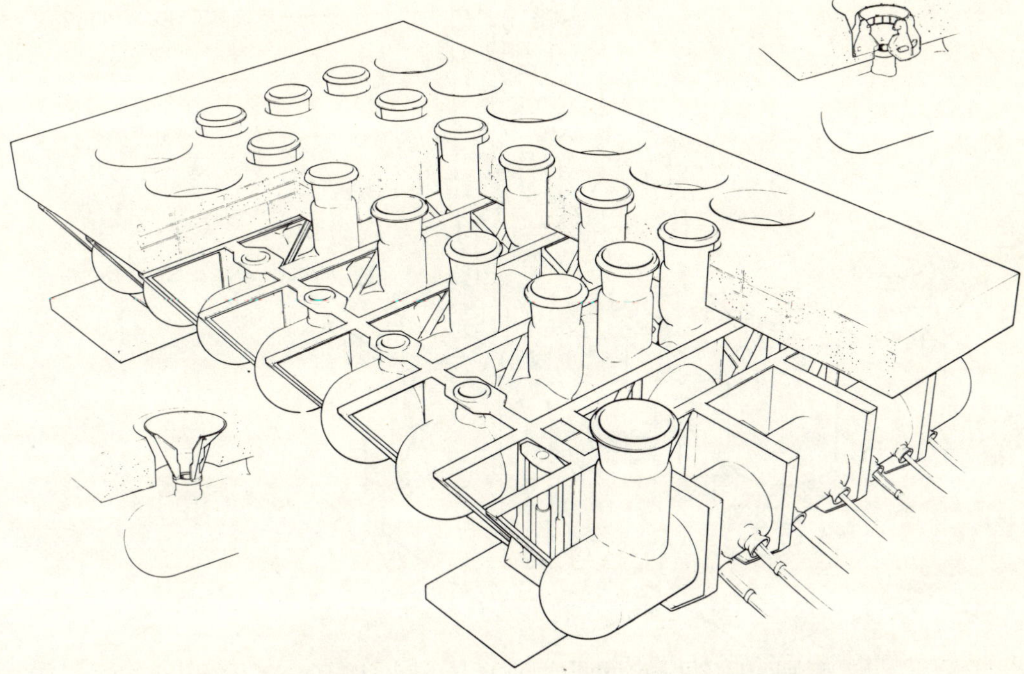

Fig. 15 Integrated Multi-well Template

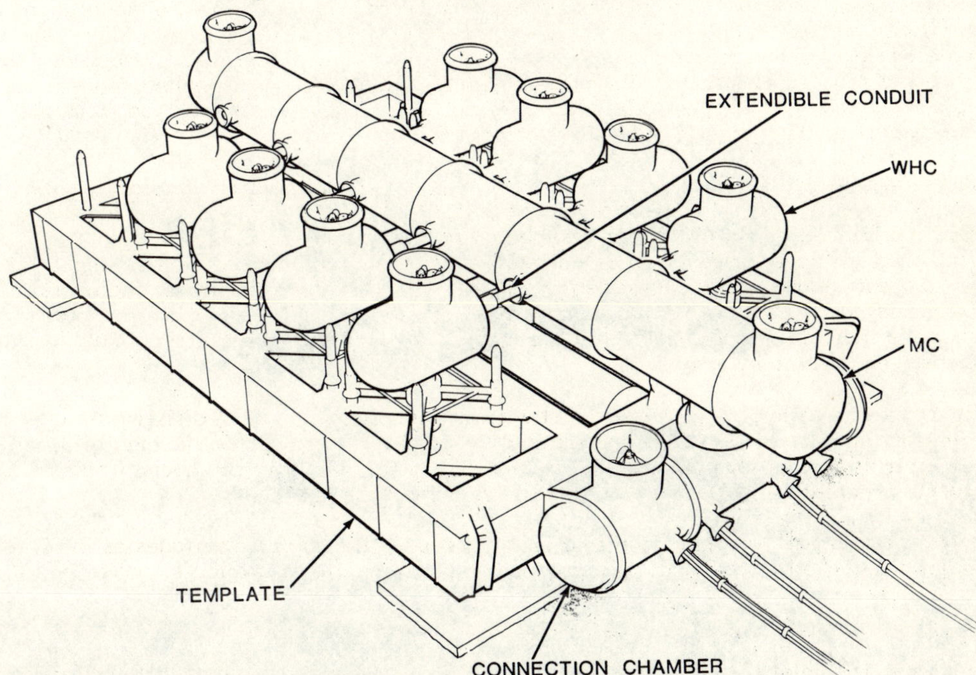

Fig. 16 Modular Multi-well Template

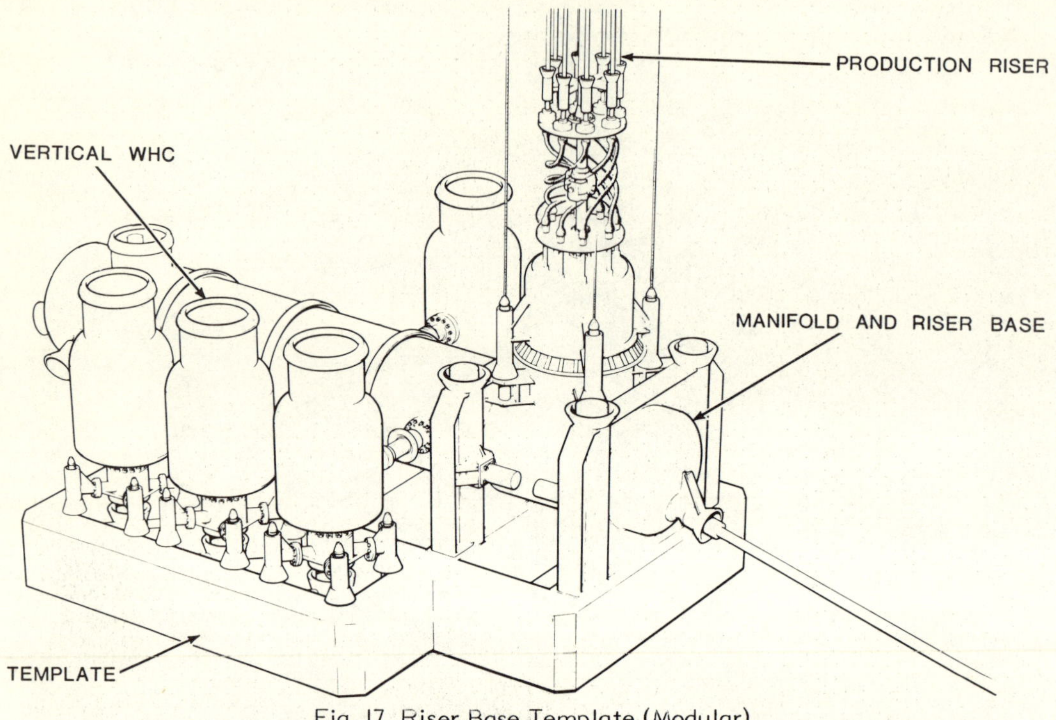

Fig. 17 Riser Base Template (Modular)

Fig. 18 Garoupa Manifold at Sea

Fig. 18) have proven the one-atmosphere manifold chamber to be a viable and practical solution to subsea oil and gas field development.

Fig. 19 Shell Manifold at Sea

These manifolds have:

- been shown to have less downtime than a platform
- demonstrated a reliable subsea pipeline connection technique
- versatility for use in satellite or template field development programs
- flexibility for resolution of subsea problems with manned intervention
- versatility in the selection of components for use subsea
- proven a subsea field completion technique which is independent of water depth.

REFERENCES

Woodlock, D.F. (1977). Subsea flow line connections in one-atmosphere chambers. Offshore Europe Conference.

Apostolidis, M; Crippa, R.A. and VanWalleghem, W. (1978). Design and fabrication of Petrobras atmospheric manifold center. ASME, Energy Technology Conference and Exhibition.

Housden, A.J.D. (1977). Shell group onstream with deepwater manifold. Petroleum Engineer.

English, J.G. (1979). Subsea production - one-atmosphere chambers. Petroleum Society of C.I.M.

WIRELINING IN ONE-ATMOSPHERE CHAMBERS

S. Cejalvo

CanOcean Resources Limited, New Westminster, BC, Canada

ABSTRACT

Present day offshore well workover techniques for wet subsea completions are either through flowline (TFL) techniques or workover rig operations. Both of these are expensive. With a dry one-atmosphere completion additional alternatives are available.

This paper describes the use of downhole workover techniques with dry one-atmosphere subsea completions. Aspects of well design, completion techniques, downhole equipment selection, and the required respective workover equipment are reviewed with respect to the dry completion.

One-atmosphere completion experience in the Gulf of Mexico and offshore Brazil has shown that subsea well servicing utilizing wireline system, polished rod tools, a surface well access system or full workover equipment, provides considerable flexibility for the resolution of downhole problems.

Emphasis of the paper is on discussion of existing equipment and operations with reference to improvements presently being considered.

KEYWORDS

Composite riser; downhaul termination frame; temporary guide base; ball joint; bumper sub; extension spool; flexible hatch seal; spool cavity; LEL (lower explosive limit).

INTRODUCTION

Subsea wirelining provides an alternative to Through Flowline (TFL) or full workover. Both wet and dry completions utilize TFL and workover systems.

In the TFL Systems (see Fig. 1 & 3), pumpdown tools are circulated through the circular flowline loops for downhole work in the tubing strings. Subsequent well maintenance is carried out from remote stations through the flowlines using pumpdown tools. TFL completions require an additional service string and "H" member as well as pumping equipment and are therefore more expensive than wireline completions.

If a major workover operation is required, a rig is brought in and a stack up of blowout preventor, wellhead extension, connectors and control lines is provided to gain access to the tubing strings (see Fig. 2 & 9). Downhole maintenance could then be performed from the rig at the surface. Workover is an expensive last resort for TFL completions and may be the only alternative for wet completions.

The one-atmosphere completion (see Fig. 3) offers the additional alternative of subsea wirelining and other features involving manned intervention of the wellhead.

CanOcean Resources[1] has developed additional methods of subsea well workover with wireline and polished rod tools and a surface well access system (SWAS). These techniques, and the tools and equipment needed for servicing subsea wells in dry one-atmosphere enclosures, are the results of continuing experience with subsea well completions. Through the years from 1966, experience has been gained not only in the development of this equipment, but also in coping with problems associated with subsea well completions. Remedial work programs necessitated by downhole problems were developed using the flexibility made possible in the dry one-atmosphere chamber. The subsea one-atmosphere wireline system has been used to service wells in the Gulf of Mexico and offshore Brazil.

The techniques and the tools and equipment for servicing subsea wells in dry one-atmosphere chambers is the subject of this article.

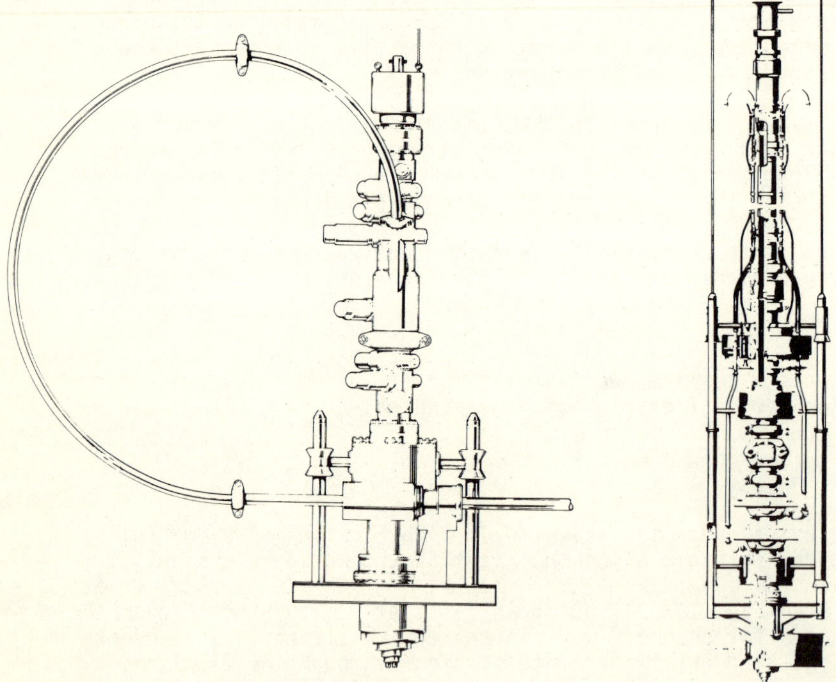

Fig. 1. Subsea TFL wet tree.

Fig. 2. Wet well workover stack-up.

---

[1](formerly Lockheed Petroleum Services Ltd., and no longer associated with the Lockheed Corporation)

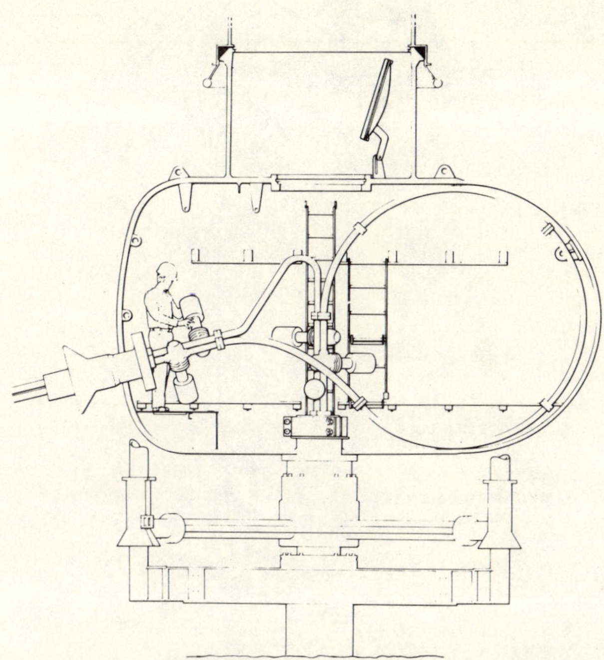

Fig. 3. Horizontal one-atmosphere wellhead cellar (TFL).

SYSTEM DESCRIPTION

One-Atmosphere Subsea Stack-Up

Manned access to the wellhead cellar is by means of a service capsule which operates from a surface support vessel moored over any subsea chamber (see Fig. 4). The capsule carries men and equipment to the subsea system and locks onto the chambers to allow the skilled, non-diver trained personnel to install and service the enclosed equipment in a shirtsleeve environment.

This service system provides access for manned intervention to perform numerous tasks including wirelining and control system and valve refurbishment or repairs on the subsea wells. Wirelining and polished rod lubricator operations are performed within a subsea stack consisting of a wellhead cellar, wireline module(s), and a capsule. The wireline module is not only used for wirelining but also provides additional space for transporting large equipment to the wellhead cellar before assembly subsea.

Refer to Fig. 5 which illustrates the subsea stack for wirelining with the assembled lubricator. Lubricator length is normally 12 feet, however, this can be extended to 19 feet with the addition of a second wireline module.

Service Capsule

Operation personnel descend to the wellhead cellar in the service capsule. When

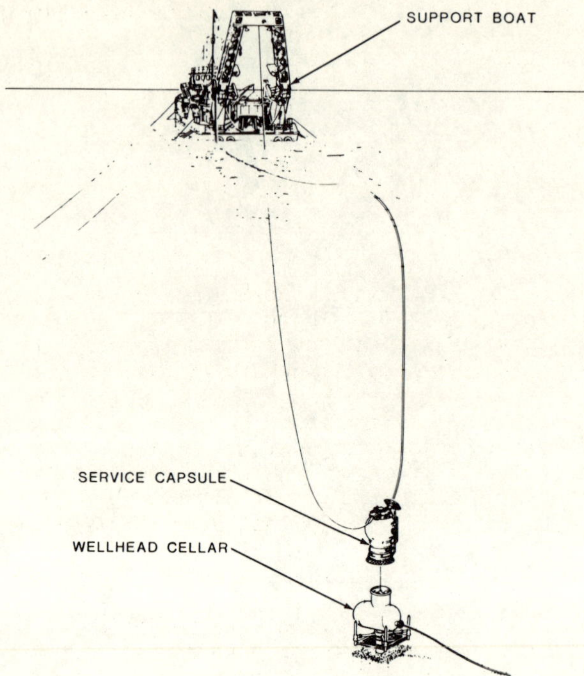

Fig. 4. Service system.

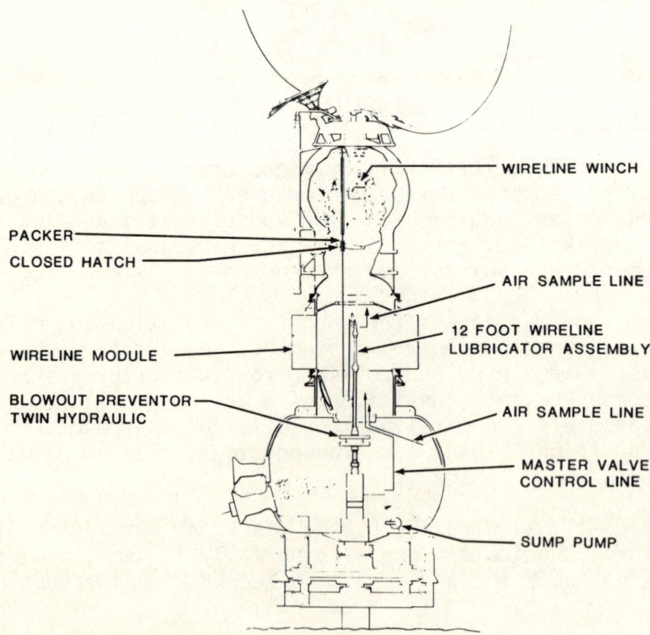

Fig. 5. Subsea light duty wireline stack-up.

inside the chamber, breathing air, power and communication are provided through an umbilical from the surface support vessel.

The service capsule also serves several wireline functions. During wireline operation, the lower hatch is closed to isolate the operation personnel from the lubricator stack and the potential liquid and gas leaks. It is in the capsule where the wireline winch and control systems are stowed. The lower hatch incorporates a removable wireline stuffing box to allow passage of a wireline.

Wireline Module

When a lubricator is needed for downhole work, a wireline module (WLM) is used. This module is an open cylinder with mating rings at the top and bottom duplicating those of the wellhead cellar and the service capsule. It is used primarily as an extension of the capsule skirt, however, it serves also as transporter for subsea equipment.

The WLM is launched using the service capsule. At the surface, the WLM is latched on to the capsule. Together they are launched, hauled down and mated to the cellar in the same manner as the capsule alone. The WLM is left mated to the WHC while the capsule ascends for a crew change or for more payload trips.

Wellhead Cellar

The wellhead cellar (WHC) is the work area in a subsea dry one-atmosphere completion. It is in this chamber where manned access to the subsea well is made possible. Although limited in space, it is ideally suited for stack-up of downhole service tools and equipment and with the WLM(s) and capsule the necessary headroom for wirelining is achieved.

Lubricator Assembly

The lubricator assembly (see Fig. 6) is adapted from a conventional dryland wireline lubricator. Its normal length is 12 feet but can be extended to 19 feet as previously noted. The lubricator is designed for 5,000 psi working pressure. Test pressure is 10,000 psi.

The major components of the lubricator assembly consist of an optional hydraulically controlled swab valve and twin blowout preventor with wireline rams, riser sections with quick disconnect couplings, and stuffing box. The swab valve would serve as the operating valve to prevent wear and damage to the master valves when wirelining. A gas collecting manifold is secured to the top of the lubricator assembly. This permits removal of gas from above the stuffing box. A wellhead adapter completes the interface connection to the Xmas tree. Vent and test ports are provided at the risers for pressure testing and flushing of the lubricator assembly.

Light Duty Wireline Winch

The light duty wireline winch is hydraulically operated and rated at 10 hp with capacity of 18,000 feet of .082 inch wireline. The winch, power unit, and controls are located in the capsule, and incorporate the usual odometer, weight indicator and guide pulleys. The wire is reeved through the wireline guard and stuffing box in the hatch penetrator to the lubricator in the stack-up. (See Fig. 5).

The light duty wireline winch is capable of setting and retrieving downhole safety valves and tubing hanger plugs at shallow depths. It could also be used

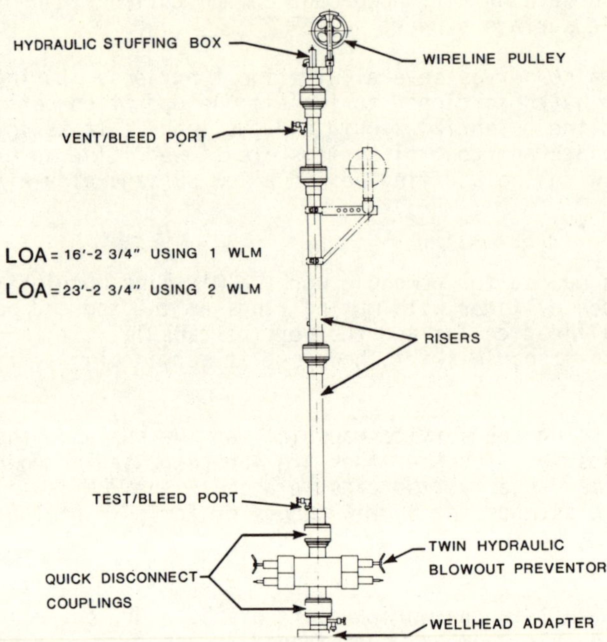

Fig. 6. Wireline lubricator assembly.

to run impression blocks, and bottom hole temperature and pressure surveys.

The subsea crew wirelining on a killed (or pressure balanced) well as in the case for Gulf of Mexico wells are free to operate inside the wellhead cellar. But they must be isolated from the wellhead in a live well situation. In this case, the crew and wireline operator remain in the service capsule with the lower hatch closed.

To prepare a well for wireline operation, the wireline equipment is lowered into the WHC using the service capsule. The Xmas tree is flushed and bled down to remove any hydrocarbon in the piping. A wellhead adaptor is fitted to the top of the Xmas tree and then the blowout preventor, risers, and stuffing box are assembled. The wireline is reeved through and connected to the tool string. Pressure testing is done to test the integrity of the lubrication assembly. On completion, the crew retire in the capsule, close the hatch and commence wireline operation. A TV camera in the lubricator stack monitors the wirelining operation to check leakage, and wireline rigging.

The light duty system described here is the second generation of subsea wireline systems. The original system did not have a stuffing box in the hatch which meant the hatch remained open during wirelining. The stuffing box was added when it became necessary to wireline on an unbalanced well.

All subsea wireline operations to date have been performed by wireline operators from Otis or Flopetrol assisted by the CanOcean Resources technicians as required.

Standard Wireline System

The third generation subsea wireline system is under development by a major wireline manufacturer for use in the WLM (see Fig. 7). It is a heavier duty wireline system with a self-contained 35 hp hydraulic power unit and is used in conjunction with the 12 or 19 foot lubricator mentioned previously.

The standard wireline system is designed to overcome two limitations of the light duty system, namely: the need to cut the wire for crew change and the insufficient power for deep well work. The winch and power unit will be mounted in the WLM and will not require the wire to enter the service capsule. Its capabilities will be temperature and pressure surveys, caliper surveys, running an impression block, sand sampling, water sampling, locate bottom, end of tubing locator, set and retrieve standing valves, open and close sliding sleeves, run collar or tool stops, swaging out tubing, mechanical perforating, set tubing pack-off, set and pull gas lift valves and dump valve, selective fishing and parafin cutting.

As in the light duty wireline system the operating crew are isolated from the lubricator stack-up by the capsule hatch.

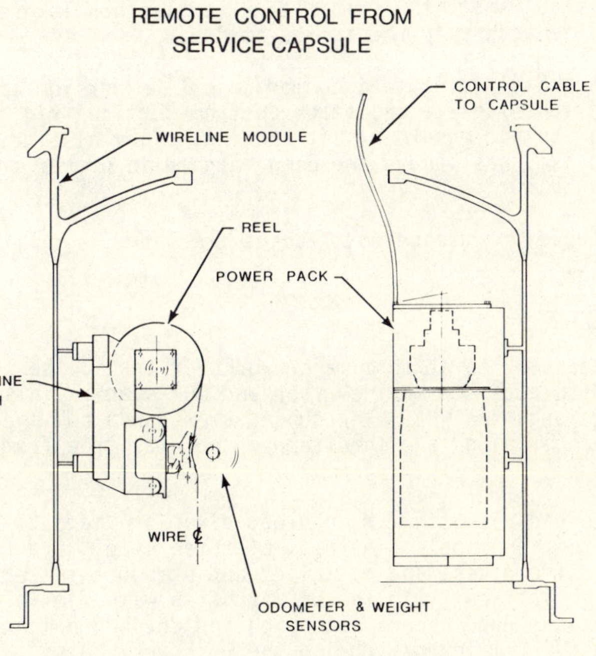

Fig. 7. Standard wireline winch.

Surface Well Access System

Because there are still a number of wireline functions which cannot be performed by either the light duty or standard duty systems, a third system is now available. A surface well access system (SWAS) (see Fig. 8) has been manufactured and will complete dry land testing in March 1980.

The SWAS provides direct access to the Xmas tree by a marine or composite riser from a surface rig. Downhole work capability is then not limited by the maximum 19 foot lubricator length, 35 hp wireline winch or to the use of a solid wire.

To prepare for surface wirelining, the wellhead extension equipment required subsea is lowered to the WHC using the service capsule. The Xmas tree is flushed and bled down to remove any hydrocarbon in the piping. Extension subs, swab and cross-over valves are then assembled on top of the Xmas tree, and pressure tested. A control package is set up and connected to the valve actuators. On completion, the crew move to the teacup and replace the existing hatch with a special hatch. This new hatch has a stab-in housing that interfaces with the SWAS riser attached to the composite riser of the surface rig. A downhaul termination frame containing a hatch housing plug, a buoy, an acoustic release and downhaul cable is installed in the teacup and completes the SWAS preparation in the WHC. The capsule then ascends and the whole service system leaves the scene.

A surface rig then moves in and the downhaul cable is acoustically called up to the surface. Using the downhaul cable as a guide, the temporary guide base is lowered to the WHC. This establishes the necessary guidepost and orientation requirements for the subsequent stab-in operations. The running tool used for this retrieves the downhaul termination frame at the same time. After the guide wires are established, the composite riser is then lowered down and stabbed in place to complete access to the tubing string.

The SWAS features a ball joint to accommodate $\pm 10^{0}$ excursion of the rig and bumper subs. The subs isolate the stack up from vertical rig movements during the running operation and ensure a soft landing. They also drive the stab-in feature after the SWAS has landed and been latched in proper orientation on the guide base.

The SWAS is adaptable for use on most subsea trees and is scheduled for use in the summer of 1980.

Major Workover System

If it becomes necessary to perform major workover tasks, the subsea well (in the WHC) is prepared to accept a marine riser and BOP stack. This allows complete workover tasks through the WHC. The subsea crew use a polished rod tool to install tubing hanger plugs and then remove the Xmas tree from the connector spool.

Using the service capsule with WLM, payload dives are made to the WHC to transport wellhead extension spools. A total of three spools are required, each one carried in successive dives, and installed and pressure tested separately for integrity. On the last entry to the WHC, a hatch with flexible seal is installed between the upper spool extension and the WHC hatch. This hatch ensures that the WHC remains dry during the workover. Also, a downhaul termination frame containing a buoy, an acoustic release, and downhaul cable reel is installed. As in SWAS, the capsule then ascends and the whole service system leaves the scene. The drill rig moves in, recalls the downhaul cable, and uses it to run a 6 foot radius guide post platform onto the teacup.

The BOP stack is fitted with a hydraulic connector compatible with the extension spool. Adapters are also used to suit the type of the BOP. When run and stabbed onto the WHC extension spools, the BOP stack and riser completes access to the tubing string from the drill rig (see Fig. 9). Workover tasks can then be performed.

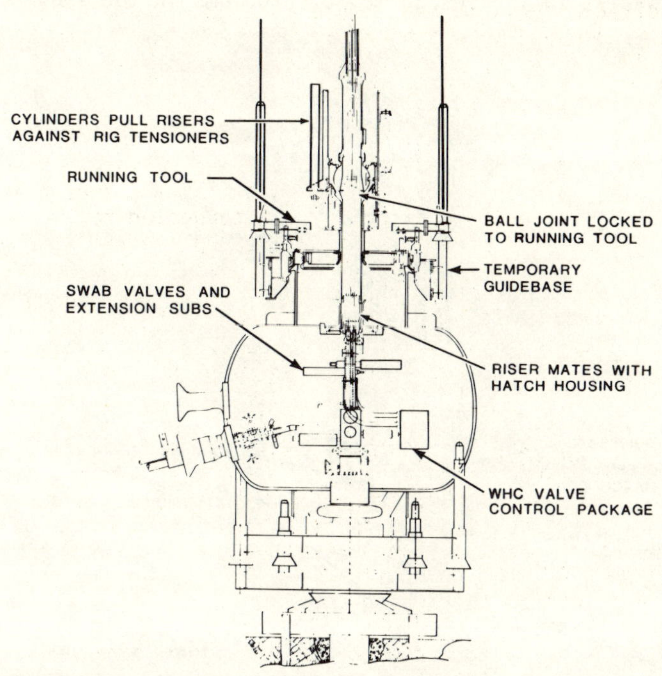

Fig. 8. Surface well access system stack-up.

On completion of workover, the well is put back into production in a reverse sequence as previously outlined.

The workover equipment was built and fully tested in 1977 and is ready for operational use in any WHC having a 13-5/8 in. or 16-3/4 in. 5000 psi completion.

The Polished Rod Machine

To reduce the risk of breaking the wireline while retrieving tubing hanger plugs and to provide a means of setting and retrieving mechanical packers, a polished rod machine (PRM) was developed (see Fig. 10).

The machine is used with 6 foot solid or hollow rods. It has a downward mechanical jarring feature and allows torque of up to 200 ft-lbs to be applied. Pulling capability is 12,000 lbs. If time was of no concern, several 6 foot rods could be used to reach down for deeper work. It has been used with special tools for spool cavity fishing which is difficult to do by wirelining. There is also a spring loaded jarring tool for upward jarring.

Subsea crew attach the PRM to the Xmas tree or spool cavity ball valve using adapters with a BOP in the stack up (see Fig. 11). Tools are attached to the bottom of the rod and succeeding 6 foot rods are connected at the top of the PRM. The stack up is pressure tested prior to performing polished rod operation.

The polished rod machine was specifically developed for use in subsea wells offshore Brazil where five wells have been serviced to remove and re-set 2 inch A-2 and 3 inch A plugs (Camco), for fishing of slugs and control line plugs, and for taking impressions of unknown configurations. The polished rod machine has proven to be versatile in the resolution of problems in the spool cavity.

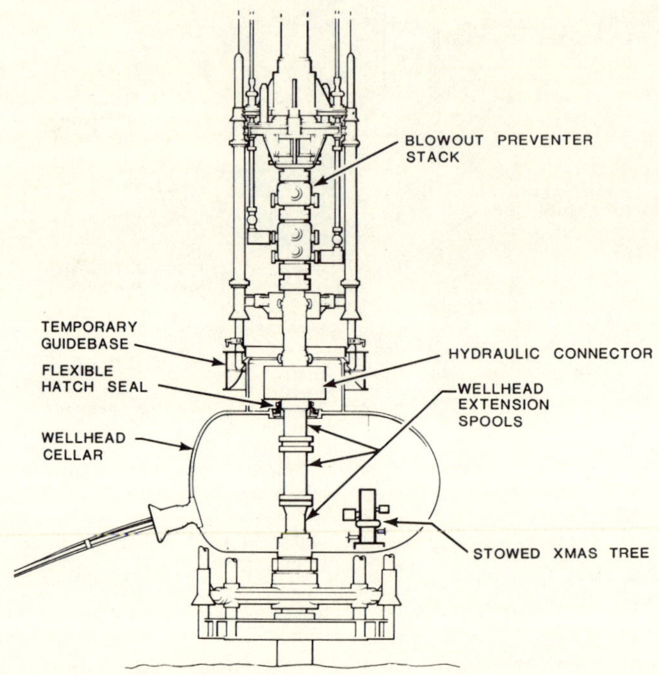

Fig. 9. Major well workover.

Downhole Equipment Selection

Subsea wells completed with a one-atmosphere system offer considerable flexibility for downhole work but some consideration must be given to downhole equipment selection to reduce subsea operating time and crew hazards.

In general, in a field where a working knowledge of TFL techniques and TFL pumpdown equipment is already available TFL completions are preferable. In this case, the tubing completion program is not influenced by the one-atmosphere well completion.

For wells to be serviced by wireline, normal tubing completion are used. The primary concept is that there must be two barriers between the live well and the subsea crew. Each of these barriers must be subject to test to confirm their integrity.

The preferred program uses a service line to allow the well to be balanced (killed) before any subsea wirelining operations begin.

If the well cannot be killed, the annulus line must have several pipe joints and two landing nipples for setting plugs and packers.

Experience with existing wells indicate that wireline retrievable downhole safety valves are not reliable.

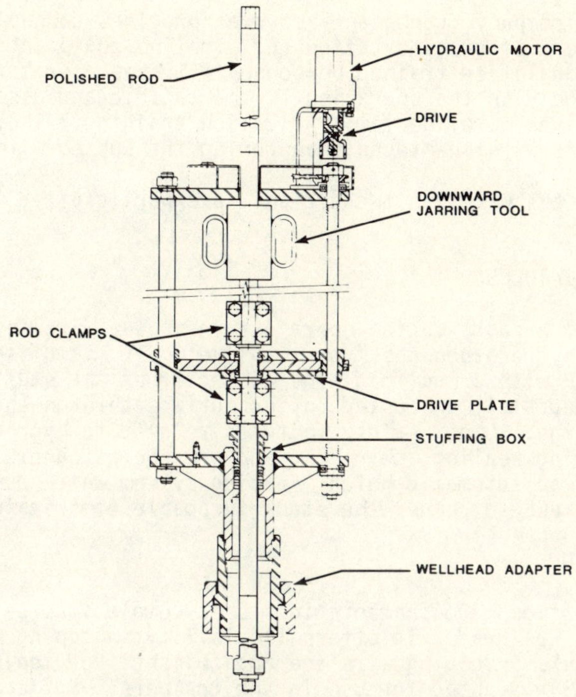

Fig. 10. Polished rod machine.

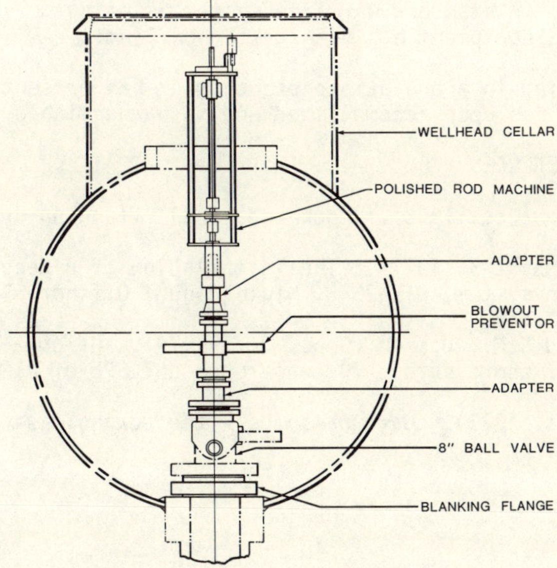

Fig. 11. Polished rod machine stack-up.

PERSONNEL

As a subsea service company, CanOcean Resources provides downhole service capability in the form of transportation and wireline equipment as previously discussed and also qualified trained personnel. Subsea operators are trained by specialists not only in the operation of the capsule and associated equipment but they are also given wireline familiarization training. They are trained to assist the customer's wireline technician during the subsea wireline work.

Both Flopetrol and Otis wireline technicians have participated in subsea wireline operations.

SAFETY FEATURES

One of the principal hazards facing operators in subsea wirelining is the possible escape of hydrocarbon gas from the live well. To offset this, the lubricator is fitted with a manifold type collector at the stuffing box. Any escaping gas is immediately exhausted to the surface through the capsule umbilical exhaust line. Other safety features related to hydrocarbon gas include sensors, double O-ring sealing, twin BOP, and pre-operation pressure testing. The sensors trigger an automatic halon inerting system which dumps freon into the stack-up at 60% LEL to render the stack incapable of sustaining an explosion.

CONCLUSION

The use of dry one-atmosphere chambers in subsea completions permits manned intervention at the wellhead. It offers flexibility in coping with downhole problems and the tasks involving wireline and polished rod tools. Dry land techniques are readily adapted for use in the chambers. Surface well servicing utilizing a drill rig for major wirelining and workover operations are also adaptable to the dry one-atmosphere subsea completions.

A light duty wireline system and a polished rod machine have been used extensively subsea already. A standard wireline system is being developed. Full workover system and SWAS equipment has been tested on land.

Subsea wirelining in a one-atmosphere chamber has presented unique technological challenges but has been accomplished and is now a viable reality.

REFERENCES

Bleakley, W.B. (Oct. 1972). Lockheed, Shell flang up subsea well. The Oil and Gas Journal, 64.

Stone, W.H. (May 2-5, 1977). The installation of a deep water dry subsea completion system, OTC-2818, Ninth Annual Offshore Technology Conference, Houston, Texas.

DeJong, J., and S.M. Adamson (Sept. 3-7, 1979). Garoupa subsea production system, Offshore Europe 79 Conference and SPE of AIME, Aberdeen, Scotland, SPE 8152.4.

Irish, J. (Sept. 1979). Garoupa snags - the downhole story. Offshore Engineer, 40.

A NEW OFFSHORE LOADING SYSTEM DIRECTLY FROM THE SEABED

S. Mathiassen and I. Øvergaard

Kongsberg Engineering A.S., Asker, Norway

ABSTRACT

A new offshore loading system.
The system consists basicly of a dynamic positioned tanker with a flexible riser wheeled up on a drum. The riser will be lowered down to the seafloor via an active heave compensated boom-arrangement. The lower end of the riser, attached with a thruster and docking package, makes it possible to mate with a seafloor terminal. This operation is fully controlled from a control room onboard the tanker.
The system is designed for operation in significant waveheights up to 10 m. The system will eliminate the need for a loading buoy or tower and improve regularity. In addition will the system require modest investments.

KEYWORDS

Offshore loading system, directly from a seafloor terminal.

INTRODUCTION

Offshore exploitation of crude requires a transportation system to bring the crude ashore, to the refineries and the market.

There are two main alternatives for offshore transportation:

a) Seafloor Pipelines

This is a safe transportation system with a weather-independant capacity and regularity.
The disadvantages are that it is expensive, and has limited depth capability. Accordingly deepwater fields, or fields far from shore or small fields that can-not justify the costs of a pipeline may be dependant on the second transportation alternative:

b) Offshore Loading

Offshore Loading today means:

1) Floating buoys/columns
2) Gravity towers

3) ALP's

Loading to tankers is performed via flexible hoses, while the tankers are moored by hawsers.

This alternative is considerably cheaper than a pipeline and requires less lead time for completion.

Operational limitations for the systems are:

a) Hawser stress limit
b) Tanker Master's concern for impact with buoy
c) Buoy mooring stress limit being reached
d) Sea-sickness of SPM operators
e) Downtime due to a) and b).

The major disadvantages is that loading is deferred at seastates generally in excess of $H_S > 6$ m.

As the problems of existing offshore loading methods were established, Kongsberg Engineering started an internal project to study feasibility of an alternative method.
Kongsberg Engineering has experience on risers, subsea equipment, dynamic positioning and automation systems, in addition to its general experience within the various diciplines.
The concept of KbE SUBLOAD was conceived under the execution of a Floating Production System project. A/S TELEPLAN has been subcontracted the development of an efficient heave-compensating system, based on their own patent pending. Their contribution of the project is greatly acknowledged.

SYSTEM DESCRIPTION

KbE SUBLOAD is based on a dynamically positioned shuttle tanker connecting a flexible riser to a seafloor terminal at the end of an export pipeline from the production platform.
The seafloor terminal may typically be located some 600 - 1000 m from the production platform, dependant upon the water depth and the lay-out of the field complex.

The main subsystems of KbE SUBLOAD are:

Shuttle Tanker equipped with Dynamic Positioning System, and thrusters for stationkeeping capability.

Heave Compensating System compensating for the vessel vertical and horizontal motions due to environmental forces. The heave compensating system will provide constant tension to the riser during the operation, and enable controlled and stable motions of the riser in the vertical direction during the connection phase.

Riser Storage & Handling system will store the flexible riser on a large-diameter storage drum during transit and will facilitate deployment/retrieval of the riser at arrival/return.

Riser & Control Cable consists of a flexible hose with tension capacity for conducting the crude from the seafloor terminal to the storage holds of the shuttle tanker. Parallel with the riser there will be a separate control cable providing power to the subsea equipment and conducting the two-way signal transmission.

Lower Riser Package will be suspended at the lower end of the riser.
It will include:

- communication module
- power distribution module
- underwater navigation package
- thruster unit
- docking/reentry system
- jack-down and locking system
- riser purge valve assembly

Seafloor Terminal will be mounted on a piled-in 30" heavy-walled extension conductor with wellhead housing at the upper end.
The terminal will incorporate pipeline tie-in system, wellhead with guide base, retrievable isolating ball valve (fail safe) with hydraulic connector beneath, and transponders for the underwater navigation package.

NORMAL OPERATING PROCEDURES

The operation of the system can be divided in serveral phases:

- First the shuttle tanker will navigate to a surface position just above or within the proximity of the seafloor terminal location. To hold the tanker within acceptable operational tolerance limits, regardless of the environmental conditions, the tanker will be equipped with DP-system and thrusters.

- When in position, the ship will be kept heading between say 0° to 20° from the incoming waves.

- After the tanker movements are checked to be within acceptable limits and location versus the seafloor terminal are accepted, the Lower Riser Package (LRP) is lifted from the stowed position on the deck and lowered on the lee side into the sea.
  This is done by means of a crane with a built-in wave-motion compensator.

- The LRP will be lowered to a preset altitude above the seafloor terminal, giving a safe margin to avoid collosion with possible subsea obstacles.

- Provided the surface position of the tanker is correct and within allowable tolerances, the LRP will be lowered the last distance until mating. This is accomplished by means of relatively wide cone guide funnels on the LRP-matching the guide posts of the seafloor terminal.

- After mating, hydraulically operated claws will secure the joint.

- Then the connector will be hydraulically jacked down, this to avoid damaging seals on the seafloor hub.

After the connection have been made and the control lines stabbed, the subsea ball valve can be opened and the loading performed.

When connected the motion-compensator will maintain constant tension in the flexible riser hose, paying out/pulling in the riser according to static vessel offsets or dynamic wave-induced motions of the ship.

The heave compensating system will be able to absorb the following max. values:

- Heave amplitudes : $\pm$ 8 m.

- Velocities up to 3,8 m/sec.

As the handling system will be located amidship, this means that the KbE SUBLOAD can be operational in significant waves up to approximately 10 m, when using a 120.000 tdw. shuttle tanker.

SYSTEM ADVANTAGES AND DISADVANTAGES

The subsystem advantages compared with existing systems are:

a) Improved regularity as the weather downtime only will be limited to the ship station keeping capability.

b) The SUBLOAD system exclude the need for a surface structure (buoy) which again eliminates collosion risks and improves safety.

c) Connecting the riser to a seafloor terminal reduces weather dependancy as the connection is moved away from the extreme wave zone.

d) The major part of the equipment is onboard the tanker, which makes it easy accessible for maintenance.
Further all equipment subject to wear on the seafloor terminal will be retrievable to surface for maintenance under normal working conditions.

e) The KbE SUBLOAD requires no personnel transprt to an offshore loading platform.

f) The system will require modest investments.

The disadvantages for the described system may be increased loading time due to limited riser dimensions.

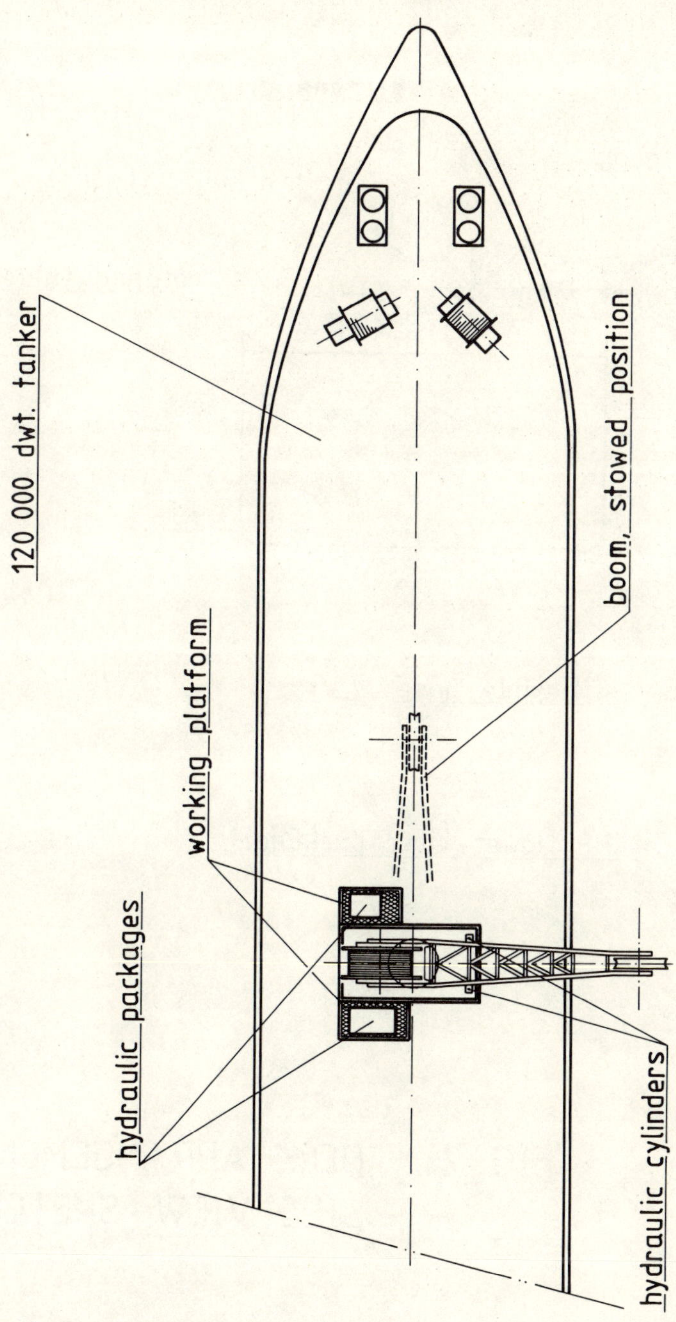

FIG. 1. DECK ARRANGEMENT
PLAN VIEW SKETCH

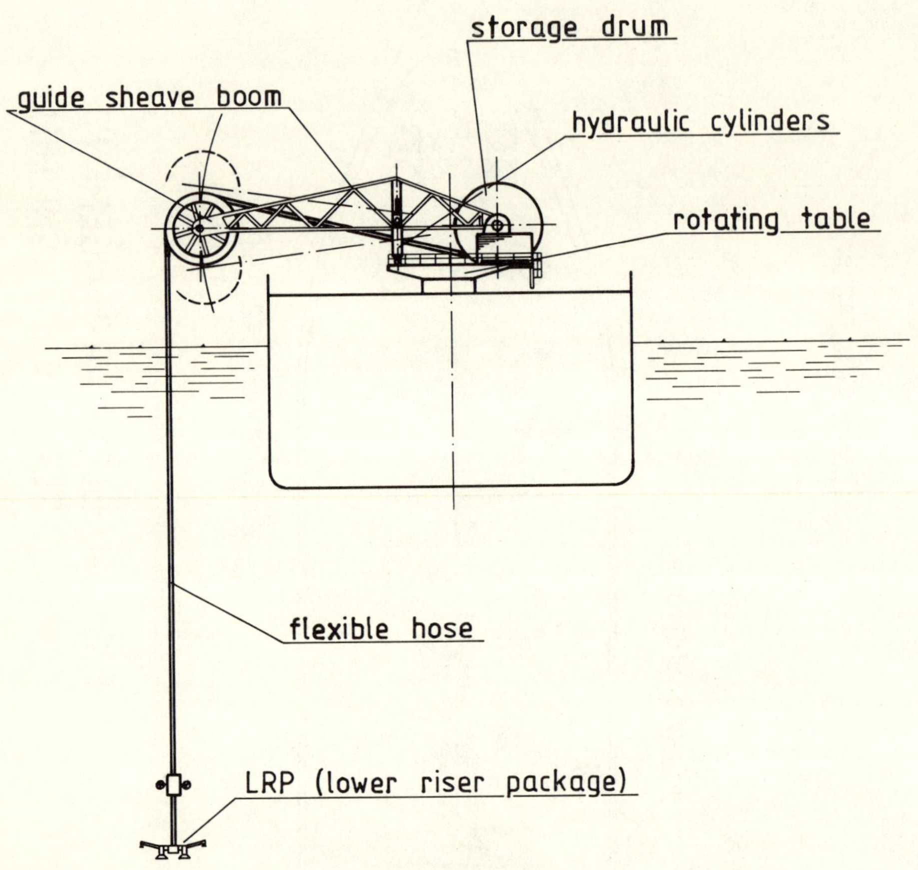

FIG. 2. DECK ARRANGEMENT SIDE VIEW SKETCH

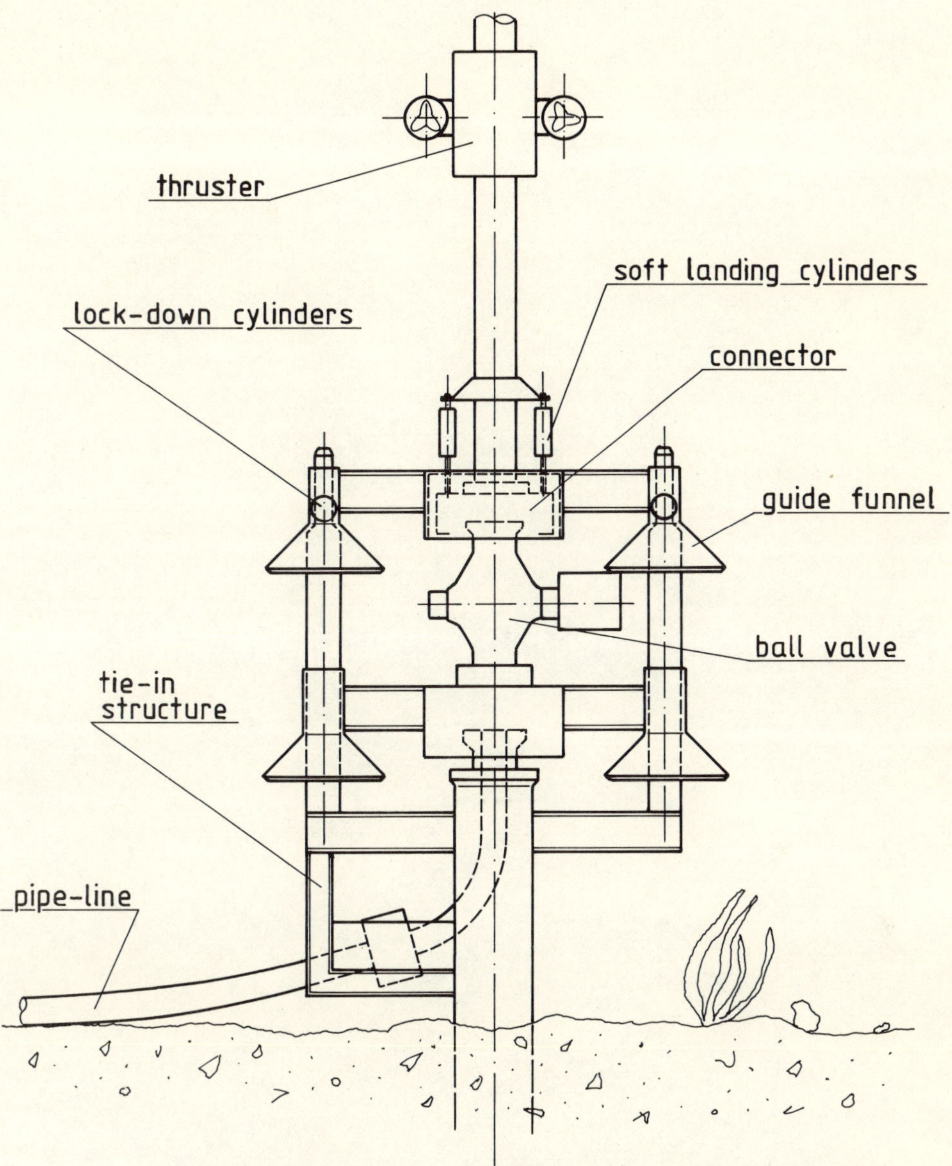

FIG. 3. SEAFLOOR TERMINAL
(WITH LRP IN MATED POSITION)

A NEW CONCEPT OF AN OFFSHORE LNG-TANKER LOADING AND UNLOADING SYSTEM

H. W. Backhaus and K. Friedrichs

LGA Gastechnik GmbH, Federal Republic of Germany

ABSTRACT

Over the last two decades, liquefied natural gas has become an increasingly important source of energy for the major industrialized areas of the world, the United States, Japan and Western Europe. The initiative to liquefy natural gas as a means of providing supplementary energy has several roots including the growing awareness of the environmental qualities of this fuel, the distaste for flaring large volumes of gas in oil producing countries and the realization that liquefied natural gas (LNG) is amongst the few proven marketable methods of transporting additional natural gas over large distances and/or oceans. The liquefaction of offshore natural gas as a method to exploit remote and marginal offshore gas deposits is basically the result of the same considerations.

While the technique of installing and operating liquefaction plants on offshore platforms today does not create major problems, the offshore loading of LNG tankers adjacent to LNG production and/or storage facilities still seems to be a risky undertaking for several reasons. In order to reduce these risks, the unloading technique should be based upon well-proven components for offshore loading procedures. The established "hose-system" comes closest to this philosophy, but it has to be adapted to the specific requirements of the ultra-cold LNG and to the fact that boil-off gas has to be returned from the loading tanker to the LNG production plant.

The proposed hose-handling arrangements do not require direct intervention by deck hands, and the "hands-off principle" is practically realized by remote-controls. The sturdy, simple and proven components of the hose system are most likely to give dependable services in the hostile marine environment. In designing the hose systems the authors always had possible break-downs in mind and they kept checking if their designs would provide fall-back features.

Depending upon the amount of LNG to be transferred, the double-hose system offers a redundancy (each single hose for full capacity) or a duplication (single hose for the half capacity). In anticipation that the manoeuvring tanker will undoubtedly be the greatest threat to the integrity of the loading (unloading) tower, a fool-proof berthing method was developed, comprising for example a remote-controlled tanker approach/mooring procedure, together with a multiple-redundancy propelling system on the tanker itself.

KEYWORDS

Underwater LNG pipeline; submerged mooring buoy; LNG hose-systems for loading and unloading operations.

INTRODUCTION

Recent months have made it abundantly clear how much the highly industrialized world depends upon non-disrupted supplies of energy, which today means the supply of crude oil and also natural gas. It is well-known that oil and gas reserves are declining at an alarming rate and this is particularly true for onshore fields. OPEC countries, furthermore, have drawn-up a new philosophy of restricted exports obviously resulting in conservation of their oil and gas. They are no longer willing to sustain "over-production" in order to accommodate the industrialized countries's welfare.

Assuming that this scenario proves to be right, the search for alternative energy resources and supplies will have to be intensified. Offshore exploration and the exploitation of oil and gas will thus experience a tremendous upturn in the forthcoming decades. Geologists estimate that nearly 50 % of the undiscovered oil and gas reserves will probably be found on continental shelves and continental slopes. Most of these areas have yet to be explored, and the prospects have never looked better for the offshore industry, a trend clearly sustained by the soaring prices for oil and gas.

But what is the current state of the art?

While the techniques of offshore exploration may be regarded as quite sophisticated the means of exploitation have in many cases yet to find optimum solutions. This is particularly true when offshore natural gas is involved. Today, the submarine pipeline is the only proven technical system for natural gas transportation from an offshore field to a gas consuming area.

Compared to pipeline facilities on the mainland, the subsea transportation of offshore natural gas by pipeline in relatively small quantities and over large distances will create even more technical and economic problems. In recent years a number of different concepts have been developed in order to liquefy offshore natural gas directly above its point of discovery. The envisaged technical solutions today prove the feasibility of offshore liquefaction, in particular if the plants are operated on platforms firmly standing on the seabed. The same, however, cannot be stated regarding the problems of transfering the low-temperature LNG from the production plant to an exporting tanker.

LNG OFFSHORE HANDLING

The transfer of liquids such as crude oil from an offshore storage tank to an exporting tanker is well established today. There is a variety of special hoses available for the loading procedure. These hoses either float on the water surface or hang between the boom of a mooring-tower and the tanker's bow. In certain weather and sea conditions, a hose system permits a large degree of tanker movements at the mooring buoy during the loading procedure. But there are considerable problems if the fluid to be conveyed is not crude oil but liquefied natural gas, which has a temperature of minus 162° C (111 K; -324° F). To overcome the obstacles of the ultra-low temperature, the transfer system has to be insulated, thus preventing the formation of ice on the outside of the hose and avoiding high boil-off rates of the cold liquid inside the transfer line.

Having accepted this fundamental requirement of insulation, one has to consider how - in an offshore environment - the connection between the LNG production platform and the LNG tanker should be realized. In principle, there are two ways of approaching the problem:

- LNG handling systems in contact with the surrounding sea-water, similar to floating crude oil hoses (Fig. 1)
- LNG loading facilities completely elevated above the sea-level, i.e. the principle of crude oil loading-towers (Fig. 2)

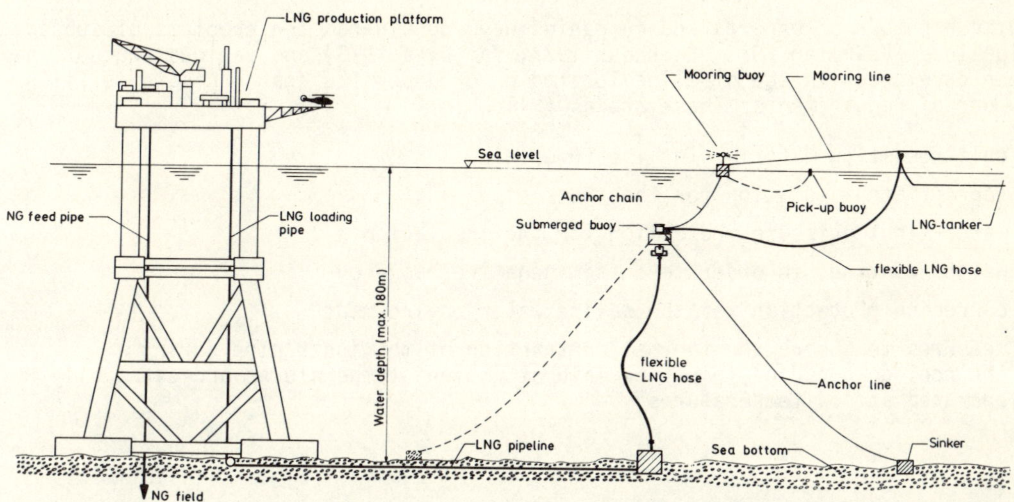

Fig. 1 Underwater LNG loading system

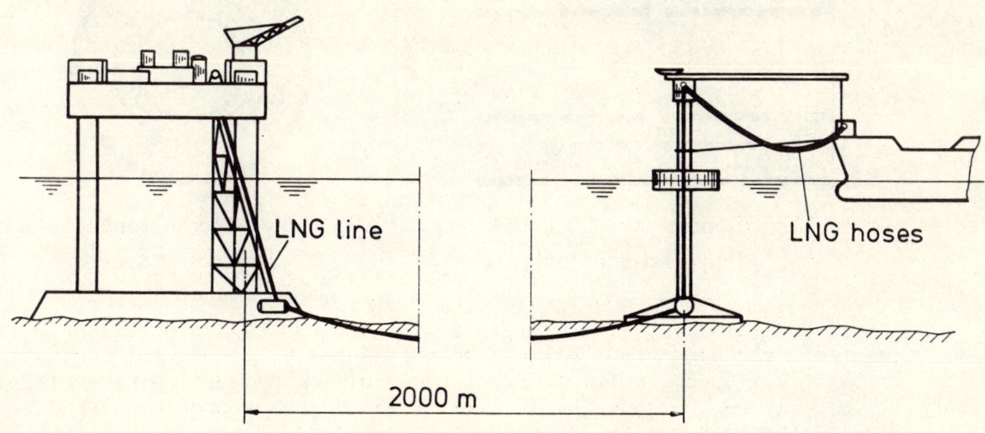

Fig. 2 Tower loading system for LNG

Regardless of the concept which finally proves to be feasible, there is a submarine LNG pipeline with the function of separating and then interconnecting two offshore structures (Fig. 1 and 2). For example, this can be the production platform and a mooring buoy (tower) and/or a separate offshore LNG storage facility. One main reason for having a certain distance between the platform and the LNG loading point is the presence of a manoeuvering tanker and the possibility of its colliding with nearby offshore structures.

LNG Underwater Pipeline

Quite a number of experts and companies have considered the problems of subsea pipelines (Veerling 1972; Backhaus 1977; Van Dyke 1978) and various concepts have been developed. An LNG pipeline located on or under the seabed has to fulfill a number of requirements. These are as follows:

- cold-resisting material for the inner pipe
- appropriate insulation for the inner pipe
- water-tight pressure protection for the insulation
- weight-coating, in order to obtain negative buoyancy
- corrosion protection for the saline water environment
- measures to absorb the thermal contraction of the inner pipe during the cool-down period (pipeline laid at ambient temperatures and eventually operated at low temperatures)

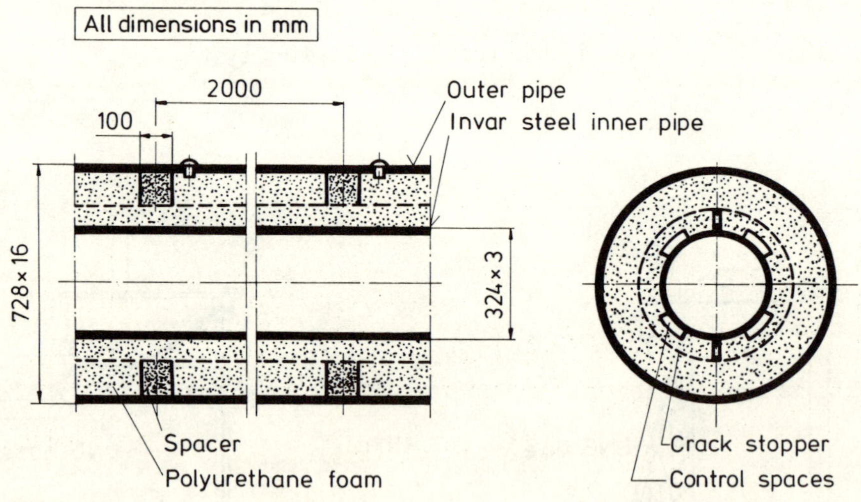

Fig. 3 Design features of underwater LNG pipelines

- provisions for sea transport and placing the complete line or parts hereof into the final offshore position

As it may be seen from Fig. 3, the submarine pipe is a coaxial construction comprising several layers, in particular the insulation. The latter cannot be exposed to water or water pressure. Therefore, the insulation has to be provided with a water-proof cover, for example a steel sleeve, whose thickness will depend upon the water depth.

The compensation for shrinkage of the cold inner-pipe versus the "warm" outside sleeve should not be based upon conventional bellow compensators, because of the difficulties to be foreseen with respect to maintenance and repairs. There are a lot of other propositions to resolve the shrinkage compensation, but the most efficient way seems to be the utilizing of invar steel (36% nickel alloy) for the inner pipe. The extremely low shrinkage values of this material (1/7 of the normal CrNi steel) makes it possible to fix the inner pipe at its end-points. For this purpose, special ports of "dry habitats" are used, in which the installation of the fixing points for the invar pipe causes no major problems.

Another problem may be seen in laying the relatively complex and sophisticated LNG pipeline in an offshore environment. In response to these pipeline properties, a new method has been developed based upon the well-known "near-bottom-tow" method using buoys for positive buoyancy and chains for negative buoyancy (Brown, 1977; Borelli, 1978; Kotani, 1979). Consequently, the LNG pipeline at its full length is manufactured in an appropriate place near the coast. The pipeline then will be provided with special buoyancy aids, which can be ballasted with a precalculated amount of seawater and also deballasted again (Fig. 4).

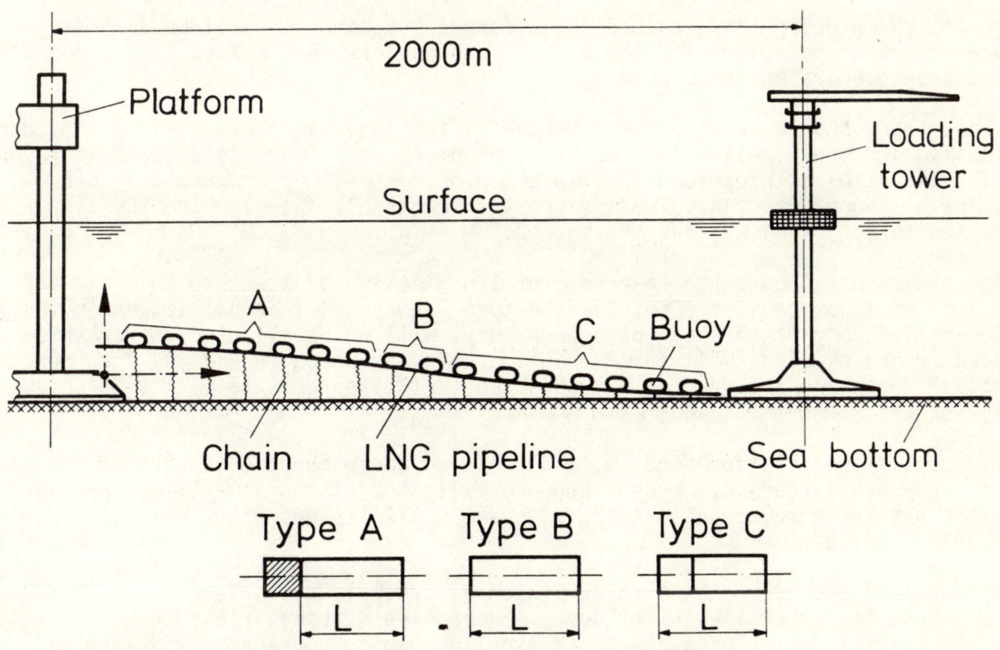

Fig. 4 Underwater LNG pipeline and buoyancy aids

If a 2,000 m LNG pipeline is to be laid, the total number of buoyancy aids is 60, which will have to be fitted to the pipeline in a certain order. Furthermore, each single buoy will be precisely precalculated, not only with respect to its overall volume, but also as to the volume, which can be ballasted (or deballasted). The partition of each buoy into ballast compartments (smaller or larger) depends upon its later position on the LNG pipeline.

As may be seen from Fig. 4, there are three types of buoyancy aids (A, B and C), having principally the same volume (defined by the length L), which provides the positive buoyancy for the near-bottom-tow operation. The type A has an additional volume filled with water right from the beginning of the towing operation. This additional ballast volume is different for each buoy of type A, with a maximum for the buoy positioned at the "platform-end" of the pipeline and a minimum for the one next to the B-type buoys.

The type C buoys have the ballasting compartment within the length L, and - similar to type A - the volume of this compartment is different for each buoy, with the maximum value at the "tower-end" of the pipeline.

The design and the arrangement of the buoyancy aids permit not only a safe towing of the LNG pipe from its fabrication area to the offshore installation point; it also provides all the necessary means for accurate and smooth positioning of the pipeline within a prescribed target-area on the seabed.

During the towing operation, the distance from the seabed is approx. 4 m, which means that the LNG pipeline is not exposed to the wave forces of the sea surface. The towing is carried out by a tug in front of the pipeline. A second tug at the pipeline-end is employed in order to prevent the pipeline from fishtailing and moving laterally.

When the LNG pipeline has reached the offshore location, it will be directed into a pre-described position with the aid of transponders on the line itself and reference-transponders on the seabed.

In case there should be an adverse weather situation, all buoys will be completely ballasted and the pipeline laid on the seabed. As soon as the sea state is favourable, the "tie-in" procedure for the LNG pipeline will be commenced by deballasting all compartments (volumes given the length L). The pipeline then is elevated to approx. 4 m above the seabed.

After having positioned the tower-end of the pipeline adjacent to the "tie-in" port of the tower-habitat (Fig. 5), the type C buoys will be ballasted and the water-filled compartments of the type A buoys will be deballasted. After this procedure the pipeline will form a configuration as shown in Fig. 4. As a next step, the pipeline's tower-end will be tied in the appropriate port and locked water-tight.

Subsequently the "platform end" will be aligned above the tie-in port of the platform's dry habitat and all buoys then are ballasted at the same time. The dry habitat has to be moved against the pipeline until its end piece has matched with the port-opening.

The described procedure has been developed in order to avoid large deflections - tolerable for other tie-in methods - and bending stresses (less than 5 % yield) for the coaxial design of a cryogenic pipeline, which combines rigid layers (inner pipe and outer sleeve) with relatively soft material (insulation).

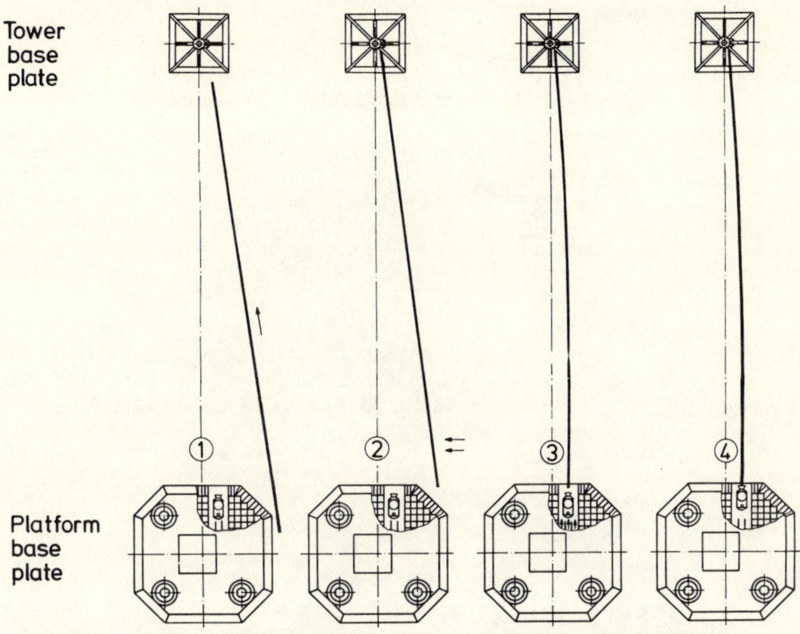

Fig. 5 Installation steps for underwater LNG pipeline

Mooring System and LNG Loading Hose

Apart from the LNG underwater pipeline, which permits the placing of the loading point at a safe distance from the LNG production platform, the offshore LNG transfer system also requires a mooring facility. This can be realized either by an underwater mooring buoy (Fig. 1) or a mooring tower (Fig. 2). The difference between the two propositions has to be seen from a safety point of view. While the underwater buoy provides a maximum safety, the mooring tower is still exposed to possible collisions with the tanker.

On the other hand, the safety aspects have to be compared with the given hose manufacturing techniques. As a matter of fact, a transfer hose which connects the loading tower with the tanker bow - above the water surface - does not require fundamental development. On the contrary, the flexible hose-system leading from an underwater buoy to a tanker still poses a few problems (Backhaus, 1977).

Underwater mooring buoy and LNG loading hose. The crux of the underwater loading system is a mooring buoy, anchored to the seabed at a depth of about 30 meters (Fig. 6). It has a rotatable top through which the LNG pipe is conducted by means of a swivel flange. The sealing principle underlying such a swivel for low-temperature liquids is established (for example in onshore loading-arms for LNG tankers).

The low temperature swivel joint is subject to wear and tear. An easy way of maintenance has to be provided, and the rotatable top shown in Fig. 6 is under atmos-

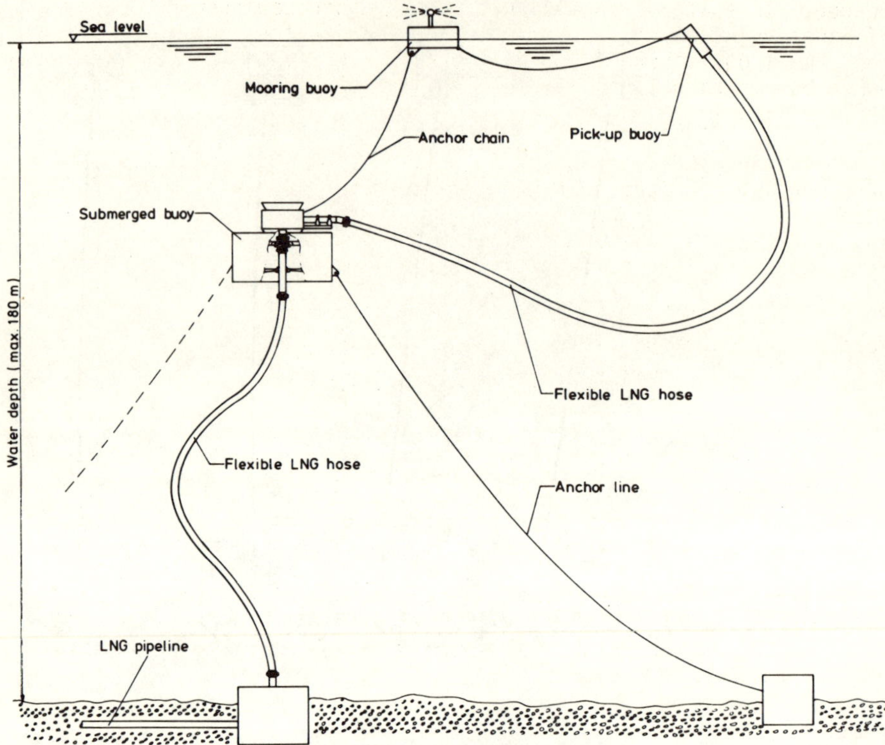

Fig. 6 Underwater mooring buoy

pheric pressure and free of water. This is obtained by a special swivel gasket based on the same design criteria as in the propeller shaft packing of sea-going vessels. The requisite maintenance and repair work is carried out by a crew, who are taken to the buoy's top in a submersible service boat.

Although the described underwater buoy has - in a model test - proved to be feasible and the manufacturing technique of the flexible subsea hose is fully developed, there is the problem of guiding the hose from the underwater buoy to the tanker. Here, the anticipated changes in buoyancy pose some problems. In order to allow a smooth configuration of the water-exposed hose, the specific gravity of the same should have a magnitude of 1 or just above 1. In this case, there will be a slightly upward bending curve (Fig. 6) until the "tanker-end" of the hose reaches the sea surface. But one has to consider, that the hose can be empty or filled with LNG and this results in a distinct gravity difference of the hose as a unit. If specific gravity of the empty hose is a little over 1, it will have a negative buoyancy when filled with LNG. That may happen after the LNG tanker has been loaded and the hose again is dropped into the water. Even with a pick-up buoy, which keeps the hose's "tanker-end" floating, the hose will eventually have a sharply U-shaped configuration, thus causing destructive bending stresses at the hose entry point in the top of the underwater buoy.

Consequently, further development work will be necessary before the submarine mooring buoy together with all its components may be regarded as technical feasible. If the boil-off gas has to be returned from the loading tanker to the LNG-plant, the transfer system requires a "return-line" and this will further complicate the concept. In spite of this situation the described concept offers the safest way of manoeuvring an LNG tanker into its loading position, even in severe weather and sea conditions. There is no danger of collision between the tanker and the submerged buoy, which is positioned considerably lower than the tanker's keel.

Tower mooring and LNG loading system. Having thoroughly evaluated various other offshore mooring and loading concepts (Backhaus, 1977; Ehret, 1979; Jones, 1979), the authors came to the conclusion that a tower mooring and a suspended hose-loading system today is the most feasible solution. Pursuant to this reflection priority has been given to the following design criteria:

- application of proven offshore systems and components
- lowest possible weight of the components to be used
- maximum free moving envelope for the loading tanker
- optimum accessibility to all parts for inspection, maintenance and repairs
- easy and - where possible - remote-controlled connecting and disconnecting coupling-procedures and
- avoidance of LNG spills on to steel structures, in particular the tanker's loading area

The hoses used for the LNG and return-gas transfer are suspended above the water surface between 2 pivoting points (Fig. 7). One is on the loading tower and the

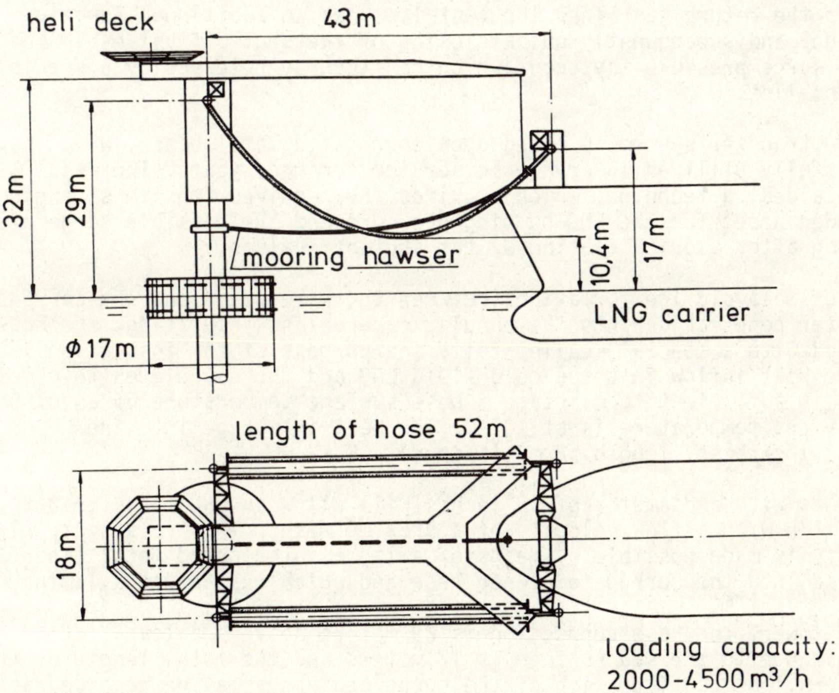

Fig. 7 LNG hose between tanker and tower

other on the tanker's bow. On either end of the LNG and the return-gas hoses (at the tower as well as at the tanker bow), there are universal cryogenic swivel joints (Fig. 8). They protect the hoses from high bending stresses, which in turn could shorten their service life or even cause damage.

Since in general there ought only to be minor bending stresses in the suspended hoses, the distance between the loading tower and the ship's bow has to be fairly large. In the case of the described concept this amounts to 43 m. In fulfilling this requirement, one achieves at the same time sufficient freedom for the tanker's movement. The obtained wide "moving-envelope" is shown in Fig. 9.

The complete transfer system consists of 2 pairs of hoses. One is utilized for the LNG transfer to the tanker and the other for returning the boil-off and flash gas from the tanker to the liquefaction plant. The LNG and the return gas hoses are conducted parallel at a distance of about 18 meters (Fig. 7). This is realized by small supporting structures on the tower as well as on the tanker's bow. This design has a supplementary feature, namely the hose connecting points can be positioned outside the tanker's deck area and all LNG spills (whether a leakage or emergency disconnection) will drop into the sea and they cannot harm at all the tanker's forecastle. When there is no LNG-tanker moored at the tower, the coupling ends of a hose-system are pulled up to the Y-shaped end of the tower boom, an operation for which the "holding ropes" are used. The coupling-end is then guided into an arresting device, thus taking away the coupling's weight from the holding-lines. The latter then can be "freely moved" through a special opening in order to commence the next hose-connecting procedure.

The residual LNG is left in the hoses in order to guarantee that they remain in a cold state. A small vent-line allows the boil-off gas to escape from the LNG hose ends to the return gas line. The vent-line has an additional function: in case of emergency and subsequently quick closing of the shut-off valves in the LNG hose, a high surge pressure may occur which can then be released by a special valve into the vent-line.

The LNG transfer system is based upon corrugated metal hoses, which have been successfully utilized in cryogenic service for many years. The metal hose is wire braided, a technique which provides the required tensile strength for the suspended hose, for the LNG-pumping pressure and the possible surge pressure occuring after a quick closing of the shut-off valves.

In order to avoid ice formation (between the wire braid and the corrugated hose) and water penetration into the annular fiber-glass insulation, the hose has to be equipped with a special sealing-foil. The purpose of the insulation is the reduction of heat inflow into the cold fluid LNG and the calculated thickness of 124 mm is sufficient to maintain a hose surface temperature of about 6° C when the ambient temperature is at 10° C. The heat inleakage into the LNG hose system (2 x 12 inch hose, length 52 m) is approx. 8 kW.

Two hoses with a diameter of 12 inches will allow an LNG transfer rate of about 4,500 m^3/h with a flow velocity of approx. 9 m/sec. This relatively high flow velocity is made possible by means of a light interlocked metal liner, which is inserted into the corrugated inner hose and which guarantees a laminar flow.

At the tower, the hose connection is 29 meters above sea-level; at the ship's bow, the distance to the sea surface is 17 meters and the total length of the hose is 52 meters. The configuration of the suspended hoses may be regarded as a catenary curve and the hose behaviour in various ship positions can be calculated with the appropriate equation. For a distance of 43 meters between tower and tanker, the calculations show that the lowest point of the hose is 10.4 meters above the sea

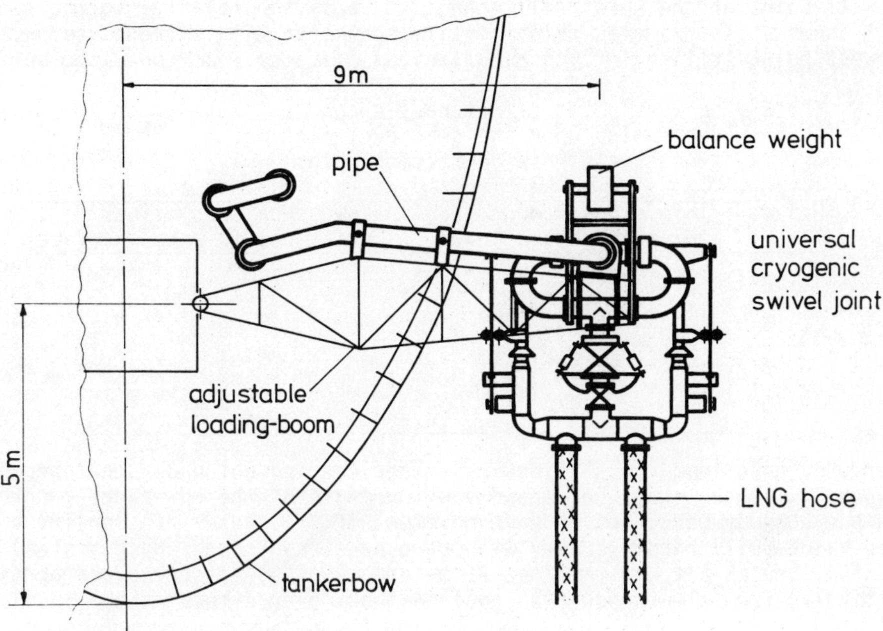

Fig. 8 Universal cryogenic swivel joint (tanker bow)

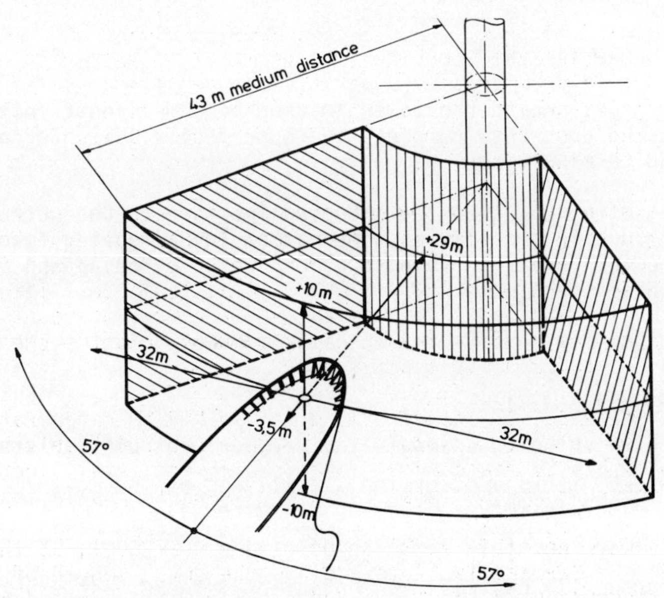

Fig. 9 Moving envelope of an LNG tanker

surface (Fig. 7). On the assumption that the tanker bow carries out heave motions, the lowest point of the hoses will change its position relative to the sea-level. Table 1 shows a few examples which confirm that even with extreme heave-values the lowest point still maintains a sufficient distance from the water surface.

TABLE 1 Lowest point of catenary-shaped LNG-hose depending upon the sea level-distance of the ship's hose connecting point

Heave	Ship's hose connecting point above sea level	Lowest catenary line point rel.to water surface
+ 4	21	12.2
- 4	13	8.5
- 8	9	6.7
- 10	7	4.8

Furthermore, hose load calculations have been carried out under the assumption of various roll, surge, sway, heave and yaw movements of the LNG tanker. For example, the ship's bow may carry out a heave movement of ± 10 meters (in practice such extreme values will hardly occur) without over-stressing the hose system. On the other hand, the ship's bow may move 32 meters (Fig. 9) to starboard (or to port respectively) from a mean position again without over-stressing the involved hoses.

Last but not least the oscillating behaviour of the system was investigated. The pendulum oscillation of the hoses (mainly influenced by the horizontal movements of the ship) has a frequency of 6.7 sec. By contrast, the ship's moving period is over 30 sec. Resonance therefore will not occur and the influence by the ship's movement on the pendulum behaviour of the hose system is negligibly small.

LNG AT-SHORE TANKER UNLOADING

The LNG underwater pipeline and the tower loading systems cannot only be used for offshore services. The described components are perfectly suitable for an unloading procedure, i.e. LNG terminalling.

Instead of using a jetty, on which the LNG pipelines are to be installed and which provides the tanker mooring facilities, one can employ a mooring tower and underwater LNG pipelines, in order to furnish a deep water berthing and unloading facility for land-based terminals.

There are many reasons for considering an "at-shore unloading" system for LNG, for example

- no need for costly harbour installations
- no need for dredging shipping channels for deep-draught LNG tankers
- freedom to locate LNG tanks and terminal facilities at the most suitable coastal place (safety and ecological aspects)
- tanker approach always possible from lee-direction
- easier tanker approach in contrast to conventional tanker manoeuvering (harbour, jetty)

At-shore unloading facilities should be usually installed in sheltered waters near the coast or in estuaries, in order to make them safe, dependable and as insensitive to weather influences as possible. But since the system described in this paper, was developed for "open-sea" employment, the above mentioned restrictions do not apply to a large extent.

The submarine LNG pipeline substitutes the costly jetty, which always would be exposed to tidal forces, currents, exceptional high waves and coastal ship's traffic (Fig. 10). The installation of the LNG pipeline for an at-shore unloading facility poses far less problems in comparison with its complete offshore counterpart. The pipeline manufactured at its offshore place merely has to be pulled into the port of the dry habitat in the tower's base-plate. The other end of the pipeline remains on the mainland, where connecting procedures can be carried out without any difficulties.

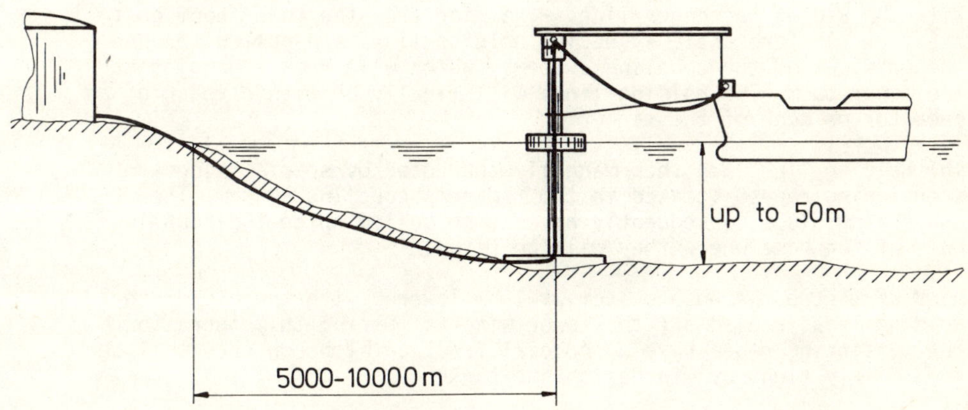

Fig. 10 Principle layout of LNG tanker unloading (at-shore)

In harbours and for approaching a jetty (mooring posts) the LNG tanker is usually berthed with the assistance of tugs. For a tower loading system it normally needs a small service boat only, helping to connect the ship's messenger to the tower's polypropylene hawser in a safe distance between tanker and tower.

The alternative to this method would be a self-mooring, pioneered by Phillips (Wheeler, 1974) or a remotely-controlled mooring (Patent pending for LGA Gastechnik and Howaldtswerke-Deutsche Werft AG). While the Phillips method requires a small boat to care for the mooring line (if the out-going tanker is not immediately followed by an incoming one), the LGA/HDW system is fully unattended. But here the tanker has to approach the mooring tower within some 40 meters. The authors, therefore, propose to furnish the LNG tanker with bow-thrusters, a variable pitch propeller and a bow-control bridge in order to increase the ease of control and the manoeuverability of the tanker. Furthermore, the LNG tanker has to have a special mooring device at her bow. In putting together all the various requirements, it is to say, that a tanker used for offshore and at-shore loading and unloading operations, is to be considered as an integral part of the whole system.

LNG LOADING AND UNLOADING PROCEDURE

The LNG shuttle tanker will be manoeuvred to the loading tower with special electronic nautical aids. As soon as the tanker has been moored, the connecting procedures for the hose system may be started.

Connecting Phase

Apart from a few manual operations, the whole connecting procedure is remotely controlled. The sequence of the connection will be supervised from a special control room, situated on the deck of the tanker's forecastle.

The two hose systems, hanging with their coupling flanges on the Y-shaped end of the tower boom, finally have to be brought to the coupling counter-systems at the ship's bow, and this is carried out as follows.

1. With the aid of messenger lines - hanging from the tower boom on to the tanker's forecastle - special holding-lines are pulled through the orifices of the coupling systems, which will remain hanging at the tower boom. The holding-lines will finally be within reach of an operator on deck of the tanker.

2. The holding lines are then manually connected by special plugs to "tensioning ropes" located in the tanker's coupling system. The connecting plugs subsequently have to be pulled up to the coupling ends of the hose where they will be fixed.

3. As a next step the hose system will be lowered with the aid of the holding-lines reeled off the tower winches. During this operation the tensioning ropes have no special function although they will be kept fairly tight by winches on the tanker.

4. When a hose system has reached a certain position opposite the ship's coupling system (Fig. 11), it is finally pulled against the ship's universal joint with the aid of the tensioning-ropes. As soon as both coupling halves are in contact, they are flanged by remote control. During this operation the two tensioning ropes will align the system even in situations where the ship is carrying out heavy movements. The alignment will improve if the distance between the two coupling halves becomes shorter.

5. As a last step the tensioning-ropes will be disconnected and the coupling system is ready for a possible case of emergency. Then, the tower winches are switched to a constant tensioning force in order to tighten the holding-lines.

The described procedure will last about 30 minutes. Subsequently the ball valves are opened by a remote controlled mechanism, and the transfer system is ready to convey the LNG from the intermediate storage to the tanker.

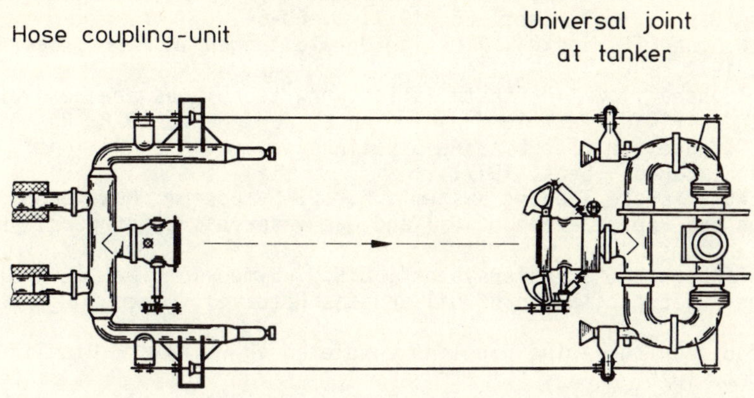

Fig. 11 Hose system connecting procedure

Disconnecting Phase and Emergency Procedures

After the LNG tanker is completely loaded, the normal disconnecting procedure of the hose systems is carried out in the reverse sequence.

In case of emergency, there is one distinct difference: The tensioning-ropes will not be reconnected to the holding-lines. After shutting off the valves, the clamps of the special coupling swing to open position. The hose-side coupling systems are immediately released and they swing away from the ship's bow. Only the holding lines (retaining rope) from the tower boom prevent the systems from dropping into the water. Finally, the hose-systems are pulled up to the Y-shaped end of the tower boom, a position from which a new connecting procedure can be started.

Adjustment of the Tower Boom

Amongst various general aspects there is the necessity to adjust the tower boom, when currents, winds or waves permanently move the tanker out of its original position and out of the large moving envelope (Fig. 9) of the hose loading system.

The tower boom adjustment will be monitored by a measuring device, actuated from the pivoting guide rollers for the holding lines. As soon as these lines and the guide rollers have reached a certain angle relative to the originally perpendicular position, a special electrical transducer will start (and switch off) the motor, thus carrying out the boom adjustment procedure.

REFERENCES

Backhaus, H. W., "Offshore LNG handling system aimed at severe conditions" The Oil and Gas Journal, Sept. 26, 1977, p. 60-64.

Backhaus, H. W., and P. Wieske, "Flüssigerdgasleitungen im Meer", Meerestechnik, mt, No. 6, 1977.

Borelli, A., "Subsea flow line installation, new techniques are ready for use" Magasine Industrie de Petrole dans le Monde, Jan./Febr. 1978, No. 490.

Brown, R.J., "Connection of submarine pipelines in arctic conditions", POAC 77, Newfoundland, Canada, Sept. 1977.

Ehret, I., "LNG Offshore loading system", C.E.C.-Symposium: New Technologies for Exploration and Exploitation of Oil and Gas Reserves, Luxembourg, April 18-20, 1979.

Jones, E., "Offshore LNG Transfersystem", C.E.C.-Symposium: New Technologies for Exploration and Exploitation of Oil and Gas Reserves, Luxembourg, April 18-20, 1979.

Kotani, A., "56-inch submarine pipeline completed in Okinawa", Pipeline and Gas Journal, Oct. 1979.

Van Dyke, B. H. and B. C. Hanley and J. T. Sparling, "Are underwater/buried LNG pipelines feasible?", Pipeline Industry, August 1978.

Veerling, C.W.N., "A submarine offshore loading line for LNG", Third International Conference on Liquefied Natural Gas, Washington, USA, Sept. 1972.

Wheeler, R. J., "A floating pipeline - Producing Ekofisk crude into tankers" Offshore Technology Conference, May 1974, Paper No. 1866.

SUBSEA PRODUCTION SYSTEMS, AN APPROACH TO SYSTEM CERTIFICATION

Ø. Hauan and T. Andersen

Det norske Veritas, Høvik, Norway

ABSTRACT

The paper will focus on Det norske Veritas' approach as to how certification of subsea production systems might be looked upon. A general acceptance philosophy which is based upon risk analysis is discussed. The need for reliability is highlighted. The paper is also discussing maintenance and inspection philosophies related to subsea production systems.

KEYWORDS

Subsea production systems, rules, certification, probabilistic approach.

INTRODUCTION

This paper presents an approach to certification of subsea production systems. The purpose of the certification system is to assure a high degree of safety and reliability of subsea production systems.

The paper starts with an introduction to certification in general. A certification scheme for subsea production systems is then outlined. As a basis for this certification scheme, Det norske Veritas is developing rules for subsea production systems. The philosophy behind these rules is discussed. As an approach to system design evaluation and approval a risk assessment methodology is proposed and discussed (Design Analysis and Loss Control Evaluation).

Some safety and reliability aspects of subsea production systems are finally discussed.

CERTIFICATION IN THE PERSPECTIVE OF OVERALL SAFETY

It is generally accepted that certain elements and degrees of risk are always connected with technological activities. The risks can never be eliminated, but safety, reliability and protection of lives, property and environment can be evaluated and controlled throughout the life cyclus of an offshore installation in order to minimize the risks.

No certification system can offer unconditional guarantees of structural integrity or operational safety. Adequate systems adopted in a proper way should, however, assure a high degree of probability that appropriate procedures, technology and methods have been utilized, and that no recognizable problems have been overlooked. An important part of the certifications programme is thus to identify and assess risks or to be able to focus the attention on critical areas.

The consequences of failure on an oil production installation may be considerable and it is quite reasonable that activities involving new technology, or where present technology is being vastly extended, create a public demand for some sort of certification or independent quality assessment to ensure that human lives, the environment and financial investments are not unduly jeopardized.

In addition to the more traditional concern of personal safety, the governments are getting more and more concerned of the impacts from delays, loss of production of whatever reason, etc., on the overall energy and econimical situation.

Fortunately, in general a mutual confidence between the Petroleum Industry and the Governments and their representatives seems to have been developed, a very important situation contributing to the establishment of sound practice in the offshore development in the North Sea.

It is also recognized that a properly prepared and implemented certification system, managed by a third party, may have many positive effects. These include the possibility of avoiding delays of the development which might be imposed by regulatory or legal actions resulting from lack of suitable standards and lack of public credibility. Properly developed criteria and specifications can aid the designer, and the system can serve as a safeguard against omission and oversight, which can be of special value in hostile environments such as the North Sea.

When talking about the overall perspective of safety one has to consider all systems building up the total installation as well as all phases throughout the life of the installation.

Unfortunately, the responsibility for the different aspects with important safety interphase has been and still is divided between different authorities. E.g. the Certificate of Fitness comprises only parts of the installation and a total safety evaluation considering the interaction of drilling/operation procedures with emergency shutdown, structural reliability and contingency planning in case of major failure, are not required.

In this connection the subsurface valve installed in the production/injection tubing is of special importance. This valve is now subject to detailed requirements in the Outer Continental Shelf of US waters and in Norwegian and Danish waters. Being controlled from the installation, this valve is obviously a natural part of the safety systems protecting the installation.

In particular, the interphase between the production drilling equipment/control system and the process system safety and emergency shutdown system, and the consequence of any fault or failure on the personal safety and installation integrity is not adequately covered within the present verification system.

Naturally, these facts are not unknown to the authorities and strong efforts are made to cope with the situation. However, it is not easy or may be even impossible to prepare regulations ahead of the enormously quick development within the offshore technology.

CERTIFICATION OF SUBSEA PRODUCTION SYSTEMS

Growing installation costs, demand for early and additional production systems, new platform concepts, oil and gas production on increased water depths and the utilization of marginal fields are all factors which have made subsea production systems more interesting in the last years.

To our knowledge there are no specific regulations, codes or standards in existence for this type of installations. As the number of such installations grows, it is, however, likely that the national authorities will pay more attention to subsea installations. Furthermore, many of these are and will be complex installations where the reliability of the system and the need for maintenance will have a great impact on the total field economy. DnV therefore intend to offer certification services for subsea production systems in order to ensure that acceptable safety and reliability requirements are met.

The following stages in the development of a subsea production system should be subject to approval and inspection under this certification scheme:

- Design premises
- Conceptual design
- Detailed design
- Fabrication/Construction
- Installation and tie-in
- Repair and maintenance

Some parts of the system may be subject to type approval instead of specific design approval and fabrication inspection of each item.

In most development concepts, the subsea production system is only a part of the total field development and any certification scheme of such systems must therefore be seen in connection with certification of the total installations.

Extent of Certification Activity

The certification activity will normally cover the following aspects:

- Environment, including environmental conditions as wave, current, sea temperature, earthquake, etc., for relevant temporary and permanent phases.

- Loads, such as environmental loads, permanent loads, transportation and installation loads, flow induced loads and accidental loads. This includes loads on structures, risers and other relevant parts.

- Foundations, including bottom topography, site geology, geotechnical properties, settlement, scouring, etc.

- Structures, including evaluation of materials used, corrosion protection, structural strength, protection against trawling, anchoring and marine operations.

- Mechanical equipment and systems. This includes evaluation of any pressure chambers, piping and manifolds, wellheads, x-mas trees and sub surface safety valves, connectors (for trees, risers, flowlines), etc.

- Control and safety systems. This includes control systems, shutdown systems, emergency power systems and the surface part of these systems.

- Electrical installations, with emphasis on personnel protection, protection against explosion (in dry chamber) and the reliability of any electrical power and electrical control system.

- Personnel safety, in particular for encapsulated systems, such as life support systems, communication, fire protection, etc.

- Intervention methods, such as service capsules, diving bells, submersibles, remote manipulators and their surface support vessels with launching/retrieval gear.

- Pipeline/production risers, i.e. both conventional pipelines and cables, as well as marine production risers to a floating facility, including surface tensioning equipment

Above is listed the most significant aspects related to subsea production systems to achieve a high level of safety and reliability.

To secure the reliability and safety of the system during all phases the subsea production system must be fully compatible with:

- The downhole well completion
- The vessels and equipment used during installation and servicing of the system

The certification scheme should therefore include a scrutiny of equipment and operational procedures intended for use in the following operations:

- Installation, retrieval and reentry.
- Well completion, workover and wireline
- Monitoring, inspection and maintenance

Rules for Subsea Production Systems

In order to have a formal basis for the evaluation of subsea production systems Veritas is developing rules for such systems. These rules are based on normally accepted requirements in the offshore industry.

The Rules will be tentative due to the relative novelty of these installations and the large variety of types. These installations differ from conventional platform installations in that the risk for operating personnel and the investment is reduced while far more emphasis has to be put on system reliability and maintenance operations. These facts led to selection of a risk analysis approach to the Rules. It is intended that this will be most pronounced in the section of the Rules dealing with mechanical and electrical equipment, control and safety systems and safety of personnel. On the other hand, structural requirements will be of conventional nature and with exception of functional loads mainly refer to DnV's Rules for the Design, Construction and Inspection of Offshore Structures.

Philosophy of Rules

In order to evaluate the vast variety of sub-sea production systems it will be necessary to establish some sort of general acceptance criteria. These criteria have also to be set to be able to compare subsea production systems with more conventional production systems and techniques and to use experience from these areas.

A general acceptance criteria can for instance be that the total risk created by an oil field development and production with subsea systems should be comparable and preferably better than for a traditional platform development. The total risk will be a weighed sum of risks for human life and health, damage to environment and major economic losses.

The different risks, i.e. different consequences and their probability for occurence vary from conventional production systems to subsea systems, between different subsea systems and for each system depending on phase (installation, operation, workover).

A total risk assessment for each system to be certified might therefore represent an ultimate goal. For practical certification purpose, however, the acceptance criteria should be expressed as more easily approvable principles and detailed functional requirements.

The specific characteristics of subsea systems are taken into account in addition to normally accepted principles and safety philosophies in the offshore industry.

Among the main principles to be applied in the rules can be mentioned:

Safety Principles

- Fail-safe closing of valves by accidental damage to system.
- By accidental damage to any part of system no hazardous situation should develop without the failure of other elements in system at the same time.
- During installation all subsea connections to hold pressure should be tested.
- During operation no single human error should result in a hazardous situation.
- During the total operational life of system any corrosion, defect on sealing elements, or damage caused by impact should be possible to detect before hazardous situation develops.

Reliability Principles

- The consequences of equipment break down (long and costly repair) justify the use of high quality components with a low failure rate.
- Only components tested for long term reliability in subsea environment should be used.
- System should be designed specially for easy installation and retrieval. Repair of one component should have little influence on the total production rate.
- The components, which by failure would cause close down of all producing wells should be specially designed for high reliability and easy repair. The number of such components should be minimized.

In many cases it may be beneficial and necessary to perform safety and reliability analysis of systems and operations to establish proper design requirements.

DESIGN ANALYSIS AND LOSS CONTROL EVALUATION

Outline of Method

A suitable method for assessing safety and reliability of the system design at a conceptual stage is outlined below. A loss control evaluation including selection of design basis accidents is an essential part of the study.

The method is basically a risk assessment approach utilizing an Event Tree technique. The principles of the analysis is illustrated in Fig. 1.

The major steps in the analysis are as follows:

- Definition of system, system limitations and system interaction
  - Hardware system
  - Environmental conditions
  - Operational interface events e.g.
    - construction
    - installation
    - drilling
    - maintenance
    - recovery and abandonment
  - Third party activities e.g.
    - fishing
    - other offshore activities

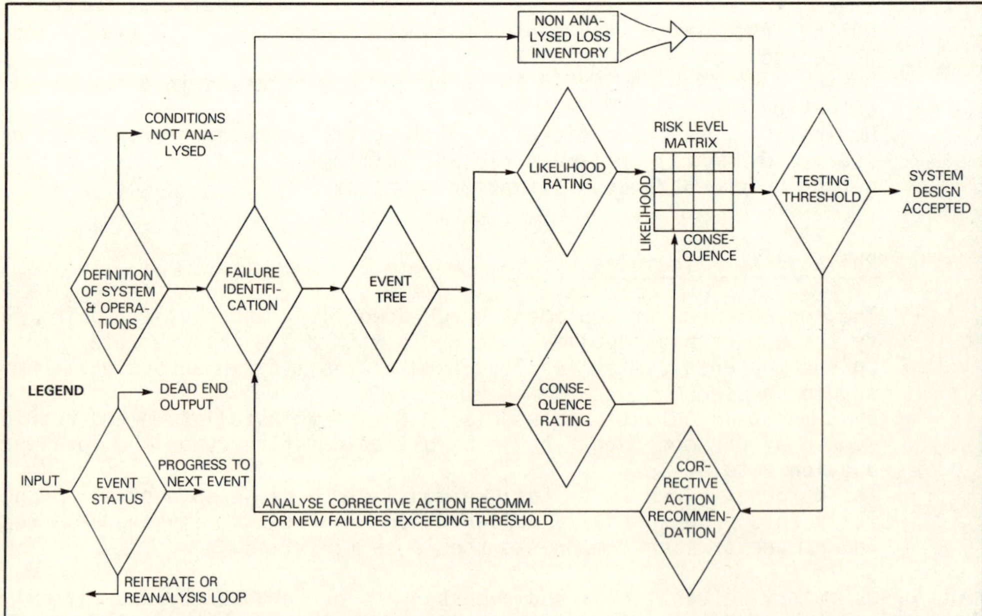

Fig. 1. RISK ASSESSMENT PROCESS

- Failure identification

  Internal and external failure initiating events (FIE), should be identified for each operational phase.

- Event Tree Analysis of the FIE's

  In the Event Tree Analysis each failure or combination of failures caused by the selected FIE's will be assigned.

  - a resulting consequence or system state
  - a likelihood of occurrence of the assigned consequence or system state.

- Risk Level Matrix Assignment

  The expected outcome of identified failures are classified into risk levels by means of a two dimensional matrix format which shows the distribution of consequences and likelihood of the failure. A risk matrix should be established for each of the following:

  - risk to human life
  - risk to environment
  - risk for material losses (structural and economical)
  - risk for loss of operation

  The risk matrix approach is not intended to be an exact mathematic process, but only qualitative in relative terms so that the more severe losses will be identified.

- Risk Level Matrix Evaluation

  The risk level matrix are compared with acceptance criteria.

- Evaluation of Corrective Actions

  For FIE's which violate the acceptance criteria the effect of corrective actions and protective measures are evaluated.

<u>Risk Level Matrix Definition</u>

<u>Consequences or system state categories.</u> The consequences of a failure initiating event in terms of system damage or threat to human safety can be defined by the state in which the system is found after the event has occurred.

The following states can be defined:

A. <u>State of serviceability (availability)</u>

   The system does its designated purpose

B. <u>Unavailable but fail to safe state</u>

   The system loose availability, but no structural damage or system damage will occur.
   Repair or maintenance is necessary to bring the system back in state

of serviceability.

C. Restricted accident state

Structural or system damage may occur, but loss of lives due to process or technology, uncontrolled pollution or major economical losses should not occur. System is repairable.

D. Catastrophe or disaster state

Total loss of system, large scale loss of lives or uncontrolled pollution have happened.

Likelihood categories. The following likelihood categories can be defined:

1. May happen once or several times every year. Will always occur or occur a high percentage of time.
2. Likely to occur one or more times during designed lifetime of system. Will occur a moderate percentage of time.
3. Reasonable probable to occur during lifetime of system. Probability to occur $\gtrsim$ 5%.
4. Occurs very seldom ($\lesssim 10^{-4}$ per year)

Risk Level Matrix Discussion

The previous definitions of states gives the different possibilities in which the system can be found. Transititons between states will take place depending on internal and external events like component failure, maloperation, load impacts etc..

A philosophy can be established which form a basis for controlling these states and transitions. By designing components for safety according to recognized codes and standards, a reasonable probability to remain in service is obtained. This is a way to control serviceability of the installation and to verify the system serviceability against most probable disturbances.

Compared to platform installations, a subsea installation is likely to result in larger production downtime and economic loss in case of loss of serviceability. This is mainly due to the more costly repair and maintenance procedures and the longer repair times which can be anticipated.

In addition to economic consequences, higher risk to human lives and pollution compared to platform installations is expected during repair and maintenance to bring the system back to serviceability. An assessment of increased efforts in terms of stringent requirements to quality control, inspection, pre-installation testing and reliability analysis will be made on component and system levels in order to increase reliability and mean time between failure for the total system.

If abnormal situations arise which makes the system unavailable, a transition to state B (see above) should prefereably take place. A usual criterion is that any single basic event which results in a loss of availability should lead to a transition to a fail-safe state (state B).

The basic undesirable event could be a failure in a component or in an operation procedure. The transition to a fail-safe state is normally obtained by fail-safe design and at least two levels of protection. The two levels of protection should be independent of each other and used to protect equipment and systems from entering an accident or disaster stage.

To control a transition to a fail-safe state if an undesirable event takes place, one way is extensive use of safety features. However, any component added to the subsea system may lead to reduced serviceability and hence increase in maintenance and repair. The obtained failure control must therefore be weighed against loss of serviceability. Compared to platform installations, a reduction in requirements to number of components installed for safety reasons might therefore be relevant and instead as mentioned increase the effort to obtain high system reliability.

A transition from A (or B) to C might take place if an accident occur. This is a controlled state where loss of lives , pollution or major economical losses should not occur, but where structural or system damage is possible. This is in contrast to state D where uncontrollable situations esist. The probability of entering these states should be decreasing from C to D. Since the consequences of such accidents may be large, they must be checked for events with small probability.

Assessments should be made to establish design events for which the system will be able to withstand without leaving state of servicecability, and design accidents for which the system should not enter catastrophe or disaster stage.

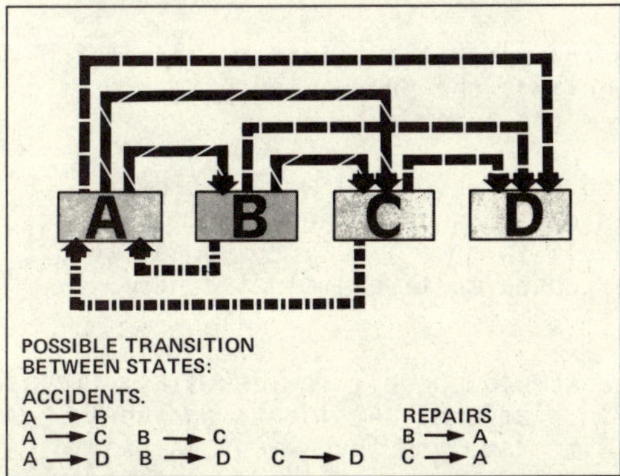

FIG. 2. POSSIBLE TRANSITION BETWEEN STATES.

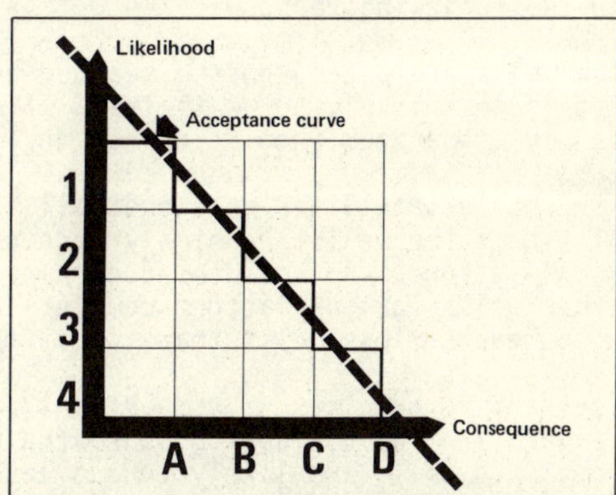

FIG. 3. POSSIBLE ACCEPTANCE LEVEL.

RISK ASPECTS

Below is found some basic ideas related to risks in connection to subsea production systems.

Installation Phase

As each part of a subsea completion system has a lower investment cost, the consequence of a single loss will be smaller than for fixed platforms.

The following accidental events can be foreseen:

- Loss of subsea template and/or manifold centre structure during tow-out and installation
- Damage to well during well completion and x-mas tree installation
- Damage to well and x-mas tree during flowline laying and tie-in
- Diver accidents during installation and hook-up of equipment

If the well is not properly killed a damage to the well and wellhead may result in oil pollution to the sea.

During installation of subsea equipment events may occur which will result in loss and founder of the vessel taking part in installation, but this is not specific for subsea completions.

Maintenance and Operation Phase

Subsea completions are likely to result in more operational nuisances and production downtime than with on-platform well completion.

Consequences to human life and material properties will normally be smaller due to the following reasons:

The wells are placed on the sea-bed away from the production platform and a blow-out is less likely to be ignited. If it is ignited, the production platform is safely separated from it or may move away from the hazardous area.

For single satellites only one well is affected in an eventual well accident. On platforms the wells are highly concentrated in clusters of 20 wells or more on each platform. An accident occurring at one well is more likely to spread to other wells for on-platform completions. One subsea multiwell cluster normally accommodates less wells than a platform.

Contrary to on-platform completions, a sub-sea completed well is less likely to be damaged from an accident onboard the production platform. Accidents on sub-sea completed wells are likely to put less human life and less investment values in danger than on-platform completed wells.

In case of a gas blow-out, or rupture of a gas riser a floating producting facility laying above the gas plume may loss much of its buoyancy.

A blow-out may occur during workover, x-mas tree retrieval or while doing some other work on the well system or x-mas tree. As such situations there will be a service vessel above the well. These vessels will often be capable of moving away from the area, before the situation becomes catastrophic.

Pollution. In case of a blow-out from a subsea completed well, the only way to stop the blow-out at the time seems to be by drilling a relief well.

The x-mas tree and the flowlines are vulnerable to external damage from fishing and anchoring. Experiences from North Sea also indicate that there may be some problems with internal corrosion and erosion in the two-phase flowlines from the wells to the processing facilities.

Small leaks may develop in the flowlines connectors, riser connectors, in seals and gaskets.

The hazard of environmental pollution therefore seems larger. Especially more small leakages and spills are likely to develop.

When subsea installations are placed in areas with large fishing activity or where much anchoring of ships may be anticipated, the installations should be protected against anchors and fishing gear.

The type of events can be summarized as follows:

- External damage to multiwell cluster resulting in:
  - multiwell blow-out
- External damage to wellhead and x-mas tree resulting in:
  - blow-out of well
  - loss of well without blow-out
  - loss of x-mass tree
- Blow-out during well maintenance (wireline, x-mas tree retrieval, workover)
- Damage to flowlines caused by
  - external force
  - corrosion/erosion

  resulting in oil spills and repair
- Damage to riser system from corrosion, environmental forces, mooring failure of floating production facility or failure of riser tensioning system.
- Damage to personnel capsule for encapsulated system
- Fire in personnel capsule for encapsulated dry system
- Entaglement of diver or submersible in subsea installation

CONCLUSION

Some of the above mentioned risk aspects are related to specific subsea production concepts, others are related to locations.

Due to above mentioned factors, and due to the fact that a rapid development of the hardware of subsea production systems is anticipated, it is not possible to cover all the risk aspects in detail in the tentative rules.

The rules have to be flexible and kept on a general level. They will aim at outlining a safety and reliability level which should be achieved and state some general principles on how to obtain this safety level.
We therefore find it prudent that a risk assessment approach is used for large parts of the rules.

OPERATIONS SYSTEM

REVIEW OF REMOTE CONTROLLED AND ATMOSPHERIC VEHICLE SYSTEMS

R. F. Wharton

Wharton Williams Limited, Underwater Engineers and Diving Contractors

ABSTRACT

This paper deals specifically with underwater operations that do not involve divers, i.e., operations that involve remote controlled and manned vehicles where the operators are at atmospheric pressure.

These vehicles can be divided into five groups:

Remote Control Eyeball Cameras

Remote Control Pipeline Inspection Vehicles

Remote Control Platform Work Vehicles

One-Manned Atmospheric Vehicles

Conventional Submersibles

The paper discusses these various groups in detail and their ability to fulfill the three basic requirements essential for machines to progressively replace man as set forth by the British Advisory Group for the Progressive Replacement of Man Underwater.

These criteria are as follows:

Hand-eye coordination

Tactile sense

Ability to react to forces

The paper also discusses the future potential for these vehicles and their ability to replace man and do meaningful work. And it attempts to reach some conclusions about the future requirements for water depths between 300-500 meters and to define areas in which technical research should be focused.

INTRODUCTION

In writing a review paper on operations systems for deep water work we are, of course, talking about diverless systems. However, before finally abandoning the diver, it is necessary to recap his limitations.

The current realistic limit of manned hyperbaric intervention is on the order of 500 meters. Divers have been deeper than this in controlled experiments, but at 500 meters the density of the breathing medium is so high as to virtually preclude bodily effort other than the act of breathing.

In practice, I do not think that divers are likely to operate regularly at 500 meters but could be used to that depth for specific tasks. Between 300-500 meters, however, I do feel that diving will become common and cost effective.

I remember in 1973, with great trepidation, carrying out the first dive at 600 feet in the North Sea. The event was attended by doctors and hyperbaric specialists and was regarded as the frontier of technology. There are now hundreds of divers in the North Sea working regularly to this depth as a completely routine operation. We must then assume that if we go deeper, divers will continue to be used down to 500 meters.

When we are talking about operations systems, we are talking about alternatives to divers.

There are a number of alternatives, but none are ideal. In trying to develop a system that can replace the diver underwater, it is necessary to define in detail the essential functions of the diver that the machine must replace. In this context, we actually mean replacing a diver to do work, as it is now quite standard to replace the diver's visual capacities with television.

In performing work, the diver has three essential abilities:

Hand-eye coordination

Tactile sense

The ability to react to forces

These essential abilities must be built into a machine before it can attempt to do work previously done by a diver.

REQUIREMENTS. THE PRESENT.

The first criterion is self-explanatory. Tactile sense is the ability of a diver to feel an object in zero visibility and recognize it. This is the most serious limitation on vehicles and effectively restricts them to operations where there is good visibility.

The ability to react to forces must not be underestimated. To every action there is an equal and oppsite reaction, and this is never more true than underwater. The diver subconsciously wraps his arms and legs around objects underwater to react to forces that he wishes to apply, such as tightening up a nut on a stud. The diver uses his body automatically provide an exactly equal and opposite reaction to the force he is applying. Currently some machines can provide limited restraint to react force, but this is usually restricted to one or two dimensions.

The problem that the oil industry faces, is that the diver is used in all phases of offshore work. He works on drilling rigs, on derrick barges, on pipelay barges, on bury barges, on maintenance work, on subsea completion work, etc. He performs an enormous variety of tasks, and there is probably no one individual who is aware of all of them.

Hitherto, there has been no real pressure on the oil industry or among the underwater contractors to provide alternative systems, although most operators and contractors now realize the coming need.

The result of this has been piecemeal development by a few forward-thinking organizations.

These developments have been contractor led rather than market led. Undoubtedly, as necessity is the mother of invention, when the real need is there, there will be a sharp increase in successful equipment development. However, we have to analyze what is available today and in which direction the industry is likely to go.

In examining the present state of the art, it is possible to divide the existing vehicles into five main groups:

Remote controlled eyeball cameras

Remote controlled pipeline inspection vehicles

Remote controlled platform work vehicles

One-manned atmospheric vehicles

Conventional submersibles

Each group will be discussed below.

The Remote Controlled Eyeball Camera

Over the past three or four years, these have become very accepted and reliable vehicles. They can be extended to carry out a limited range of tasks, such as still photography and cathodic protection readings, but their basic function is to act as a self-propelled, remote television camera.

Remote Controlled Pipeline Inspection Vehicles

While not as advanced as the eyeball vehicles, these can now be regarded as current technology. They have the ability to swim buried pipeline routes and to measure depths of bury and cathodic protection potential.

They are also able to cover substantial distances in one day of survey and have very rapidly displaced the manned submersible in this mode during the last two seasons.

Remote Controlled Platform Work Vehicles

This is a very new area of technology in which no successful vehicles have yet been produced. This is because to build a remote-controlled platform work vehicle one has to build into it the three previously mentioned criteria, not one of which has as yet been correctly reproduced by machine. This is, however, the most interesting and rewarding area for development work.

One-Man Atmospheric Vehicle

The one-manned atmospheric vehicle comes halfway between the diver and the remote controlled vehicle. Its biggest negative factor is that it still has a man inside and, therefore, must provide some element of risk to the operator which is not present when using remote controlled vehicles. However, the one-man atmospheric vehicle is likely to provide coverage for the interim period of five to ten years while sucessful platform work vehicles are developed. The greatest advantage of the one-man atmospheric vehicle is the presence of an intelligent operator at the site, who has a sense of spatial perception which cannot be reproduced in the controls of a remote controlled vehicle. This class of vehicle will be described in more detail later in this paper.

Conventional Manned Submersibles

The conventional manned submersible is already an accepted part of the offshore business and may be regarded as a classic technology. The utilization of this vehicle has been stretched substantially over the last five years, and its particular advantage is its payload-carrying capacity which is absent from most remote controlled vehicles. Its main disadvantage is its size and complicated support equipment required. It is interesting to note that the main market for conventional submersibles has historically been pipeline survey work. This sector of work is now being substantially taken over by remote controlled vehicles.

FUTURE INTENSIVE DEVELOPMENT

The purpose of the above listings is to highlight the areas for future intensive development. The ultimate aim is to develop a successful remote controlled platform work vehicle which can reproduce the principal function of the diver:

Hand-eye coordination

Tactile sense

Ability to react to forces

This utopian development is still in the future. In the interim period, one manned atmospheric vehicles will undoubtedly be developed. If we can learn to accomplish all required functions with the manned vehicles, then the task of removing the operator and controlling the vehicle remotely will be relatively simple. To attempt immediate development of unmanned work vehicles, however, would require considerably more time.

We are, without doubt, facing a major development problem. It is apparent that many manufacturers have failed to analyze in detail the exact requirements for diverless systems and are all busily engaged in building a better "mousetrap". Most manufacturers are developing systems with existing offshore structures and pipelines in mind. It must be recognized that it is virtually an impossibility to build remote controlled equipment that can perform all the repair functions on a present day North Sea platform. At some stage, there must be a major and fundamental cooperation between the designers of offshore installations and the designers of remote controlled vehicles. There are already encouraging signs of a tentative move in this direction in regards to the design of subsea drilling equipment enabling remote controlled and atmospheric vehicles to support deep water drilling operations.

Another major problem facing the designers of remote controlled vehicles is that of identifying a standard chassis. There is already a proliferation of different chassis types and undoubtedly a great deal of designers' efforts in imagination and money is being applied to building a new chassis for each vehicle developed.

If we look at the future developments of remote controlled vehicle systems over the next few years, some obvious trends are already visible.

Firstly, we can expect to see technical refinements and lowering of costs for eyeball remote controlled television systems.

We will see improvements of reliability and capability in pipeline inspection vehicles.

We will see a relative static development in conventional submersibles with all attention being paid to peripheral systems to utilize their payload capacity.

We will see a lot of money spent and wasted in trying to produce viable platform work vehicles.

Finally, we will see a considerable technical move forward in atmospheric systems, since we are not yet ready to dispense with an operator.

A number of atmospheric vehicles have been produced, ranging from self-propelled atmospheric bells carrying a two-man crew to the one-man atmospheric vehicles of the JIM, WASP, MANTIS and SPIDER types.

ONE DEVELOPMENT: THE SPIDER

My own company has put a considerable amount of effort into the development of the SPIDER, which is by far the most technically advanced of the one-manned atmospheric vehicles and is now commercially operational. When developing this vehicle, it was imperative that the following criteria be met:

Maximum operator safety under all conditons.

Powerful free swimming mid-water capability.

Sophisticated launch and recovery system.

High quality arm functions

Ability to react forces.

Spare power and payload.

Dealing first with the area of safety, we endeavored to develop a vehicle that could be used independently in an area where deep diving back-up is within 48 hours travel time away. The SPIDER has, in addition to its hard wire communications through the umbilical, a through-water communications system. In the event of the umbilical becoming snagged, the operator can release the umbilical and still maintain through-water communications. Having released the umbilical, he has forty minutes autonomy at full power provided by onboard batteries.

He also has the ability to release ballast to provide substantial buoyancy to return the vehicle to the surface.

In the event that the vehicle surfaces taking on water, it is equipped with an explosively fired buoyancy collar which is controlled by the operator from inside the vehicle. Additionally, the operator is able to remove the main dome himself from inside the vehicle.

The vehicle is equipped with two suction pads which are fail-safe. Their suction is derived from pumping a flow of water across the pads, and a power failure results in immediate loss of suction.

We provided the vehicle with hydraulically-powered hands, to provide powered opening and closing, and the ability to spin the hands continously in either direction. This gives rise to the slight possible danger of a power failure occurring when one of the hands is gripping a structure underwater. Should this happen, the operator is able to sheer off the hand using a hydraulic hand pump inside the vehicle without effecting the pressure integrity of the vehicle.

The SPIDER is built entirely of glass-reinforced plastic and is Lloyd's classified to 2,000 ft. It has a sophisticated handling system fitted with 2,000 ft. of umbilical.

The SPIDER is still a long way from replacing the diver but, compared with the previous generation of atmospheric vehicles, it represents a substantial jump in technology.

A realistic rating of the SPIDER's ability to meet our three specifications for replacing the diver is as follows:

Hand-eye coordination - 80%

Tactile sense - 10%

Ability to react forces - 30%

It can be seen that the industry still has a long way to go before it can reliably dispense with the diver, but the SPIDER represents a dramatic step in that direction. It is able to effectively provide the support required on deep exploration work and can carry out a substantial range of non-destructive testing inspection tasks.

INSPECTION OF PIPELINES, RISERS, STEEL OR CONCRETE STRUCTURES. A COMPREHENSIVE RANGE OF EQUIPMENT TO BE OPERATED FROM MANNED OR UNMANNED SUBMERSIBLES

K. Larsen* and Y. Durand**

*Kvaerner InterSub A/S, 3401 Lier, Norway
**InterSub Développement, Rungis, France

ABSTRACT

During the past few years, all the major operators of manned and unmanned submersibles, particularly InterSub, have developed a number of new inspection devices and techniques to provide efficient service to obtain reliable information on the condition of the most important underwater installations.

Pipelines: InterSub has developed and operates a broad range of equipment to track buried pipelines, measure the depth of cover, establish longitudinal and transverse profiles, monitor the cathodic protection and condition of anodes, etc. In addition, this equipment is now fully automatic, and is installed on an unmanned vehicle (TROV) operated from a dynamic positioned vessel and this delivers the resulting pipelogging data in real time.

Steel platforms: Manned lock-out submersibles constitute the safest and most efficient package to perform many inspection tasks on steel platforms. Special diver heating and breathing systems have been developed to extend the bottom operational time. A broad range of mechanical tools and an accoustic weld inspection system (AWIS) have also been developed and tested, to detect cracks and most important, defects in node welds. The system does not rely on diver expertise to provide reliable results, as sensor data are processed and displayed by computer. The system is based on the properties of acoustical holography, and uses a flexible matrix of electronically focused ultrasonic transducers.

Marine risers: Internal corrosion of marine risers is a major problem that has to be monitored frequently and accurately. A special version of the AWIS has been tested on pieces of corroded risers. The service is operational. Surface of corroded area can be evaluated to an accuracy of 5%, and the depth of corrosion and average reduction in thickness measured to better than 1 mm. Data processing and interpretation is performed by computer.

Concrete platforms: An ingenious system has been designed to detect cracks in concrete by measuring the electric current produced by the corrosion of steel reinforcing and is currently being tested at sea. A doppler sonar navigation system is used to fix the exact position of defects.

KEYWORDS

Underwater inspection, real time pipelogging, acoustical weld inspection, internal corrosion, accurate navigation.

INTRODUCTION

The inspection of offshore installations and structures is now of critical importance as many platforms and pipelines in the North Sea are nearing full production. The adverse weather conditions of the area have subjected offshore structures to high fatigue rates and corrosion factors which make it essential for the offshore operator to know and understand what is happening to their platforms and to be able to take any necessary remedial action in time. To this end inspection techniques have been developed to provide Inspection and Corrosion Engineers with definitive data to enable them to make the necessary decisions to maintain the structure in a fully operational state.

The equipment necessary to undertake the required inspections has been developed with an eye to the future so that when the water depth increases to such an extent that divers and manned submersibles are no longer cost effective then the equipment can be easily adapted for the remote unmanned vehicle, which has already been achieved for pipeline inspection. The inspection techniques of pipelines, risers and platforms are discussed to show how a system´s approach has radically reduced the time required for inspection, increased the amount of data, and now provides hard copy for all inspection results.

PIPELINE INSPECTION - MANNED SUBMERSIBLES

Most pipelines in the North Sea have been inspected by a submersible which can provide a fully documented permanent record of the pipeline to include: Video-recordings, Current Density, Cathodic Potential, As-laid, As-trenched and As-buried conditions (Larsen, 1979). This gives the pipeline inspection engineer most of the information required to analyse the status and life of the pipeline.

The support ship should be capable of operating two survey submersibles enabling a 24 hour operation. The ship should be fitted with thrusters to safely operate in close vicinity of structures and barges and also have the space to house the men and equipment to process all data and drawings on site. The general configuration of a manned submersible pipeline inspection spread is shown in Fig. 1.

To undertake this survey the full capabilities of the observation submersible (Fig. 2) are realised. The following equipment is fitted in the submersible and kept operational:

- SHORT BASE NAVIGATION to position the submersible on the seabed interfaced through the computer with the surface navigation system to fix the geographical location in real time.

- PIPETRACKER to permit the submersible to locate and follow a buried pipeline to to a depth of burial of 3 m ±10 cm depending on the size of the pipeline. The system induces eddy currents in the metallic mass of the pipeline and measures the resulting field which gives the location of the pipeline both verticacally and laterally (Durand and Stankoff, 1978).

- CONTINUOUS SEABED PROFILER which combines an Echo Sounder and a Pressure Sensor, uses the surface sea level to give a continuous seabed profile related to absolute depth.

- TRENCH PROFILER to produce a profile of the seabed every 25 m at a normal survey speed of about one knot. The 140° sector scan cycle lasts about 60 s which includes data recording. The effective scan width is about 20 m with the profiler 4 m above the seabed.

- CURRENT DENSITY measures the D.C. Currents produced by the anodes which vary between 25 mA/M^2 and 500 mA/m^2. This gives a curve showing current densities along the pipeline whether it is buried or not (Bournat and Stankoff, 1979).

- CATHODIC POTENTIAL to provide potential readings of the pipeline giving the cathodic protection level being afforded.

- DATA ACQUISITION UNIT to record all information on magnetic tape for processing on board the mothership after the dive.

- VIDEO RECORDED WITH TIME to produce a continuous video tape of the trench and pipeline when visible ensuring that all information is visually recorded for future reference.

- STILL CAMERA to take colour photographs of any interesting areas: damage, debris, etc.

- LEAK DETECTION which is not yet fully proven at sea includes a fluorometer leak detector system which continuously measures the hydrocarbon content of the seawater. The system combined with an acoustic hydrophone to monitor the noise level allows very small leaks to be accurately pin-pointed.

The presentation of data scales, etc., will be chosen in consultation with the clients, pipeline engineers and surveyors to provide optimum information and details taking into consideration the number of charts to cover the area, cost and time.

PIPELINE INSPECTION - UNMANNED VEHICLE

Advances in technology have made it possible to integrate all inspection systems into a remote, unmanned tethered inspection vehicle for pipeline inspections.

An anchor handling vessel was converted to be fully dynamic positioned and to operate both an unmanned tethered underwater vehicle (Fig. 3) and a manned submersible.

The basic unmanned tethered vehicle has been specially designed (Durand and Le-Bouteiller, 1980) to carry the sensors for pipeline inspection (Fig. 4).

To carry out a full inspection the following equipment is mounted on the unmanned tethered vehicle:

- SHORT BASE NAVIGATION SYSTEM is an acoustic system which provides an accurate position of the vehicle in real time, relative to the support ship axis. The accuracy of the system is proportional to the depth of water and the horizontal range:

  Better than 2% of water depth within a horizontal range of 2 x (water depth).
  Better than 1% of water depth within a horizontal range of 1/2 x (water depth).

  The sea state and structure of the water will also affect the accuracy of the equipment. The short base system is interfaced with a surface navigation system through a computer with a DP3 plotter giving the coordinates of the vehicle in real time. The greatest expected error of the combined systems will statistically be:

2.6 x (sum of maximum errors of both systems)
where 6 is the standard deviation.

Any Surface Navigation System can be used such as Decca Pulse 8, Hi-Fix, Argo, Trisponder or Syledis to name a few.

- DOPPLER SONAR continuously measures the speed of the vehicle in relation to the longitudinal and transversal axis. This measurement when integrated with time thus enables the distance run by the vehicle to be known with great accuracy in relation to a specific reference point.

- OPTICAL SYSTEMS are fitted on the vehicle. A Sub Sea Systems CM-8 Newvicon camera performs two functions. It will allow the surface controller to see where the vehicle is going and also provide a visual record of the inspection. For detailed photographic work of specific areas of interest colour stills are taken using a Benthos 372 still camera.

Equipment earlier described for the manned submersible as pipetracker, continuous seabed profiler, trench profiler, current density, cathodic protential and leak detection systems are also fitted to the unmanned vehicle.

All information is by the various sensors continuously transmitted to the computers on the surface. This information coming from the vehicle is collected at the same time as the other relevant data on the surface (short base, surface navigation) by the first computer which provides in real time a plot of the vehicle´s position, thus allowing the route to be plotted point by point.

This information is then passed to the second computer which processes in real time, the data from the different sensors, calibrates them, corrects and interrelates them if necessary and finally plots them in the form of curves as a function of the distance run.

The Support Ship positions herself in the immediate vicinity over the pipeline to be surveyed and launches the tethered vehicle. The vehicle is then dived to the seabed and controlled via the umbilical. When in position over the pipeline all systems are tested and recorders checked to ensure no malfunctions.

The vehicle is now electronically locked onto the pipeline via the pipetracker coupled to an automatic pilot which ensures that the vehicle follows along the pipeline and directly over it. The active SIMRAD short base navigation system interfaced with the G.E.C. dynamic positioning ensures that the ship remains directly over the underwater vehicle. Offsets can be fed into the system to allow for umbilical drag.

Figure 5 shows a print out of the data as provided by the plotter connected to the computer. The example is relevant both for manned and unmanned pipeline inspection.

From bottom to top the curves represent successively:

- The respective positions of the pipeline and the bottom of the trench, which allows a rapid evaluation of the areas where the pipeline is covered (shaded) and those where it is visible on the seabed or in the trench.

- The height of the covering which is positive if the pipeline is covered (shaded) and negative if the pipeline is exposed. In the case where the height of the "negative" covering is greater than the diameter of the pipeline, this indicates that the pipeline is suspended and is no longer resting on the bottom of the trench.

- The route followed by the submersible in relation to the ais of the pipeline (lateral distance).

- The section of the trench, perpendicular to the axis of the pipeline, shown at intervals of the order of 30 to 40 m.

- The measurement of current density where each large peak represents an anode, from which the value of the current passed is shown above. (This has required integrating the peak as a function of distance, this calculation being done after the initial plot).

It should be noted that the horizontal scale represents the extended length of the pipeline.

These examples represent the basic elements of the composite report that will be passed to the client following the inspection operations.

RISER INSPECTION

Monitoring and maintenance of marine risers is essential as any damage to the risers can result in a partial or total shut down of production, with significant economical consequences.

Significant damage can manifest itself as:

- A geometrical deformation, easily identified by a visual survey.

- Some external corrosion evaluated through visual surveys and potential measurements.

- Internal corrosion or erosion frequently localized (pitting).

- Fatigue cracks.

The acoustic imaging system presented is based on acoustical holography and has shown major advantages over all other known system.

- Automation: once installed on the area selcted by the diver, the sensor scans electronically a complete volume of the metal.

- Efficiency: the dive time is only used for data acquisition. Interpretation and reporting are performed off-dive, via a computer processing of the digital magnetic recording of the acoustic images. A conventional cleaning only is required.

- Technical quality: the system provides a high resolution acoustic image of the metal volume, through adaptive electronic focusing.

The maintenance engineer will be able to determine objectively the maximum operating pressure of the riser, to verify the efficiency of the corrosion inhibitors and to plan in advance the replacement of damaged sections when necessary.

The principle of operation is presented in Fig. 6. The acoustic sensor consists in a flexible matrix of 160 acoustic transducers, arranged into 5 rows of 32 elements, and conformed to the surface of the riser. To inspect one point of the volume, twelve elements among 32 in a given row are simultaneously activated with twelve bursts of r.f. power (frequency of r.f. signal: 2 MHz), but properly phased with respect to an internal reference clock.

As shown in Fig. 7 and 8, the zone of interest can be located either before or after a reflection of the beam on the bottom of the riser wall, depending on the type of inspection required.

The present acoustical imaging system (Stankoff, 1980; and Stankoff, Guenon and Thomas, 1980) is operated from a diver lockout submersible. The pilot, the diving supervisor and the system operator dive in the one atmosphere compartment, with a wide viewport allowing direct control of the operation. The divers are in lockout compartments allowing intervention at the specified depth of inspection.

In the past, such vehicles have suffered from limited autonomy due to electric power and gas storage limitations. Recently the power capabilities have been significantly extended with the introduction of efficient heat storage units (a small molten salt storage unit providing up to 30 kWh), independent of the submersible battery, and allowing heat for two divers and the diving compartment for 4 hours. The gas capabilities have also been extended ten times with closed circuit breathing systems and nowadays, the lockout submersible can compete with conventional diving bells.

The lockout submersible can operate on the seabed as shown on Fig. 9, to inspect riser bends, pipelines or platforms nodes.

The acoustical imaging system consists of:

- The acoustic module generating the phase reference signals and the addresses corresponding to the inspection of a given point of the volume image.

- The display processor including the memory for storage of the image volume, and appropriate electronics to build up a video image of its projections.

- The display module consisting of a video monitor and remote controls.

- The digital recorder and its interface for recording the acoustic image.

These components are connected to the sensor via a 16 m umbilical and diver pack which contains the transmitters and the receivers feeding the 160 transducers. The acoustic sensor can be installed in two types of housings depending on the type of inspection. For diameters greater than 25 cm and unrestricted access, which is the case of most riser inspection, the acoustic sensor is installed in a light aluminium frame, with adjustable stubs to take into account the curvature of the riser and secured to the riser with a strap (Fig. 9). This allows either inner corrosion evaluation or crack detection. For large diameters, restricted access or inspection across the cylinder axis, another type of mechanical frame should be used.

A typical riser inspection falls into three steps:

- After positioning of the lockout submersible the diver cleans the riser area to be inspected with a rotating brush. The degree of cleanliness required by the system is not as critical as it is for magnetic particle inspections.

- The diver marks the surface to be inspected in order to show clearly the successive locations of the sensor positioned (the size of a location is 100 mm x 100 mm in order to allow overlapping between successive images).

- The diver positions the sensor on a reference location and acquires four separate acoustic images to ensure redundancy. If satisfactory, the acoustic images are recorded by the cassette recorder and the diver proceeds to the next location.

Although the acoustic image is acquired in 4 s most of the time is spent installing the sensor. The average time required to inspect a location is 2 min. In more complex configurations, such as weld inspections on platform nodes, the time per location could be in the order of 5 min.

To simulate a riser inspection (Fig. 10), the test piece was marked with paint at the locations (zones 1 to 9) where the sensor should be applied over the defective area. Repeated inspections showed that a diver without special training, can perform such inspection within 12 and 18 min.

Partial results (relative to zone I to 3) are shown on Fig. 10 (the depth resolution is 1 mm in this case): defect A is well shown with the proper depth, but defect B is oversized due to the lack of longitudinal resolution.

A detailed comparison and measurement was performed on a North Sea piece of riser between reference mechanical measurements and the acoustic data, both relative to 85 well identified pits. The results were found as:

- 91% of the defects were well represented
- 100% of the defects were partially or totally imaged
- The average error on depth evaluation was 0.42 mm
- The fluctuations were 1.1 mm in good agreement with the depth resolution.

The Fig. 11 shows the extent of the corrosion as a function of the corrosion depth on the four different zones where acoustical images have been recorded. The curves exhibit high similarity indicating as homogenous corrosion stage on an average point of view. As these curves seem to characterize the corrosion level at a given place, several parameters were extracted to summarize the situation:

- The depth Dm corresponding to a corrosion extent half of the total corroded area.

Although the quantification occurs through 1 mm steps, this value can be determined more accurately and appeared very reproducible during repeated inspections (within 0.2 mm).

- The depth Dmax. corresponding to a corrosion extent equal to 5% of the total corroded area. Such a depth will represent the depth of the most severe pits.

- The total extent of corrosion.

- A more comprehensive representation of the corrosion patterns can be obtained if the defects only deeper than a given depth are represented on a corrosion map.

In order to calculate the pressure reduction inside the riser, the maintenance engineer needs an estimate of the depth and size of the major pits. From a conservative point of view, the depth Dmax. and the the constructed maps at different cuts can be used to evaluate the pressure reduction inside the riser.

If the corroded zone is inspected at regular intervals the evolution of the corrosion can be monitored accurately through the behaviour of the above parameters and the date of replacement of the riser forecast with some certainty.

It has been concluded that the system can provide the following information of major interest:

- Full coverage of the area surveyed without dead zones.
- Localisation of the corroded parts can be achieved within ± 5 mm.
- The depth of individual defects is evaluated with ± 1 mm and the corroded area within ± 50%.

In the case of crack detection and evaluation, it has been shown offshore that surface cracks larger than 10 mm and deeper than 1.5 mm are clearly represented.

The system has indicated to have the following advantages for offshore operations:

- The time spent by the diver to inspect a riser is approximately 15 min per 0.1 m surveyed.
- The system is especially easy to operate in the case of a riser inspection and does not require any special training from the diver.
- The inspection is performed from the outside of the riser, thus avoiding a shut down of production.
- The dive time is essentially dedicated to data acquisition. The analysis is performed off-dive by a computer providing the reporting documents.

INSPECTION OF STEEL STRUCTURES

The system described for inspection of marine risers has also been modified for use on steel platforms enabling weld inspections to be performed (Stankoff, 1980).

The equipment provides both an optical view of the weld surface and an acoustic image of the volume of the weld.

The typical operational configuaration employ a lockout submersible attached to the platform with the diver at the weld inspection site (Fig. 12). The external weld surface is cleaned and reference marks inserted to provide accurate identification of the inspection area as for the riser inspection. The entire weld volume is inspected and the data recorded in the submersible via digital and/or video tape.

The system has been designed to perform inspections on steel bracings with diameter greater than 1 m and with thicknesses between 2.5 cm and 7.6 cm. The relative angle between two bracings cannot be less than 30° or greater than 150° to allow allow complete inspection of the weld zone. The area of each focal plane explored for a given position of the sensor, is 15.0 x 11.1 cm. These specifications allow inspections on the majority of configurations on North Sea platforms.

The present principle chosen to inspect the underwater welds features the following advantages:

- Adaptability to almost all welds likely to be met on offshore platforms through a flexible sensor.

- Systematic and reproducible exploration of the whole weld volume through electronic scanning, that does not require any action from the diver.

- High signal to noise ratio due to the focusing of the shear waves and holographic detection.

- High operational efficiency due to the speed of acquisition of the volume image.

- Interpretation, report, and filing can be achieved outside the operational time, on the raw data themselves, which can be stored on magnetic tape under digital format.

INSPECTION OF CONCRETE STRUCTURES

A system has been developed that allows detection and location of corrosion and cathodic protection currents, especially those D.C. currents due to the contact between free seawater and the reinforcement. These D.C. currents reveal dangerous cracks or abnormal porosity leading to concrete breakup in the long time.

The operations are conducted with a standard current density sensor and a gyro doppler navigator installed on a manned or unmanned submersible. The data is recorded or transmitted to the surface while the vehicle is following a "grid patch" that covers the whole surface to be inspected. Data processing is performed on the support ship through a computer and produces charts of the current density measurement as a function of the position. Abnormal peaks are interpreted as significant damage and a closer survey of the suspected areas can be decided upon after this general survey, but only at the specified locations, thus saving time.

An estimate of the crack size can be obtained in some cases, due to the accuracy of the navigation system used. The integration of the recorded current density can also provide an estimate of the amount of iron lost every year providing the crack remains constant.

The fundamental phenomenon allowing crack detection over large concrete areas was discovered during the first experimental and commercial surveys of underwater pipelines with the current density equipment (Bournat and Stankoff, 1979).

The principle underlying the identification of cracks over large concrete areas relies on the assumption that a significant crack should reach the metallic reinforcement and create some corrosion inside the concrete. As massive concrete is quite resistive, the D.C. currents due to the corrosion process will be concentrated within the cracks and provide a means to identify them even if not visible, through an electrical measurement. The higher the current output, the more severe is the corrosion level and the consequences of the damage.

Experimental laboratory measurements were performed for confirmation (Bournat, 1980; and Bournat, Stankoff and Auboiroux, 1980).

On the first example the distance from the sensor to the plate was 10 cm. Current peaks up to 10 mA/m^2 were noticed and corresponded to the concrete plate edges where the reinforcement was nearly apparent, and to the middle of the plate where the cracks were more visible.

On the second example the distance from the sensor to the plate was 40 cm. Current peaks were altered down to 2 mA/m^2, but was still visible.

The current density equipment is self-calibrating and has the following main specifications:

- Sensitivity : M 100 A/m^2
- Noise level : M 300 A/m^2
- Range: 0.1 to 2000 mA/m^2
- Measurement rate: 3 per s
- Size : Ø : 430 mm, length: 500 mm

Consequently, an underwater vehicle (Fig. 3) carrying such a D.C. current density sensor close enough to the concrete area will be able to detect significant cracks reaching the metallic reinforcement even if fouling is covering the crack outlet, thus avoiding the constraint of cleaning the concrete surface.

When the reinforcement is interconnected, the sensitivity of the method can be increased by the use of an A.C. generator which induces A.C. currents in the reinforcement and through the walls. In this case, the sensor should be modified to take A.C. current into account. Nevertheless, self-corrosion of steel bars, or D.C. cathodic protection are delivering currents which can be measured and used for crack detection.

The efficiency of such survey will depend mainly on the ability of the vehicle to follow a regular grid path over large concrete areas. This requires a high resolution navigation system, able to position the vehicle on the concrete surface in real time.

In order to assist the pilot, simplified models of the surfaces to be inspected are taken, in which the true surfaces are inscribed. These surfaces are developed in terms of depth and bearing. Examples are given in Fig. 14: cylindrical shapes, as tanks are developed in a rectangular shape. Steps between bearings are determined by the radius of the cylinder and by the maximum distance between two measurements, which should be around 50 cm.

In the same way, legs are taken as conical models and are developed as described on the lower part of the figure. Vertical distance and bearing are kept when going from the model to the true surface, which allows an easy positioning of the vehicle.

As standard navigation systems such as long base or short base systems or even inertial navigation systems are not usable on large concrete surfaces, a new system has been developed using a Doppler Sonar.

The Doppler head senses the speed in four directions. Its main feature is that it is mounted horizontally on the vehicle so that the ultrasonic beams are reflected by the concrete wall. Two measured velocities are in a vertical plane, the two others are in a horizontal plane. They are corrected for pitch and roll by an attitude reference system. When operating, the true position in the reference surface is computed and given in real time to the pilot. Discrepancies from the preset path are also given as an aid for maneuver.

Extensive sea trials were performed in 1979 on North Sea platforms, where both the current density sensor and the doppler navigation system were evaluated for surveys on large concrete areas.

A manned submersible was used and several surveys were performed on concrete legs or tanks with the submersible facing the concrete wall. In order to protect the equipment and the transparent viewport from damage, three arms terminated by wheels was installed on the guards and the submersible was rolled up and down on the concrete wall.

When the wheels were applied to the concrete, the distance between the current density sensor and the wall was approximately 40 cm.

A detailed analysis was undertaken to evaluate the accuracy of the navigation system from several tests carried out. An evaluation of the system was then made only on the vertical displacement, by comparison of the doppler information with a high accuracy pressure sensor whose resolution is better than 1 cm in this application.

Two types of errors have been taken into account being the closing error and the scale factor.

The average closing error was found to be 4 cm, which represents 2% of the vertical displacement (22 meters) with the fluctuations amounting to ± 12 cm. These numbers determine the reproducibility of the navigation system along the vertical axis.

The scale factor error was evaluated through comparison of the modules of the vertical displacements performed during the test. The total error was 32 cm for a 72 m vertical distance covered and the resulting scale factor error is 4%. The fluctuations were ± 11 cm.

The overall accuracy can be estimated from the fluctuations of the closing error, which are predominant for short trips, and the scale factor error, becoming predominant for long term travels (higher than 30 m). The fluctuations of the closing errors and of the vertical distance measurements were found nearly identical, thus indicating neither short term correlations, nor long term drift.

Considering that the accuracy for horizontal displacement is suffering the same defects as for vertical one, the overall accuracy of the equipment can then be evaluated to 9% plus 20 cm of random noise.

Damage producing current densities higher than 5 mA/m^2 was easily detected during the test survey.

In case of a regular survey of a concrete wall, the submersible will be rolled up and down at different bearings. The corresponding display is shown on Fig. 15 where a flat representation of the survey is given on the right side of the figure, and the current density data is given as a function of depth on the left side of it.

The signal to noise ratio will be improved to make sure the small current density peaks noticed on the log correspond really to corrosion activity. On the other hand, knowledge of really significant damage is still not sufficient to determine the level of current corresponding to alarming degradations.

Based on the detection of corrosion currents, the method is only sensitive to damage really dangerous to the concrete integrity, that is to say, the cracks which reach the metallic reinforcement network and leading to concrete breakup.

This inspection method is assumed to be very effective as it is still fully operational even when marine growth covers the concrete surface, thus avoiding expensive cleaning programs prior to the usual visual inspections.

CONCLUSION

Reliable instruments do now exist enabling a complete subsea survey of pipelines, steel and concrete structures to be undertaken in a combination from a manned or unmanned vehicle providing the inspection engineer with the required information on site. It is however essential that the inspections are carried out regularly enabling variations and irregularities in conditions to be detected at the earliest stage. This again should be important in reducing the cost of maintenance.

REFERENCES

Bournat, J. P., (1980). Inspection of concrete platforms: Crack detection by current density measurements. Proc. Oceanology International, Brighton.

Bournat, J. P. and A. Stankoff (1979). Cathodic protection measurements and corrosion control of pipelines by underwater vehicles. Proc. Offshore Technology Conference, Houston, 3590.

Bournat, J. P., A. Stankoff and M. Auboiroux (1980). Inspection of concrete platforms: Crack detection by current density measurements. Proc. Offshore Technology Conference, Houston, 3765.

Durand, Y. and D. LeBouteiller (1980). Automatic pipeline inspection system using an unmanned submersible and a DP surface support vessel. Proc. Oceanology International, Brighton.

Durand, Y. and A. Stankoff (1978). Inspection of buried pipelines by submersibles pipe-tracking and pipe-logging instrumentation. Proc. Offshore Technology Conference, Houston, 3071.

Larsen, K., (1979). Experiences with offshore instrumentations. Underwater vehicles - manned. Proc. Norwegian Society of Chartered Engineers, Jæren.

Stankoff, A., (1980). Evaluation of internal corrosion and cracks on marine risers by divers using acoustic holography techniques. Proc. Oceanology International, Brighton.

Stankoff, A., Y. Guenon and Y. Thomas (1980). Evaluation of internal corrosion on marine risers by divers using acoustical holography techniques. Proc. Offshore Technology Conference, Houston, 3894.

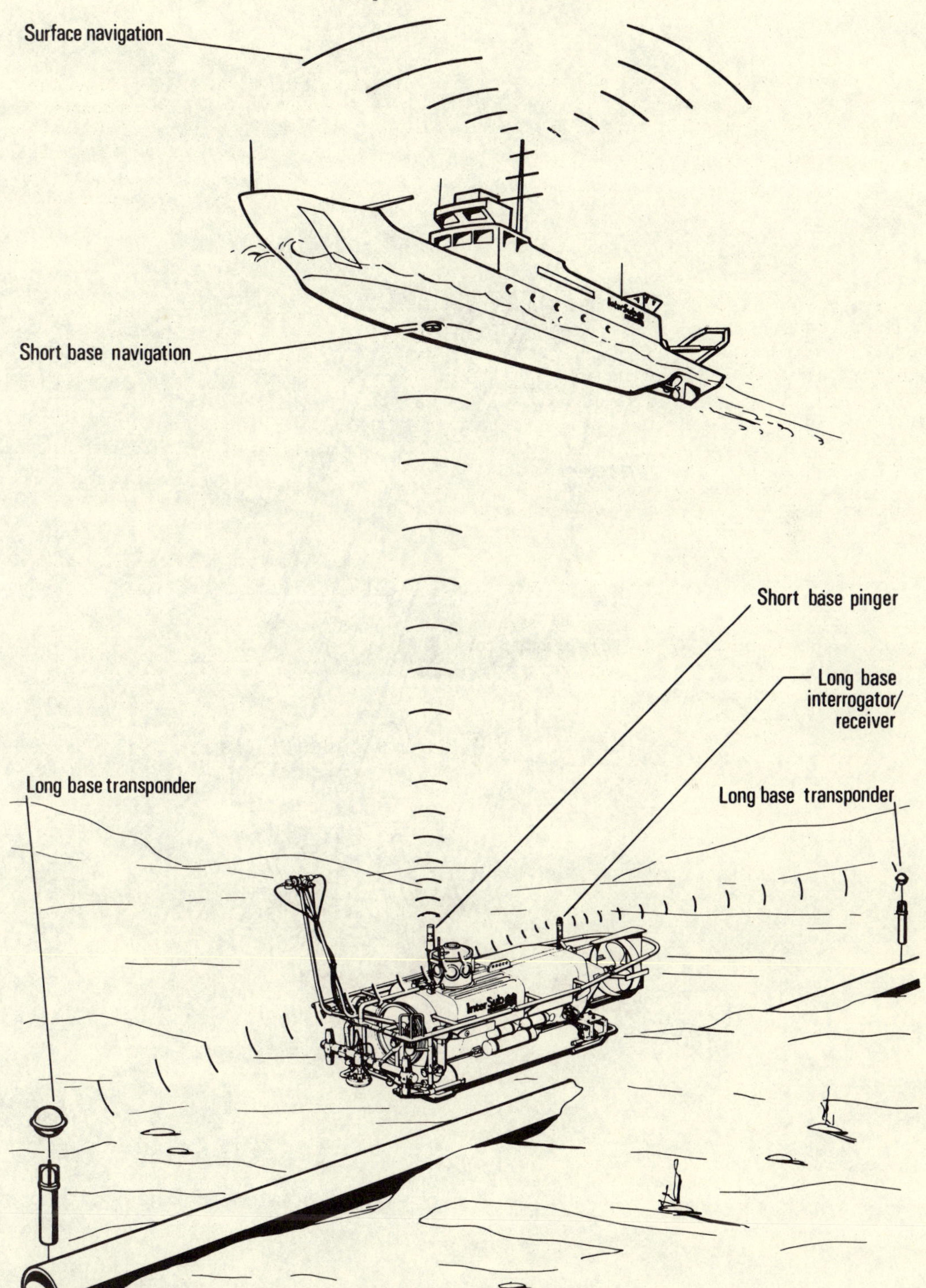

Fig. 1: Configuration of manned submersible pipeline inspection spread.

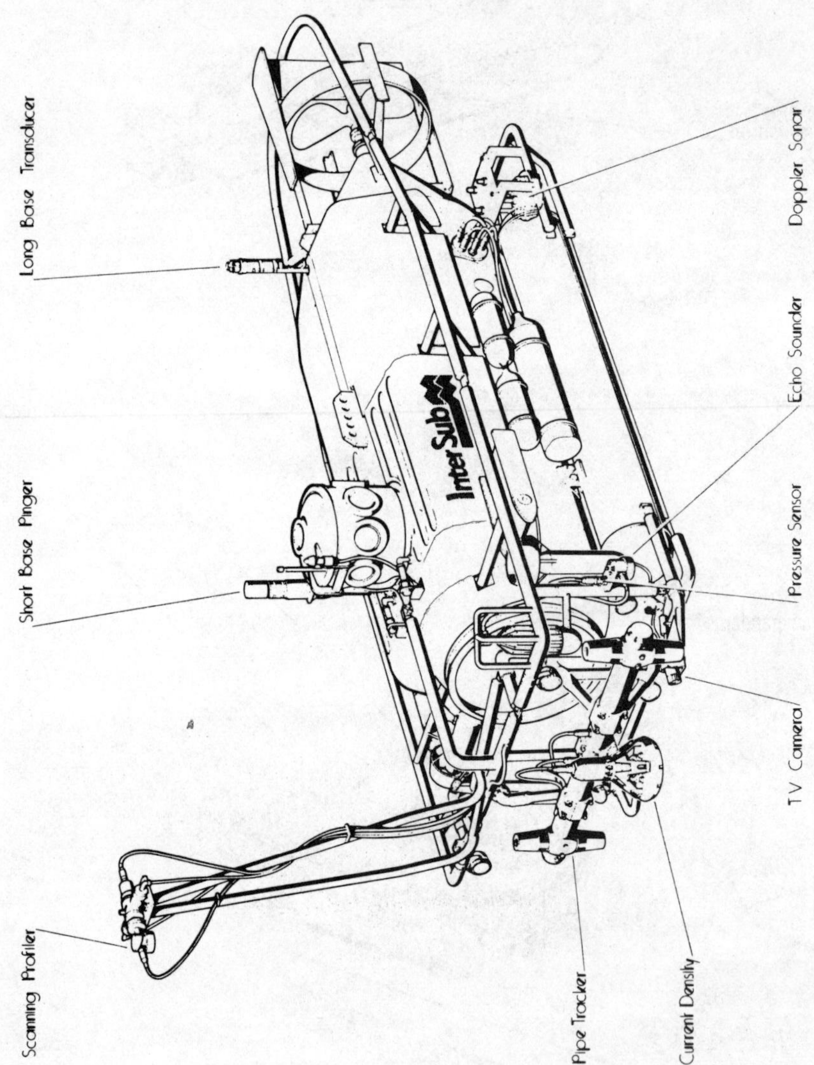

Fig. 2: Layout of submersible pipeline inspection equipment.

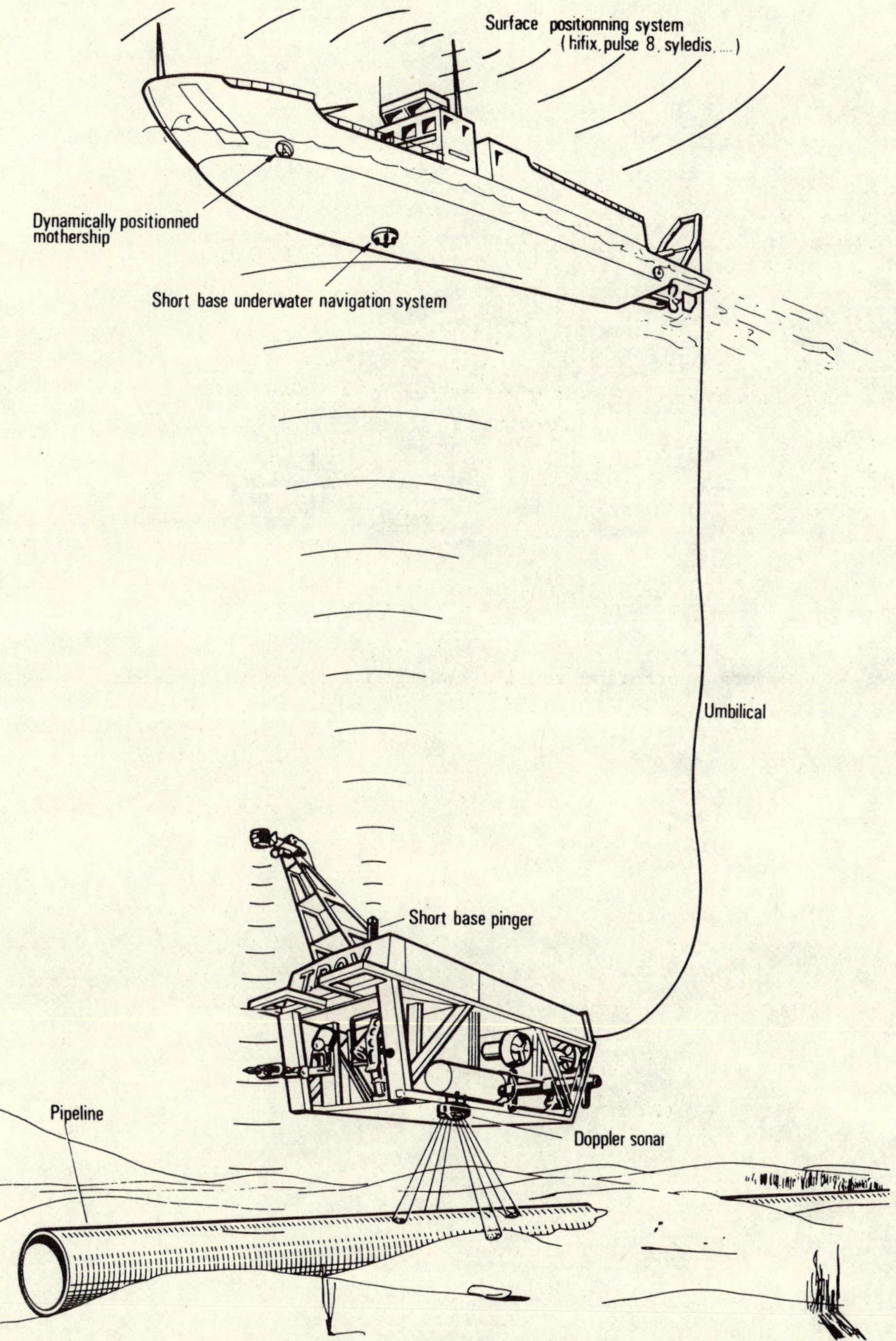

Fig. 3: Configuration of automatic unmanned vehicle pipelogging spread.

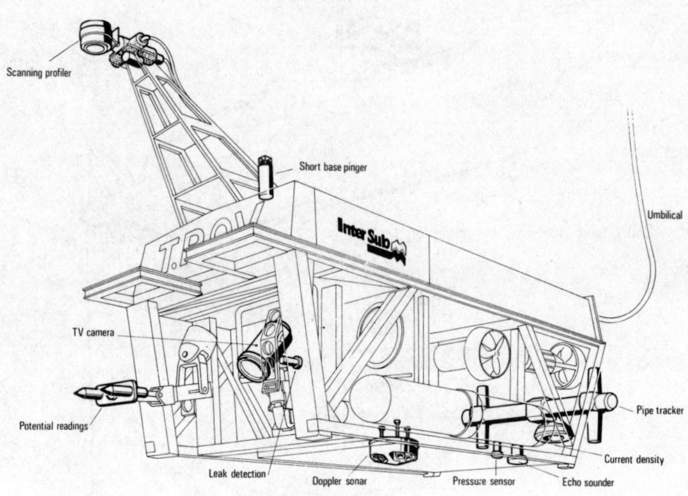

Fig. 4 : Remote controlled vehicle equipped for pipeline inspection surveys.

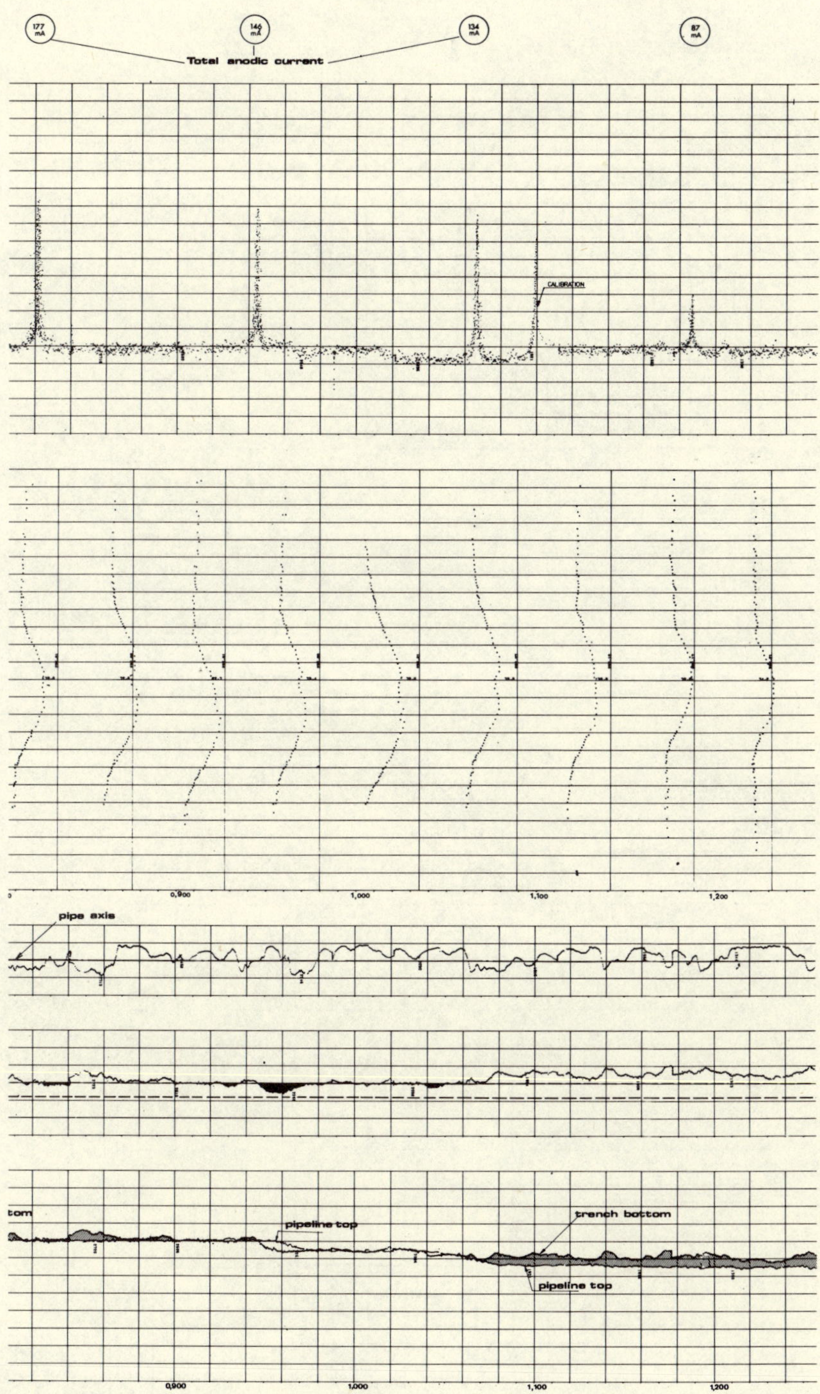

Fig. 5: Example, processed pipeline logging charts.

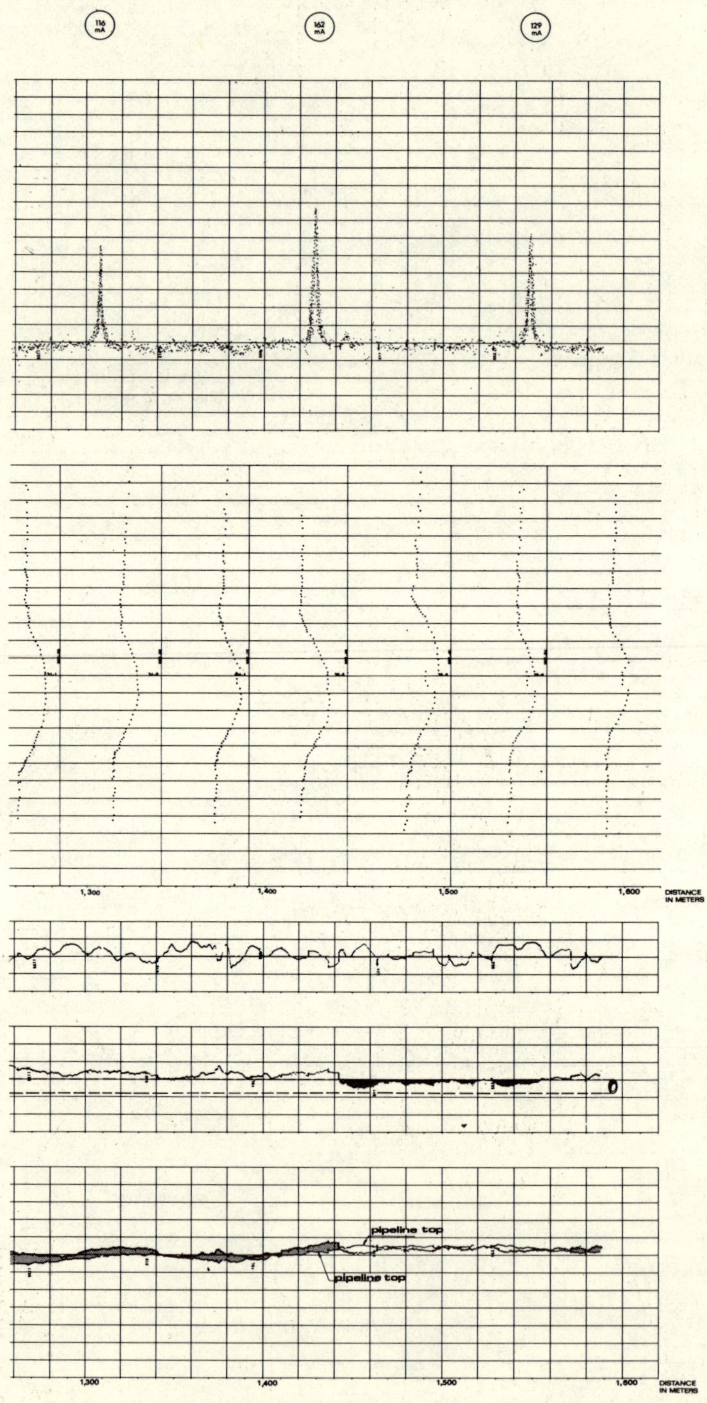

Fig. 5 : Continued.

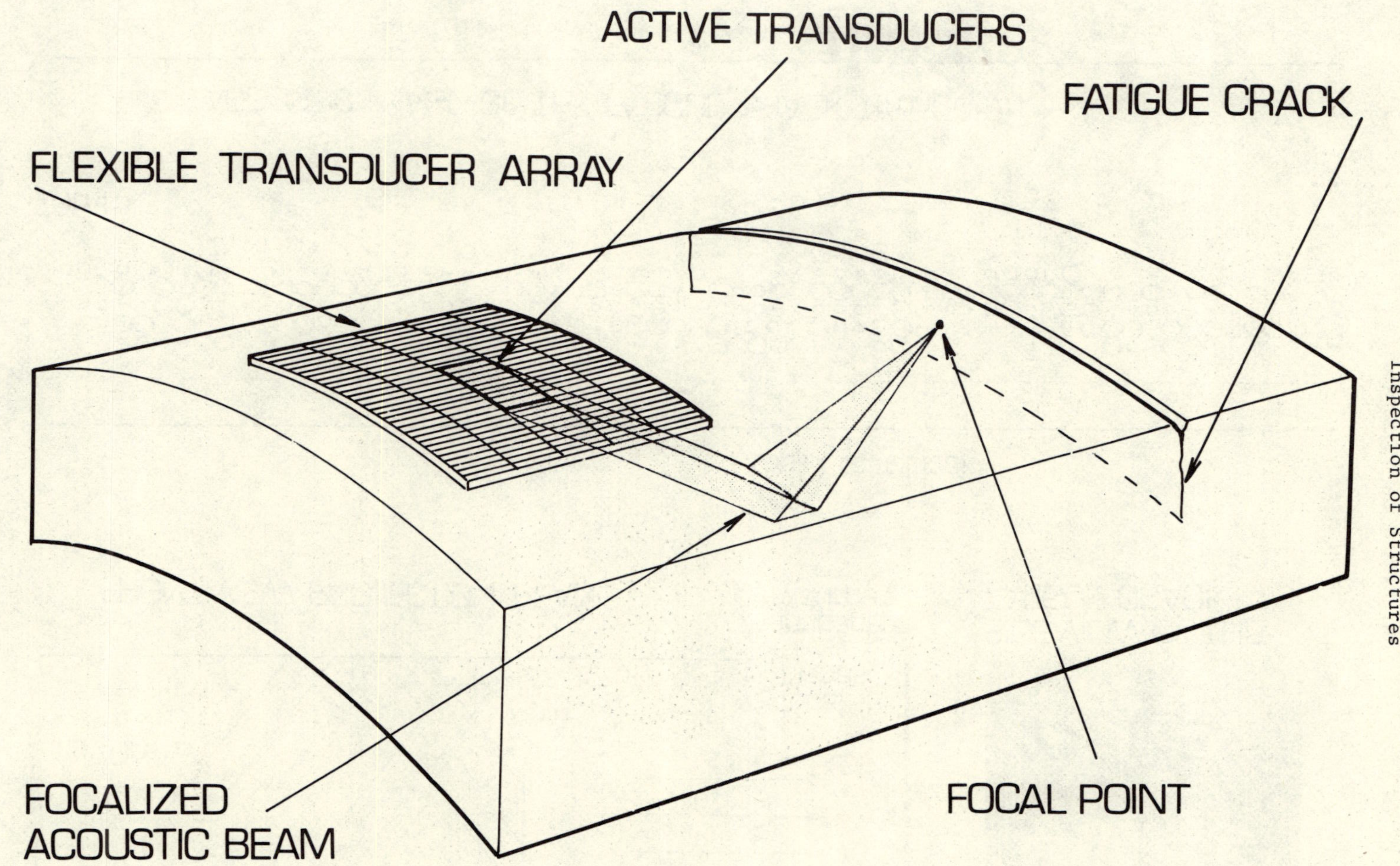

Fig. 6: Physical principle of the acoustic imaging.

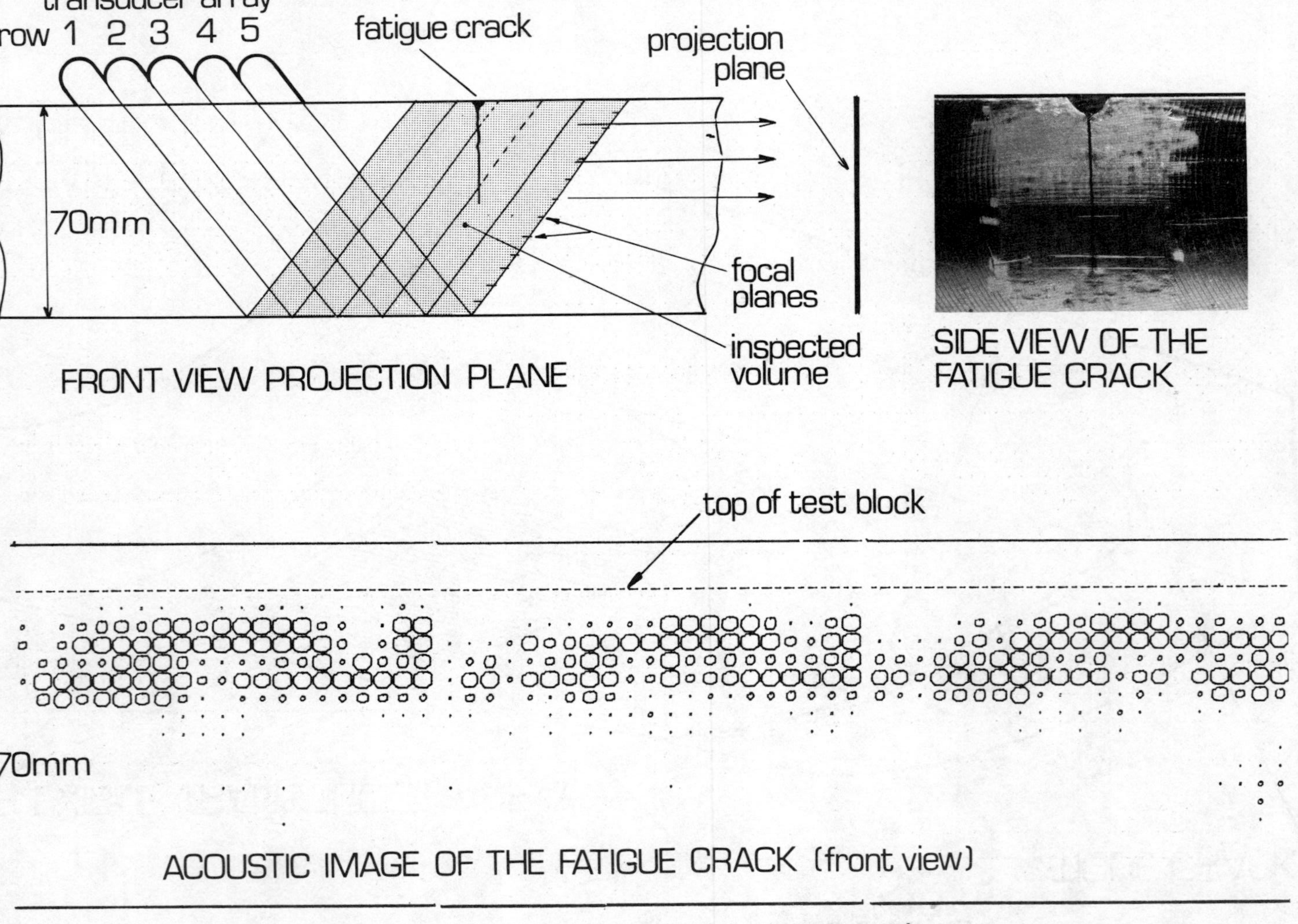

Fig. 7: Detection and evaluation of a fatigue crack.

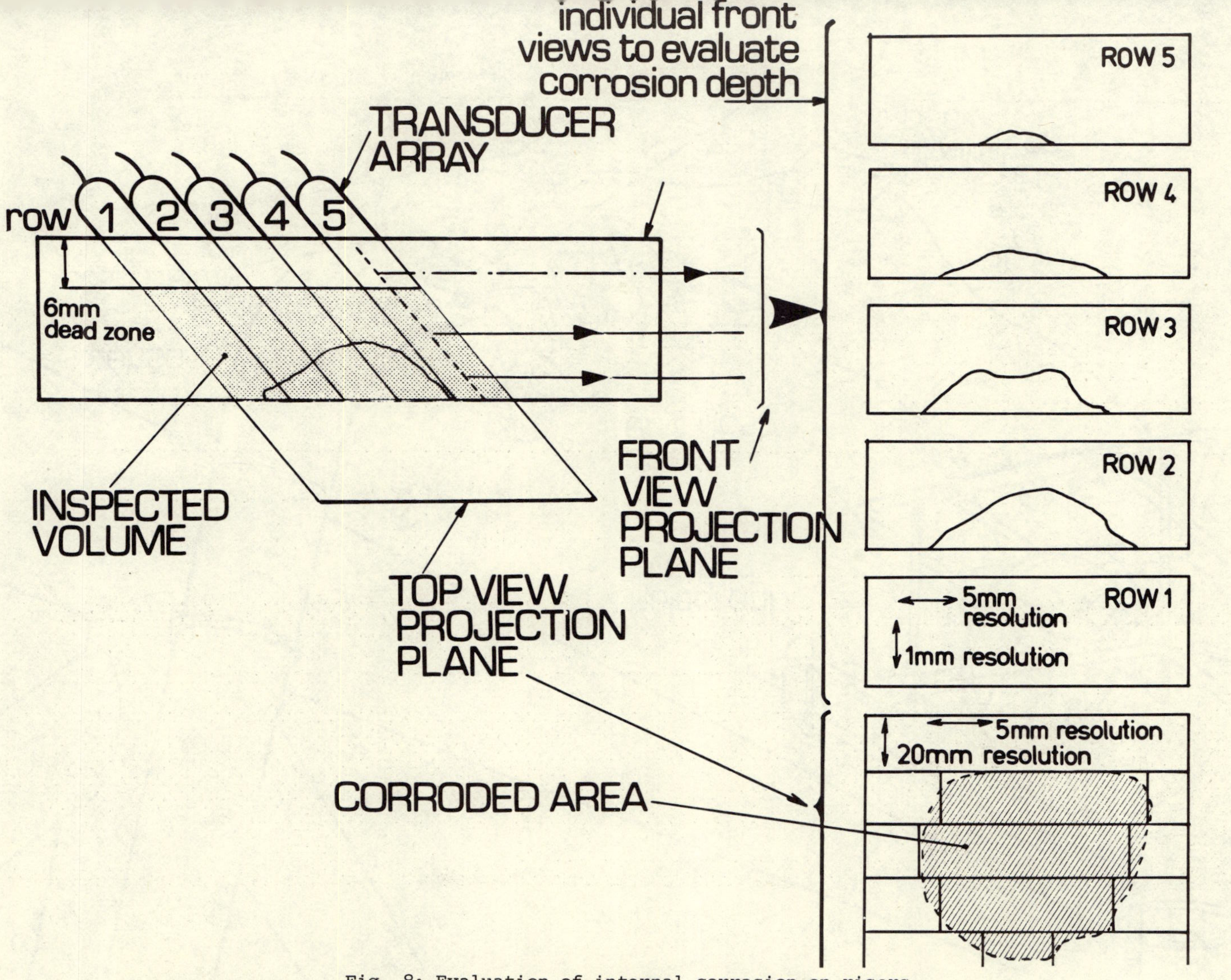

Fig. 8: Evaluation of internal corrosion on risers.

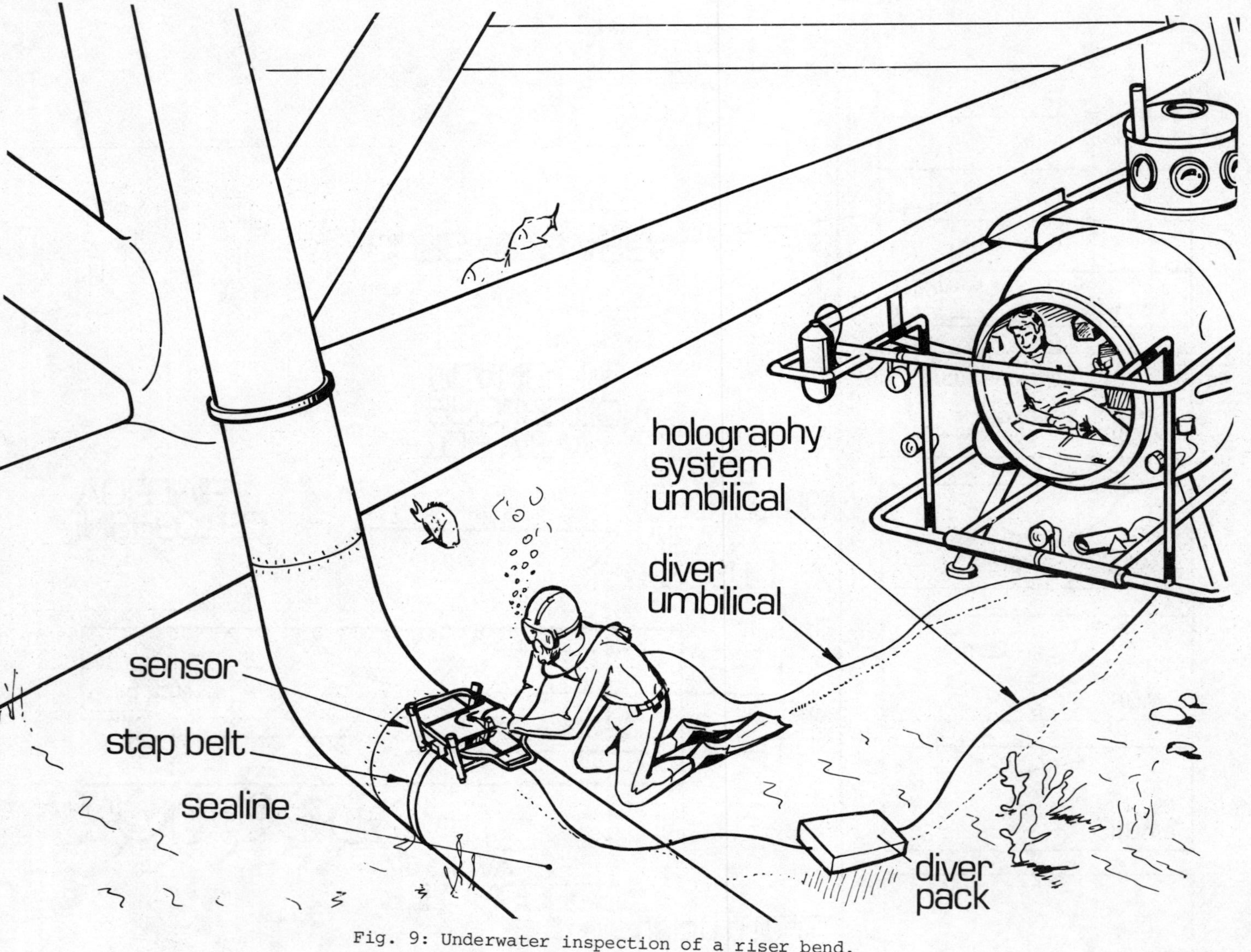

Fig. 9: Underwater inspection of a riser bend.

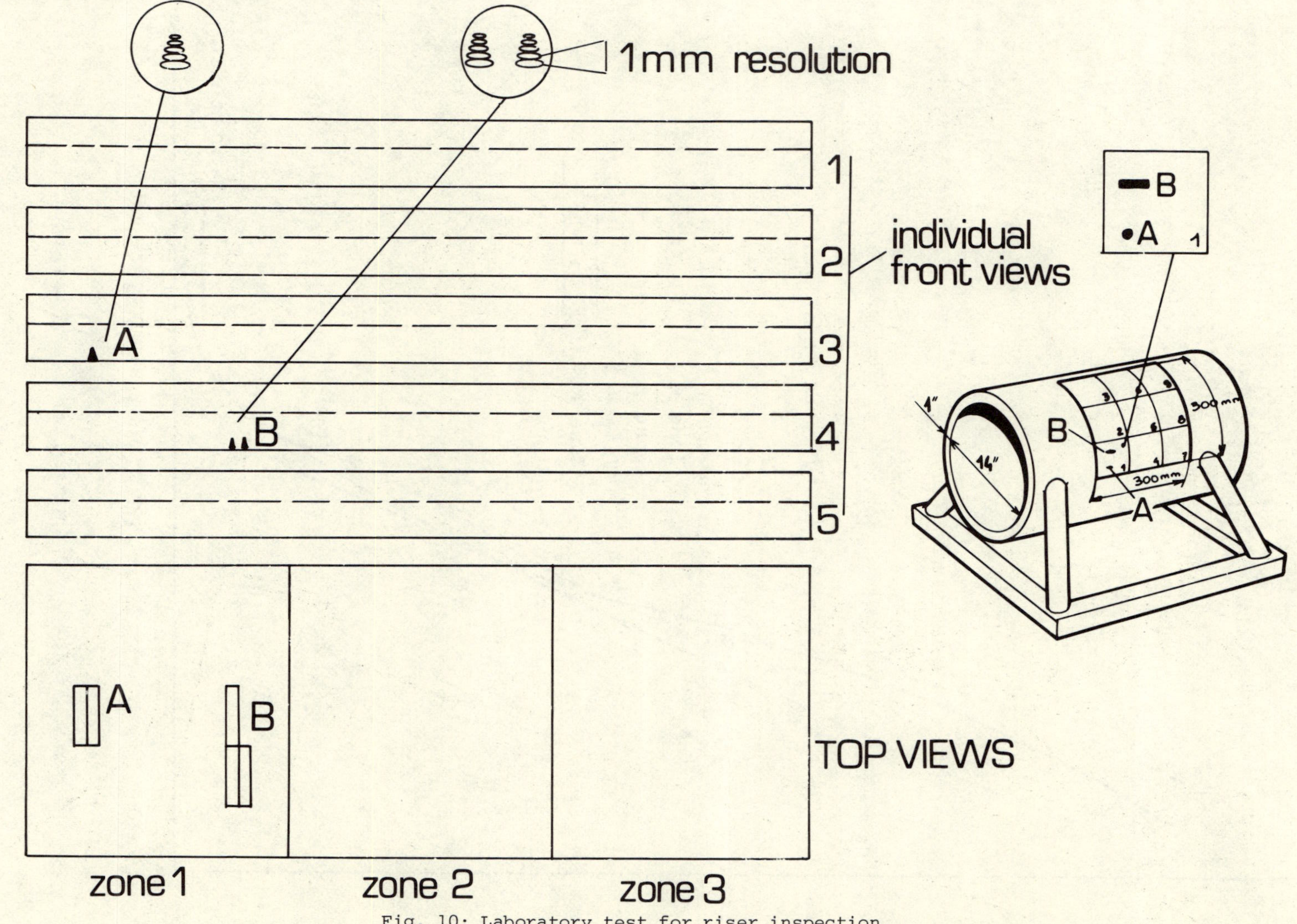

Fig. 10: Laboratory test for riser inspection.

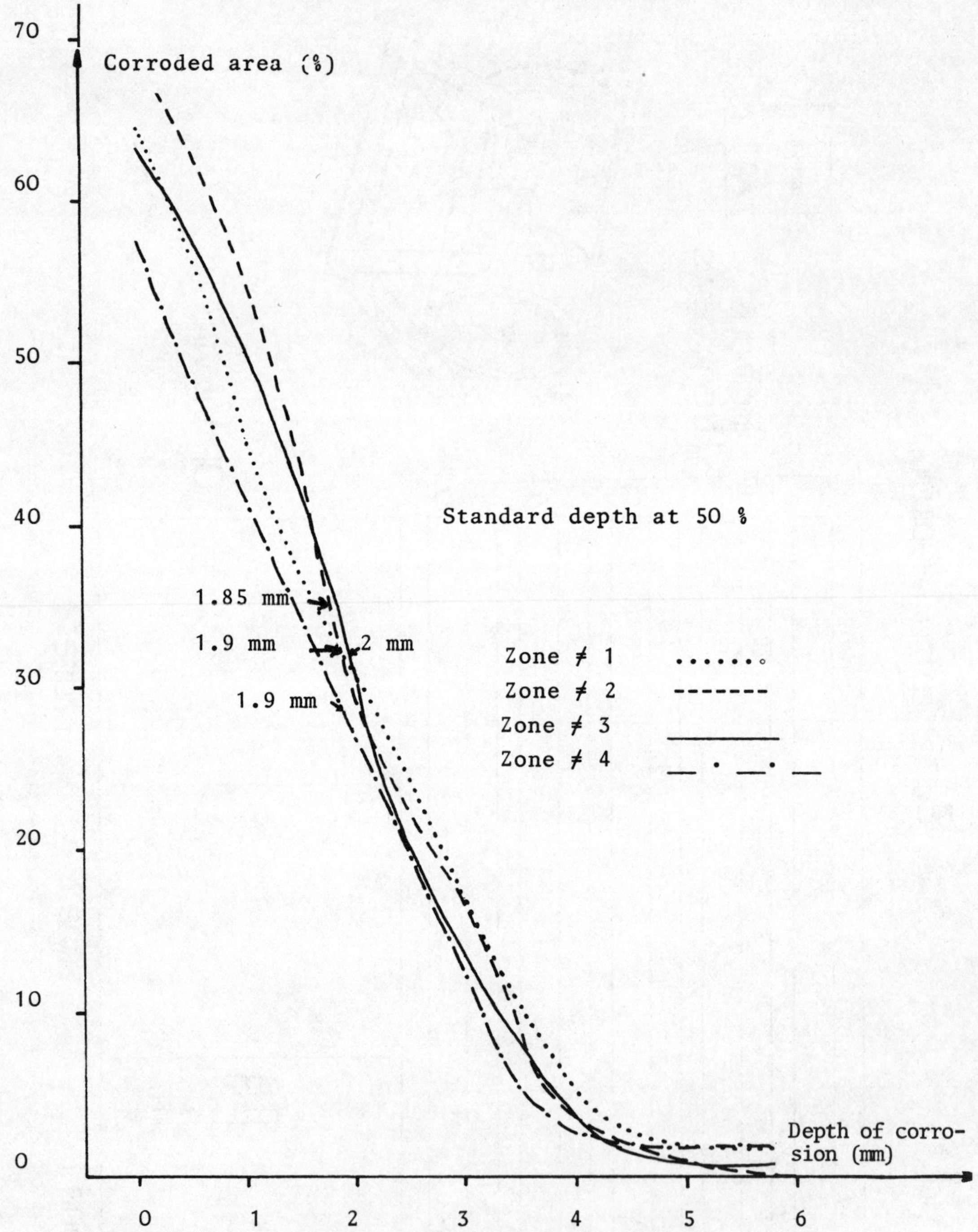

Fig. 11. Corrosion extent as a function of its depth.

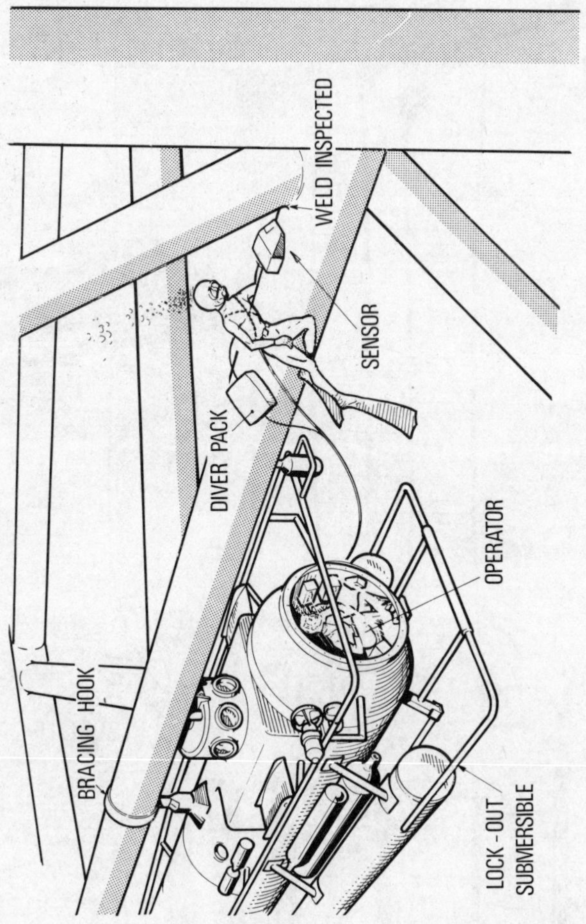

Fig. 12: Midwater inspection of a platform node.

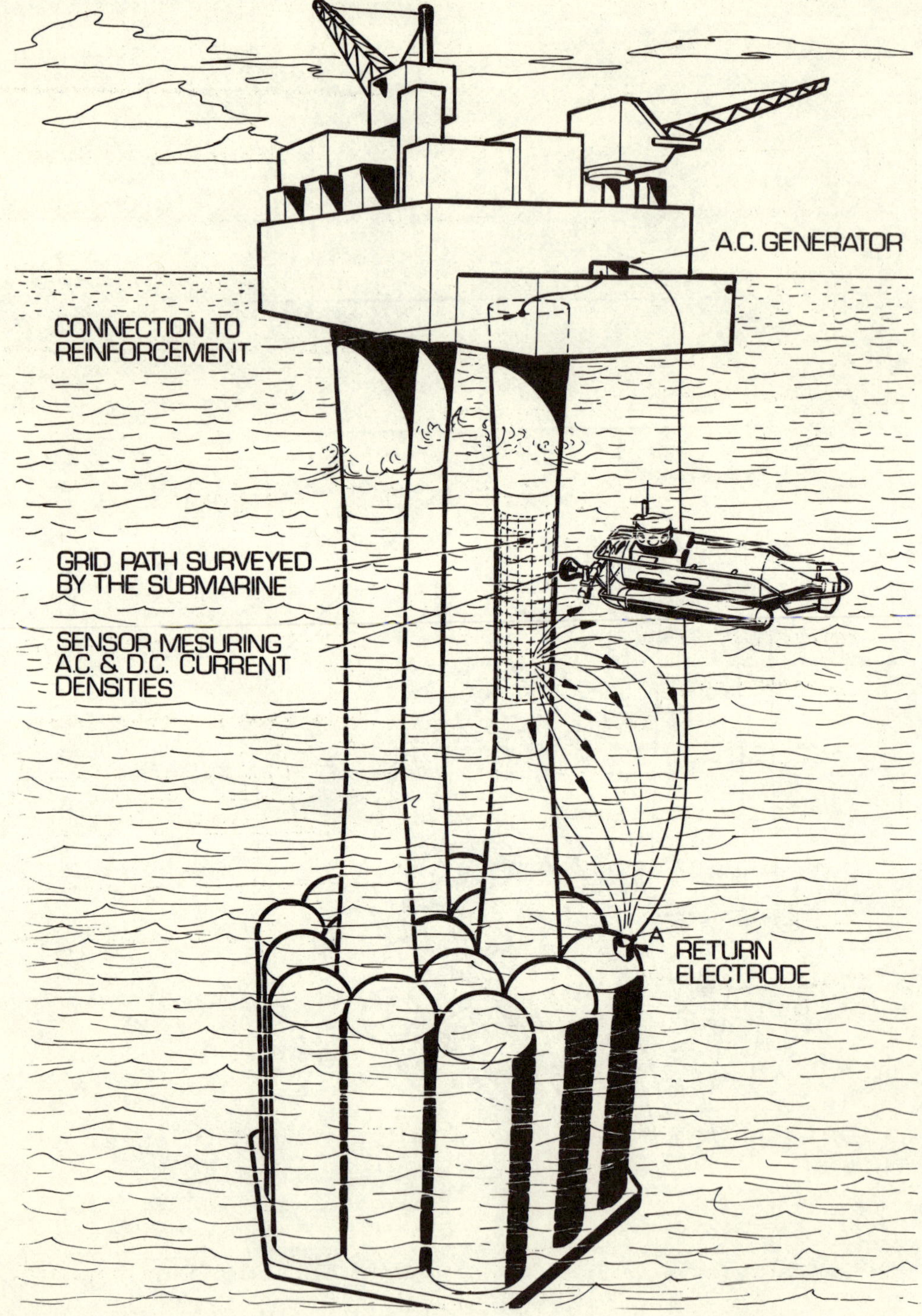

Fig. 13: Inspection of concrete platform by detection of corrosion currents. Operating principle.

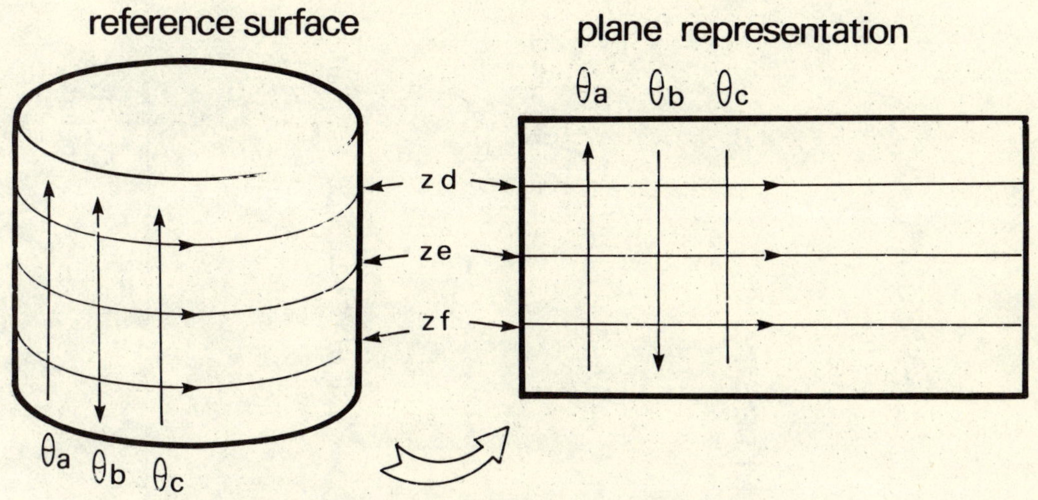

CYLINDRICAL MODEL (tank)

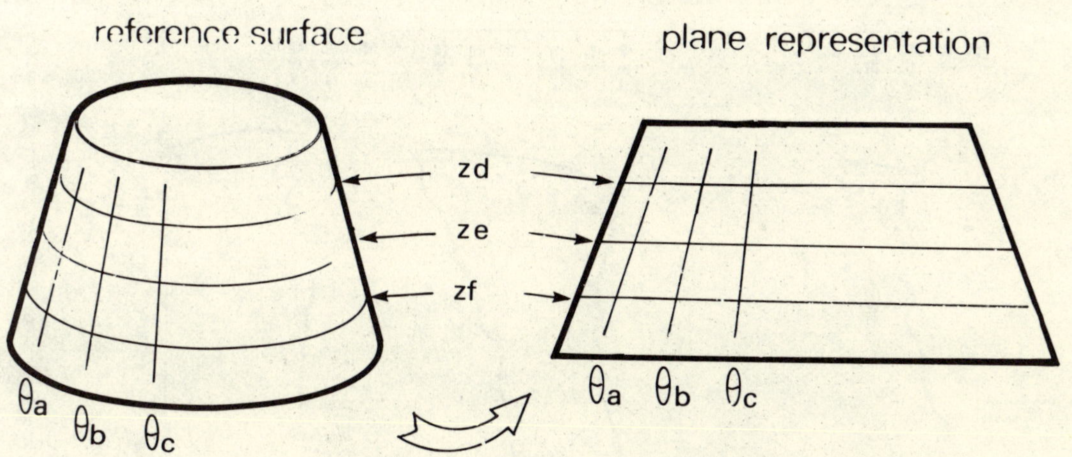

CONICAL MODEL (platform leg)

Fig.14: Examples of depth and bearing models of surfaces to be inspected.

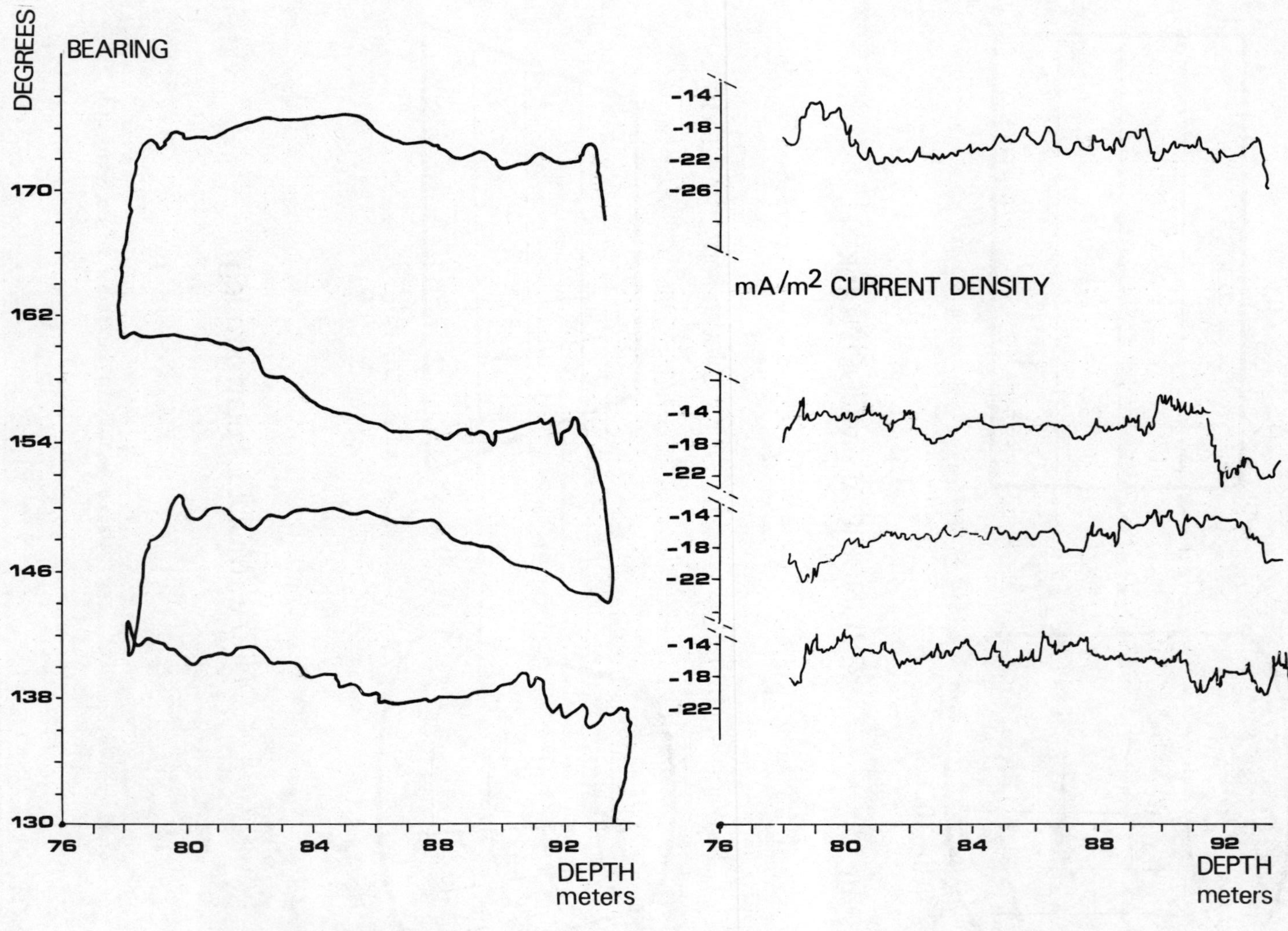

Fig.15: Model of current density display for systematic surveys.

THE DEVELOPMENT OF A MICROPROCESSOR CONTROLLED DEPTH INDEPENDENT POWER GENERATION SYSTEM

R. V. Thompson, M. R. O. Hargreaves and A. Fowler

Department of Marine Engineering, University of Newcastle upon Tyne, UK

ABSTRACT

This paper reports on the findings of an extensive evaluation of underwater power systems with particular emphasis on the 5 to 50 kw power range of medium length mission capability. The results point strongly to the adoption of a closed cycle diesel engine, on a basis of low relative cost and power to weight/volume ratios.

An experimental prototype unit using recirculated nitrogen and a chemical exhaust scrubber to achieve depth independence, has been constructed and operated at Newcastle University. The paper describes progress made in defining control algorithms using simulation techniques with the objective of implementing microprocessor control of the prototype unit.

KEYWORDS

Sub sea power, Energy conversion, Depth independence, Closed cycle diesel, Hybrid-simulation, Microprocessor control.

INTRODUCTION

Recent surveys commissioned to identify the future world requirements for underwater power sources point to an expanding world market with a greater emphasis on operational depths in excess of 200 metres.

In the majority of applications the mission duration is of the order of 10 hours or less. The power source has to be reliable and is required to utilise compact fuel or energy sources with easy and safe disposal of any effluent in submarine situations.

Present day systems,(the most common of which is the lead acid battery), have very low energy to weight and volume ratios which inhibit power output and endurance.

For Man to extend the current range of both the diver and submersible, a power source is required which offers increased shaft output and endurance capability over conventional systems.

This paper presents the findings of an extensive evaluation of units capable of satisfying the stated demand. The results point strongly to the adoption of the closed cycle diesel.

The investigation has examined available energy forms and energy conversion systems. In the selection of the most suitable combination of energy source and conversion systems, particular attention has been paid to such factors as cost, weight, volume, safety and state of development.

The latter part of the paper describes the development of a prototype unit at Newcastle University. A conventional diesel engine has been operated in the closed cycle configuration under steady state laboratory conditions, therby demonstrating the feasibility of the concept. Current research effort focusses upon the development of a high response microprocessor based control system, which will provide a capability for practical dynamic loading of the engine.

The design and development procedure features extensive use of hybrid computer simulation techniques. The presentation includes a description of the system model and a sample of the analysed results.

MARKETS FOR UNDERWATER POWER PACKS

A market survey, carried out by the Economic Planning Department (1973), identified a world requirement for over 400 underwater power units in the range of 10 - 100 kw by 1982. A similar survey by the Science Research Council (1976) agreed with these findings and highlighted a need for a reliable power source in the range 0-100 kw using compact fuel or energy sources. The following industries are involved.

Civil Engineering and Minerals Surveying

Civil engineering requirements represent one of the most important opportunities for the exploitation of underwater power. Applications include seabed corers (used in the survey phase of mineral extraction) and underwater power packs for tools. The development of more sophisticated underwater machinery will lead to a demand for large powers (eg. bulldozer systems of about 125 kw).

Research

Sub-sea habitats are required for observation purposes and for long term projects carried out at one site on the sea bed. The living and working facilities are for periods of up to a month. This would include emergency power requirements for periods when the habitat or its occupants could not be brought to the surface. They are likely to be required to operate at continental shelf depths ranging from 28 to 200m.

Offshore Oil and Gas Industry

An increase in operational depth of gas and oil retrival projects has led to a revival of interest in reliable compact power sources. The prime areas would be in pipeline burying and seabed trenching,wellhead servicing and pipeline and structure inspection. Indeed a high speed submersible, which could operate from a shore base without the need for a mother ship would be of great use in these latter areas.

SUITABLE ENERGY FORMS

Any power system needs a source of energy which can be converted into useful mechanical or electrical output.

The energy source comprises the biggest percentage of the total weight and volume of the power system. Its consideration is therefore critical for the proper selection of submersibles and portable power packs.

Amongst other considerations energy sources for diver support systems and submersibles must be easily replenishable, depth independent, simple to control and safe to operate, whereas for seabed operation they must exhibit long endurance and be capable of operating with minimum maintenance. Logistics are also important and include both the problems of fuel and oxidiser replenishment to permanent sub sea habitats as well as the weight and volume limits applicable to mobile systems.

Four energy storage systems have been considered in this study, namely:

i) Chemical energy systems;
ii) Thermal energy systems;
iii) Mechanical storage systems;
iv) Isotope energy systems.

A comparison of these four groups has been made using a derived figure of merit. This figure is based on numerical data wherever possible and where this was not possible, a qualitative assessment of the relevant factors has been applied.

Chemical Energy Systems

The choice of fuels and oxidants is usually a compromise between the requirements for a given application and the fuel/oxidiser characteristics. Various desirable features of the fuel/oxidant combinations (eg. non toxicity, high calorific value per unit weight, low cost) have been carefully compiled from various sources (Ayres and McKenna, 1971; Guthrie, Davies and Moore, 1973; Macnair, 1975), and applied to a full spectrum of fuels, oxidants and mono propellants. An overall summary of qualitative and quantitative data is given in Table 1, which includes information on the following factors:-

i) Power to weight ratio
ii) Cost of fuel and oxidiser for a given output
iii) Storage factor - the means and cost of storage
iv) Replenishability factor - describing the logistics of supply and availability of the fuel
v) Exhaust processing - the relative ease of disposal or recycling of exhaust gas
vi) Safety factor - describing the reactant or exhaust product toxicity and the hazardous nature of the reactants

Each of the above factors has been rated according to the following five point scale:-

1.5 - Excellent
1.25 - Good
1.0 - Average
0.75 - Fair
0.5 - Poor

TABLE I Comparison of Selected Chemical Energy Combinations

Energy Source	Energy to Weight Ratio	Specific Cost	Storage	Replenish-ability	Exhaust Process-ing	Safety	Figure of Merit
Butane	1.0	1.0	1.0	1.0	1.0	1.0	1.0
Diesel Oil	1.0	1.0	1.25	1.25	1.0	1.25	2.0
Gasoline	1.0	1.0	1.25	1.25	1.0	1.0	1.6
Kerosine	1.0	1.0	1.25	1.25	1.0	1.25	2.0
Methane (1)	1.0	0.75	0.5	0.75	1.0	1.0	0.3
Propane	1.0	1.0	1.0	1.0	1.0	1.0	1.0
Ammonia (2)	0.75	1.25	0.75	1.0	0.75	0.5	0.3
Isopropyl nitrate	0.5	0.5	1.25	1.25	0.5	1.25	0.2
Hydrazine (2)	0.75	0.5	1.0	0.75	0.75	0.5	0.1
Hydrogen (1)	1.5	0.5	0.5	0.75	1.5	0.75	0.3

(1) Cryogenic storage
(2) Toxic

An overall figure of merit is included in Table I and is the product of all these factors. The table reflects the exceptional heat output per kg for hydrogen (nearly 3 times better than most hydrocarbon fuels) and the inferiority of ammonia. However, on a cost basis hydrogen is nearly seventeen times more expensive than diesel oil whilst ammonia is one third cheaper.

The need for cryogenic storage for both methane and hydrogen is indicated by the low rating given for storage and replenishability whilst the superiority of diesel, gasoline and kerosene is clearly reflected in this area. These hydrocarbon fuels, which can be stored in rubber bags fitted externally to the pressure vessel, have little effect on the buoyancy of the vessel whereas special measures would have to be taken for those fuels which are stored cryogenically or under pressure. Exhaust processing tends to favour hydrogen since the products of combustion (water vapour) can be condensed easily; unlike ammonia and hydrazine which both produce nitrogen gas. Indeed both these latter fuels are toxic and thus are given a low safety rating.

The figure of merit (the product of the individual factors), illustrates the superior qualities of diesel oil and kerosene.

The careful selection of the oxidant is even more important than that of the fuel since typical values of the oxidant/fuel ratio are of the order of 3.5 to 1. Of those available, most have not been considered since they tend to be highly corrosive, toxic, expensive or react explosively with fuels (e.g. fluorine). Of those remaining, oxygen which is cheap and readily available in large quantities in both the gaseous and liquid state, is the obvious choice. Its main disadvantage is that it has to be stored in containers which are heavy and under cryogenic conditions expensive. Its main rival, hydrogen peroxide, can be stored in collapsable bags but is about four times more expensive by weight. Of this, only about one half (44%) of the weight of HTP can be converted to useful oxygen. In addition care is required in handling HTP to avoid contamination of equipment as it can become violently unstable.

Thermal Energy Storage

Of the many possibilities for storing heat in a material, the most important properties for practical applications are the change of phase between liquid and solid state and the change in temperature in a solid phase. Relatively few materials are available which have a high enough heat capacity and/or heat of fusion, in the temperature range of interest, to be considered as an energy source for underwater applications. The most suitable are either molten salts, for which the major proportion of the energy is associated with a change in state, (Table 2), or metallic oxides which store energy entirely as sensible heat.

TABLE 2 Physical Characteristics and Costs of Materials with High Heats of Fusion

Compound	Formula	Specific Gravity (a)	Melting Temp. °C	Heat of Fusion kwh/kg	£/kWhr
Lithium Hydride	LiH	0.78	448	0.71	300
Lithium Flouride (b)	LiF	2.64	847.78	0.29	25
Lithium Hydroxide	LiOH	1.41	473.33	0.24	6

(a) In solid state
(b) Shrinks by half as it freezes

Molten salts present problems with containment in metal canisters and this in turn tends to place an upper limit on the melting point of 1000°C. In addition there is the tendency to form voids in the material during rapid recharging which has an adverse effect on the rate of heat release. They do have, however, the attraction of releasing the majority of their thermal energy at constant temperature.

Metal oxides, on the other hand, present less problems with containment and can thus be stored at much higher temperatures, although they do experience a continuous decrease in temperature as energy is removed.

It should be borne in mind that thermal storage has a limited useful life of only a few days due to heat leakage.

The most attractive thermal storage medium is lithium fluoride which is some six times cheaper than lithium hydride, and chemically more stable than lithium hydroxide.

Mechanical Energy Sources

Mechanical energy storage offers one of the least attractive solutions. Energy storage in deformable solid bodies (eg. springs) or by compressed gases yields an extremely poor specific energy significantly less than for the conventional lead acid battery (0.025 kWh/kg). The only exception is the flywheel for which a high potential energy density has been claimed (Lawson, 1971). Like thermal energy storage, mechanical storage has a limited life.

Radioisotopes and Nuclear Reactors

Nuclear and isotope power systems have been extensively applied to space and submarine applications and in some cases to marine installations such as buoys. The favourable characteristics of nuclear energy sources which make them well suited for use in the ocean environment are the long maintenance free life and the ability to operate for long periods without refuelling.

The characteristics (including cost) of various isotopes, were examined by Guthrie. Of those, strontium 90 and ceasium 137, both beta emitters, were found to be relatively cheap. To put these costs in perspective, it is necessary to consider the overall or capital cost of the energy source including encapsulation. These have been estimated to be a minimum of £30,000 for a 10 kW strontium 90 unit in 1973. Making allowances for inflation this figure is probably of the order of £100,000 today.

Thus a small nuclear power plant based on isotopes may not be impressive from an initial cost or power to weight ratio point of view but it would certainly be attractive if a particular misssion requires high energy density and long endurance.

There are a number of nuclear reactor systems which could be utilised in underwater systems, but in general they are extremely expensive and operate at higher power levels, well in excess of 100 kW. Guthrie claims that £90,000 (£270,000 allowing for inflation) for 100kW output could be expected. Certainly the social consequences of a widespread use of nuclear fuels by commercial organisations would need to be given careful consideration.

ENERGY CONVERSION SYSTEMS

Systems in which the initial state is chemical, thermal or nuclear energy and the final state is mechanical shaft power or electrical power are discussed below.

Conversion processes have been categorised as direct energy conversion systems, and thermodynamic systems (incorporating both internal and external combustion).

Direct conversion systems

1. Thermoelectric
2. Thermionic
3. Thermophotovoltaic
4. Electrochemical conversion (fuel cells and storage batteries)

Thermodynamic systems

1. Stirling cycle
2. Rankine cycle
3. Brayton cycle
4. Closed diesel engine cycle

A parametric comparison of these systems has been undertaken using such factors as specific weight, specific volume, thermal efficiency, specific cost and state of development. The basic principles involved in underwater application of these systems have also been considered whilst the results of the evaluation are presented in the following sub-sections.

After a careful study of the various power systems suitable for underwater application, it was found that no single concept offers an optimal solution with respect to all the desired characteristics. The selection of the power system is strongly influenced by the mission power profile and is mainly a question of assessing trade offs between performance and cost.

Direct Energy Conversion

Direct energy conversion implies the conversion of chemical, nuclear, radioisotope and thermal energies into electrical energy without the use of mechanical elements such as rotating or reciprocating machines. Electrical energy is attractive since it is a versatile form of energy from the point of view of transmission, conversion and control.

Thermoelectric, thermionic and thermophotovoltaic devices which convert heat directly to electrical energy, were found to be unattractive. The overall efficiencies are less than 10% (Ayres and McKenna, 1971), and at the same time the power to weight ratio is less than or no better than the conventional lead acid battery.

Only the secondary storage batteries and fuel cells offer an attractive solution.

Secondary storage batteries. Electrochemical conversion systems, in the form of conventional batteries, are at present widely used to power submersibles. They present a reliable, quiet and highly engineered solution ot the problems of underwater exploration. However, they have an energy to weight and volume ratio of 0.025 kWh/kg and 50 kWh/m^3 respectively, (Rogers and others, 1978, Criddle and others, 1978). Thus a mission of 10 hours duration needing a peak guaranteed power of 50kW would require a power pack weighing 20 tonnes and occupying a volume of 10.0m^3. Clearly this is not a very practical solution.

The current cost of a battery system has been estimated to be £150/kWh, as long as the recharge rate is of a reasonable duration.

There are a number of other battery systems available. Of these only silver zinc cells have been used in submersible operation to date (e.g. Aliminant). In comparison to the lead acid battery, the energy to weight ratio shows an improvement of up to 4 times whilst the power to weight ratio shows a sevenfold improvement, (48 to 340 W/kg.) The improved energy to weight ratio is, however, offset to some extent by the higher material costs (£1000/kWhr.). It also has a major shortcoming in that the plates require replacement after about 100 charge/discharge cycles, as compared to at least 1000 cycles for an equivalent lead acid battery.

Fuel cells. A considerable amount of effort has been devoted to the development of fuel cells for underwater applications. This work has indicated the superiority of cryogenic hydrogen and oxygen as the fuel/oxidant combination. As well as being noted for their silence of operation, fuel cells are particularly attractive as they have a high efficiency (55 to 60% for H_2 - O_2 cells) and thus a low specific fuel consumption. In addition their compactness and light weight make them superior to batteries for longer mission profiles.

The main disadvantage of the fuel cell is its cost. The specific cost of the cell has ranged from £3000 to £24000 per kw for land and space applications respectively; this expense being largely attributable to the high cost of the catalyst (platinum or palladium) used.

Thermodynamic Conversion Systems

Of the various reciprocating and rotating dynamic converters, the following were selected as being most suited to underwater application:-

i) Iso-propyl nitrate turbine (Overy and Rayner, 1971)
ii) Ricardo closed cycle diesel (Puttick, 1971)
iii) Psychrodiesel (Hoffman and others, 1970)
iv) Depth independent closed cycle diesel
v) Stirling engine (rhombic drive) (Guthrie and others, 1973).

vi) Closed Brayton cycle systems (Rackley, 1969)
vii) Rankine cycles (Guthrie and others, 1973)

These systems have been compared empirically using a derived figure of merit (Table 3) which is based on estimates of the capital and running costs, the state of development, reliability, availability and finally the technical suitability for underwater application. The method suggested by Wimmer has been used to highlight the most suitable systems but using the same rating system as that adopted in Table 1. Development costs, used in the derivation, are based on estimates made by Guthrie, Davies and Moore (1973) with an allowance for inflation.

TABLE 3 A Comparison of Selected Dynamic Converters (0 to 100kw)

Converter	Development Cost Factor[1]	Running Cost Factor	Development Availability Factor	Reliability	Underwater Technical Suitability	Figure of Merit[3]
1 IPN Turbine	1.5	0.75	1.25	1.25	0.75	1.3
2 Ricardo Recycle Diesel	1.5	1.0	1.25	1.25	1.25	2.9
3 Psychrodiesel	1.0	1.0	1.25	1.25	1.25	2.0
4 Depth Independent Recycle Diesel	1.0	1.25	0.75	1.25	1.25	1.5
5 Stirling (Rhombic)	0.75	1.25	0.75	1.0	1.25	0.9
6 Closed Brayton Cycle	0.5	0.75	0.75	0.75	1.0	0.2
7 Rankine Cycles	0.5	0.75	0.75	1.0	0.75	0.2

1. Development cost factor proportional to reciprocal of development cost.
2. Running cost factor includes an allowance for fuel consumption and maintenance.
3. Figure of merit is product of individual factors.

Two of the systems gave poor ratings. Both the Rankine cycle and the closed Brayton cycle, incur a high captial cost factor reflecting the work required to develop an engine of a suitable size. The costs were estimated at up to £1.5 million

For a reciprocating steam engine, the cost disadvantage was compounded by potentially low overall efficiency (approximately 10%) whilst the steam turbine would only be feasible for the higher part of the power range if the efficiency is to be maintained. Finally, the closed cycle Brayton unit suffers currently from a very limited application; only a few development engines are in existance in the 50 kW range, and these are not directed towards subsea application.

The comparison has resulted in the closed cycle diesel engine configurations showing a consistantly good rating. The iso-propyl nitrate turbine and the Stirling engine also show potential. These three basic schemes are discussed below.

Iso-propyl nitrate turbine. This turbine, developed by Plessey and described by Hoffman, Rudnicki and Williams, offers one of the better means of converting energy into useful work. It makes use of the monopropellant iso-propyl nitrate as the energy source. Achievable values for the overall system energy to weight and volume ratios are claimed to be 0.28 kWh/kg and 250 kWh/m^3 respectively, whilst the specific cost is thought to be in the region of £750/kWh for a 50 kW unit having a medium length mission capability. The fuel was, however, found to have a specific cost over 10 times greater than that for diesel/LOX, whilst its density was less than a quarter of that for the same fuel oxidant combination. These disadvantages are compounded by the inherent depth limitation as the system exhausts into the surrounding sea. Indeed the presence of nitrogen in the exhaust raises the question of whether depth independence can be achieved for the open cycle system.

The Stirling engine. Of the various types of Stirling engine, the rhombic drive is in the most advanced state of development for application in the power range of interest. As with all externally heated engines any type of matched heat source can be used whether conventional or unconventional. Thermal energy storage using lithium fluoride is ideally suited.

When hydrogen is used as the fuel high efficiencies have been achieved, (up to 38%) and this is combined with a completely balanced and consequently vibration free engine. Detailed examination of the Stirling engine for underwater application had indicated a power to weight ratio of 0.08 kW/kg and a power to volume ratio of 44 kW/m^3 for the engine alone. For a 150 kw unit the production costs have been estimated at £80,000 per unit for a production run of 30 units. (Guthrie, Davies and Moore, 1973). This figure would not be much reduced for a 50 kw unit. However, the Stirling engine has three definite disadvantages, namely:-

1. A seal is required to separate the working volume from the exterior whilst allowing the transmission of power. This seal is known to be a basic weakness in the design, as it has a very limited life.
2. The mechanism for regulating the output of the engine is complex requiring both pressure and temperature regulation.
3. In spite of a considerable research effort over nearly 40 years, only United Stirling are known to be interested in the production of these engines. At the present stage only 50 pre-production units have been produced by them. Thus an early adoption of this engine for underwater application would inevitably mean the introduction of an untried technology with all the associated problems.

The Stirling engine is therefore not thought to be a suitable candidate for subsea application in the immediate future. Design developments and/or large scale production would, however, favour its adoption.

The closed cycle diesel engine. Two closed cycle diesel engines have successfully been demonstrated in recent years, and both adopt a different approach to closed cycle operation:-

1. Ricardo closed cycle diesel engine - carbon dioxide, used as the diluent, is enriched with oxygen prior to combustion. Following combustion, a fixed quantity of the exhaust has to be pumped overboard by compressor. The scheme

has two major drawbacks; firstly it is depth limited and secondly the selected diluent leads to a considerable loss in efficiency. The energy to weight and energy to volume are claimed to be 0.086 kWhr/kg and 95 kWhr/m^3 respectively using liquid oxygen as oxidiser for a 30 kW unit.

2. Psychrodiesel - steam is used as the diluent in this scheme. The presence of a 'lung' ensures a humidified oxygen supply to the engine but instead of pumping excess carbon dioxide overboard it is removed chemically. Whilst this method leads to depth independence, the 'lung' adds an extra degree of complexity and there is an additional need to ensure that the charge temperature is maintained above $100^{\circ}C$ to avoid condensation. The energy to weight and energy to volume are claimed to be 0.21 kWhr/kg and 150 kWhr/m^3 respectively when using liquid oxygen as oxidant and lithium hydroxide as absorber.

In an attempt to overcome some of the problems inherent in these schemes, a depth independent closed cycle diesel unit is proposed which makes use of nitrogen as diluent and chemically absorbs the spent exhaust gases. The scheme thus achieves depth independence without the added complexity of the psychrodiesel.

Based on the results of the psychodiesel and the Ricardo unit, it is estimated that a 50 kW plant with a 10 hour endurance (at peak demand) will have a power to weight ratio of 0.1 kWh/kg and a power to volume ratio of 100 kWh/m^3. Ricardo estimated that they could supply their plant, excluding fuel storage and pressure vessel for £5000 per unit in 1973. Thus a conservative estimate for the specific cost of a complete depth independent closed cycle diesel power pack (or nitro-diesel) has been calculated to be £180/kWhr for the above mission.

SELECTION OF THE OVERALL POWER PACK

The choice of power pack, comprising the energy source, an energy conversion system and the associated ancillaries depends to a large extent on the mission length. This study has concentrated on medium length missions of up to 10 hours with an energy requirement of 500 kWhrs. at 200m.

In comparing different power packs particular attention has been paid to the following factors:-

i) energy to weight ratio (kWh/kg)
ii) energy to volume ratio (kWh/m^3)
iii) fuel cost per dive (£/kWhr)
iv) specific cost index

The approach to the calculation of the first three factors makes allowances for the depth of operation (and thus containment of the energy convertor and fuel and oxidant) as well as the weight and volume of the engine, auxillaries, fuel and oxidant. In calculating the weight of fuel and oxidant the efficiency of the convertor was taken into consideration and, where appropriate, an enrichment oxygen supply was included. The approach to the specific cost determination encompasses initial cost, maintenance and depreciation (1000 dives). The results of this exercise when applied to the systems offering the greatest potential, are given in Table 4.

TABLE 4 A Comparison of Underwater Power Systems Based on 500 kWhr. Mission With A Guaranteed Continuous Output of 50 kW at 200m

Power Systems	Energy to Weight Ratio (kWhr/kg)	Energy to Volume Ratio (kWhr/m^3)	Fuel Cost per Dive (£/kWhr.)	Specific[1] Cost (£/kWhr.)
Stirling Engine	0.11	90	1	200
Recycle Diesel	0.11	100	1	180
IPN Turbine	0.20	160	15	250
Fuel Cell	0.1	80	3	2000
Lead Acid Battery[2]	0.025	50	0.1	750
Silver Zinc Battery[3]	0.09	150	0.1	50000

1. Assumes a total life of 1000 dives.
2. Assumes a battery life of 200 charge/recharge cycles.
3. Assumes a battery life of 20 charge/recharge cycles.

The superiority of the IPN turbine in terms of weight and volume is clearly evident although its specific cost is nearly 40% greater than the closed cycle diesel scheme. This factor combined with its limited operating depth (max. 50m) makes it an unsuitable candidate.

The Stirling engine shows no clear advantage in terms of volume although again the specific cost is greater than that for the closed cycle diesel scheme because of the higher cost of the prime mover.

The results for fuel cells and batteries indicate heavy, costly and in some cases, large power packs. The high efficiency of the fuel cell (up to 60%), and thus the low specific fuel consumption, has a more pronounced effect during longer missions.

Closed cycle diesel schemes provide the cheapest solution whilst at the same time providing a unit of small proportions and low weight. This approach is thought to offer the greatest potential for medium duration missions for the immediate future. However, should the Stirling engine become more widely accepted (and thus cheaper to produce) it could well become the more suitable candidate in view of its ability to burn a wide range of fuels.

CLOSED CYCLE NITRO-DIESEL TEST RIG

The potential offered by the depth independent closed cycle diesel engine inspired the construction of an experimental prototype unit in the Department of Marine Engineering at Newcastle University.

A photograph of the original prototype test rig is illustrated in Fig. 1.

Fig. 1 Original prototype unit

The prototype unit recirculates engine exhaust via a heat exchanger, scrubber, separator and oxygen injection chamber. Recirculated potassium hydroxide in liquid form, removes carbon dioxide, sulphur dioxide and oxides of nitrogen during the scrubbing process.

Nitrogen is used as diluent for the recycled gases, in place of carbon dioxide, thereby maintaining high cycle efficiency and facilitating conventional start up and shut down procedures. An additional advantage is offered in applications where changeover from open to closed cycle operation is required with the engine running under load.

Operational Experience with the Prototype Nitro-Diesel Cycle

The initial phase of the closed cycle trials has been confined to steady load situations. The oxygen supply during these tests was regulated by a manually controlled valve. This technique was not entirely successful, even under steady state operating conditions. Considerable deviations in oxygen concentration were found to occur and this in turn adversely affected engine performance.

The scrubber was observed to function in a satisfactory manner, reducing CO_2 concentrations to a minimal level without requirement for precise regulation of KOH flow rates. In this respect, the system displays a degree of inherent regulation.

Redesigned Closed Cycle Test Rig

The original prototype unit and loading circuit comprised a relatively crude assembly which was established on a limited budget in order to assess the viability of the nitrodiesel concept. The test rig has now been redesigned with the following objectives:-

i) A decrease in engine size to reduce operating costs incurred by KOH consumption.

ii) Substitution of a high response loading system capable of providing a complete range of dynamic loads.

iii) Installation of a new engine in order to objectively assess the effects of closed cycle operation over prolonged periods.

iv) Improved instrumentation and control features to facilitate dynamic engine trials.

Diesel engine selection . To be representative of a practical commercial application a multi cylinder engine of 10-100 kW is required. A Perkins three cylinder engine with rotary distributor pump and direct injection, running at a nominal 1500 rpm has been selected as a basis for the continuing experimental programme. The decision to run the engine at constant speed is based on the assumption that practical applications will almost certainly be confined to direct drives for generators or hydraulic pumps.

Dynamic loading system. The requirement for high response dynamic loading effectively restricts the choice to eddy current dynamometers, electrical generators, friction brakes and hydrostatics. Versatility and cost advantages favour the hydrostatic load, and a suitable system has been designed using standardised components of equipment.

The engine shaft is rigidly coupled to a conventional rotary vane pump through a strain gauge type torque transducer. This arrangement provides the capacity for safe torque transmission up to 150% of the rated engine torque of 160 Nm. The pump delivery pressure is set by a high response pilot actuated relief valve which has a response time of 15ms.and is electrically operated by a signal derived from a closed loop proportional controller. This system ensures that measured load torque closely follows the demanded load torque irrespective of variations in other system parameters.

A water cooled tube and shell heat exchanger is used to cool the oil and the hydraulic circuit incorporates conventional filter, strainer, oil reservoir and pressure gauge as illustrated in Figure 2.

Exhaust gas scrubber. The exhaust gas is cooled to near ambient conditions before entering the scrubber unit. Water vapour in the exhaust is condensed and drained, and the temperature is reduced to a degree which is compatible with chemical adsorption.

The cooled gas and the potassium hydroxide (KOH) are recirculated through a vertical column containing pawl rings, where direct contact between the carbon dioxide gas and liquid KOH is established over a large surface area. The operating pressure depends upon the relative concentrations (partial pressures) of the gases contained within the closed circuit, whilst the flow of KOH is processor controlled.

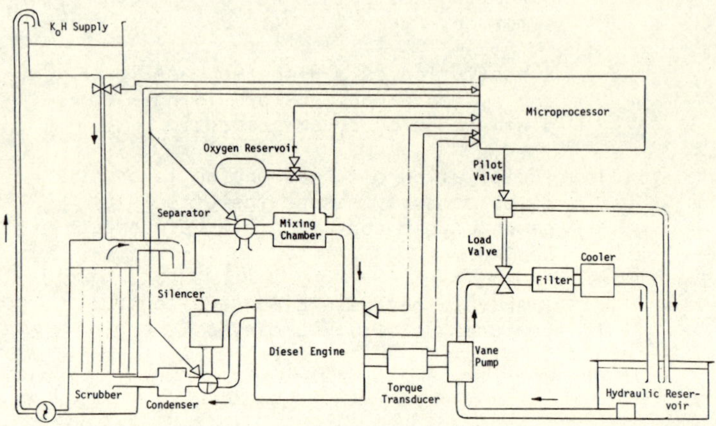

Fig. 2 Schematic of prototype unit

Instrumentation. The instrumentation for the fully developed test rig will centre upon the microprocessor unit. During the initial phases of development several control functions will be undertaken by a Miniac analogue computer, with transient data recordings presented on an eight channel chart recorder. The loading system will also be controlled by the analogue computer.

Engine speed is measured by a digital pulse counter with frequency to voltage conversion. Oxygen and CO_2 concentrations are monitored using a Westinghouse oxygen analyser mini probe (time constant 1.5 secs.) and an Irga non-dispersive gas analyser respectively. Oxygen flowrate is controlled by a pneumatically actuated control valve with linearised characteristic.

It is intended to instrument an engine cylinder with a Kistler mini probe pressure transducer, to assess changes in combustion characteristics following variation in manifold gas concentrations. Additional flow and temperature measurement in the ancillary cooling systems provides steady state energy balance data which is useful for applications in which the engine is used as a total energy source (heat and power).

Anticipated Development Problems

Diesel engine technology is extremely well developed and it is reasonable to expect safe and reliable engine performance if conditions experienced during closed cycle operation are maintained at or near to the conventional open cycle situation.

Careful attention to the following potential problem areas, will underlay the research programme.

i) Excessive oxygen concentrations in the engine cylinder may lead to problems such as:-

Burning of piston lubricating film with subsequent danger of seizure,

Overheating of piston crown or exhaust valves,

Generation of excessively high cylinder pressures due to accelerated combustion.

ii) High oxygen levels may produce adverse metalurgical effects.

iii) Low oxygen concentrations during transients will impair combustion efficiency and may lead to instability or stall.

iv) Response rates of commercially available oxygen sensors may limit rate of response of the oxygen control loop.

v) Nett pressure gradient between exhaust and inlet manifolds must not exceed the prescribed limits.

vi) Provision may be required for limited nitrogen replenishment due to formation of NOx.

MICROPROCESSOR BASED CONTROL SYSTEM

Operating experience with the existing closed cycle test rig indicates that research effort should be concentrated on the development of a high response control system capable of maintaining safe and reliable engine performance under all conditions of dynamic and steady state loading. Progress towards a practical nitrodiesel cycle is effectively halted until this objective has been satisfied.

Definition of Control Functions

i) The control system must provide the capacity for operation of a conventional naturally aspirated diesel engine in a closed cycle configuration, without risk to engine components, and without inhibiting engine performance.

ii) The engine must be capable of operating in either open or closed cycle configuration with a facility for smooth changeover between operational modes while the engine is on load.

iii) Automatic start up and shut down sequences, and automatic surveillance of principle engine and scrubber parameters.

iv) Provision of adequate stability margins whilst optimising transient response during dynamic loading situations.

It is anticipated that the control system will require a degree of sophistication due to the interaction between the various control loops. An additional problem is predicted by the preliminary linear analysis which indicates that the physical system exhibits a degree of intrinsic positive feedback in respect of the oxygen summations at the mixing chamber.

Finally, the existence of substantial transport delays associated with the loop time of the recycled gas stream and the practical limitations on response time for commercially available oxygen sensors, confirms the belief that the control problem is a formidable one.

Establishment of Microprocessor Algorithm

The present state of development of microprocessor technology provides an interesting option for solution of the previously stated control problems in terms of optimal cost, size and flexibility.

The interaction of the microprocessor with the closed cycle rig is illustrated in the schematic figure 2. Several channels of digital to analogue, analogue to digital and digital to digital conversion provide communication between the sensing elements, central processor and power actuators. The prototype justifies a programmable unit with access to the instantaneous values of the variables listed below.

Measured variables:	Engine Speed. Oxygen concentration at engine inlet or exhaust. Fuel flow rate. Carbon Dioxide concentration after scrubber. Exhaust temperature. KOH flowrate. Oxygen regulator position.
Controlled variables:	Oxygen regulator actuator. KOH flow rate controller, Engine governor set point or fuel metering valve actuator. Changeover valve positions.

It is envisaged that a production unit would operate under the control of a dedicated processor, thereby substantially reducing cost and complexity.

Hybrid computer simulation as a design tool. In order to formulate the microprocessor algorithms, detailed knowledge of the physical system dynamics is required. In an area where past experience is virtually non existent, computer simulation becomes indispensable. A preliminary simulation is now in progress in conjunction with the construction of the prototype unit and loading circuit. Data from the model is used to optimise control parameters prior to implementation on the prototype rig.

Theoretical and experimental understanding of system performance will be developed in parallel. Simulation will reduce the number of problems encountered during development of the closed cycle rig, and it is intended that at the end of the exercise a fully validated mathematical model will become available, to supplement the experimental data.

A significant feature of the hybrid computer implementation is the relative ease with which it can be interfaced to system hardware. During the final phase of microprocessor development, the physical control system will be interfaced to the hybrid model of the engine and scrubber (minus model controllers) for an extensive evaluation of performance prior to actual on engine trials.

HYBRID COMPUTER MODEL OF NITRO-DIESEL CYCLE

Hybrid computer programming is consolidated so as to take full advantage of the individual merits of the respective digital and analogue processors. The equations governing the generation and removal of constituent gases in the engine and scrubber are non-linear and contain complex interactions requiring multiplication, division and logical decision making. These functions are ideally suited to digital processing.

System dynamics (with the exception of transport delays) are executed by the analogue processor, which readily performs real time integration and produces convenient analogue signals for output recording. Variations in control parameters are generally made at the analogue console, while changes in fundamental system parameters are facilitated digitally at the keyboard.

The philosophy which has been adopted in mathematically representing the system is to develop a basically simple model whilst clearly defining all the limiting assumptions. Added degrees of complexity will be incorporated if and when required.

A description of the subsystem models which have been developed and combined to establish a first phase simulation is briefly presented in the text, together with a statement of the principal assumptions which have been made.

Diesel Engine Model

The diesel engine model is categorised into torque production (engine and governor dynamics) and gas flowrate (thermodynamic) sub sytems. Equations are omitted in the interest of brevity.

Torque generation model. The simplified model which is implemented entirely by analogue techniques assumes the diesel engine to be a torque generator of constant gain with terms for coulomb friction and speed dependent viscous friction, representing mechanical losses.

The governor is represented by a constant gain factor acting through a first order lag. Suitable limits represent non-linearities implied by fuel rack stops. Speed error is detected by comparison with a set point demand.

The torque gain is obtained from the engine operating characteristic and the total mechanical losses are based on a value of 15% at full load. The governor gain and time constants are derived from manufacturers' data.

The algebraic sum of developed engine torque and load torque is integrated through the engine rotational inertia to provide instantaneous speed.

Since the engine is naturally aspirated it is assumed that torque production is instantaneous following a change in fuel metering valve position. This assumption also implies that the concentration of oxygen at inlet to the engine remains high enough to ensure normal combustion (a basic requirement of the oxygen control system).

Gas generation subsystem model. The engine is considered to behave as a continuous gas generator, drawing a steady stream of 'air' the mass flow rate of which is assumed to be dependent on engine speed, charge density and volumetric capacity (assuming constant volumetric efficiency of 85%).

The products of combustion are calculated from the instantaneous value of engine fuel valve setting, assuming perfect combustion to CO_2 and H_2O. The fuel is assumed to be 87% carbon, 13% hydrogen, and all gas flowrates are calculated on a mass flow basis.

The constituent concentrations in the exhaust pipe to the scrubber are based on the calculated flowrates and are considered to be independent of the residual mass concentrations.

Ancilliary Models

The gas recirculation system is identified by interacting volumetric components in which transport delays, mass transfer and mass accumulation effects occur.

Scrubber/condenser subsystem model. Extraction of H_2O in an exhaust cooler is assumed to be perfect. The rate of extraction of CO_2 is assumed to be proportional to the flow rate of KOH to the scrubber. The CO_2 level controller and scrubber dynamics are incorporated by analogue techniques and are represented by a simple integral term for the purpose of preliminary modelling, thereby reflecting to some extent the tendency to inherent regulation which is observed to exist within the scrubber.

Mixing chamber subsystem model. It is envisaged that variation in the dimensions of the mixing chamber will effect the rate of change of oxygen concentration at inlet to the engine. Selection of mixing chamber volume therefore consitutes an unknown system variable which is subject to optimisation by simulation.

Changes of gas concentrations are therefore calculated from consideration of the constituent masses resident within the chamber in conjunction with the mass flow rates entering and leaving the chamber.

Mass accumulation and relative density. Relative density of the total gas content is assumed to be proportional to the instantaneous total mass of gas contained in the system. Total rate of mass accumulation is calculated from the algebraic sum of mass flow rates to and from the total system. The rate of mass accumulation between the main system components is then calculated by proportion.

A significant dynamic feature of the physical system is the transport delay associated with the loop time required for the recirculation of exhaust gas (which is in the order of seconds). The delay effects are consolidated into a single parameter which delays concentration changes between the scrubber and the mixing chamber.

Oxygen Control Sub-System Model

The control system for the oxygen replenishment is patched on the analogue computer facilitating instant changes to control configuration and variation of parameter settings.

The oxygen control system provides an option for various P + I controller configurations operating in a negative feedback loop sensing oxygen concentration at inlet to the engine. The following control options may be evaluated using the simple model.

i) Variation in the P + I Confirguration including. use of integrator authority limits.

ii) Variation in capacity of oxygen supply valve.

iii) Variations in oxygen sensor response and valve actuator time constant.

iv) Effectiveness of open loop, feed forward term activated from engine fuel setting.

v) Effectiveness of an exhaust oxygen monitoring system with infered inlet manifold concentrations.

MODEL RESULTS AND DISCUSSION

The current series of computer simulated trials is restricted to step changes in engine load between 0 - 75% of full rated capacity. This is considered to be a severe transient application which will provide an objective insight into typical dynamic behaviour of the total system in response to large signal disturbances. Extensive testing with generalised loading systems including frequency response trials will be conducted in due course, with the objective of fully defining the system dynamic characteristics.

The present discussion is limited to an appraisal of the results obtained, whilst assessing the influence of the following parameters.

i) Changes in the volumes of the scrubber, mixing chamber and associated pipe work.

ii) Changes in the transport delay associated with the recycled gas flow.

iii) Changes in the oxygen controller gain constants with oxygen concentration at the inlet manifold as the controlled variable.

iv) An assessment of the proposed fuel flow feed forward anticipatory control system.

Steady State Computer Results

The model is required to provide accurate steady state data with complete compatibility between all system parameters. For example, the sum of all concentrations should always total unity, and the algebraic sum of flow into and out from the system should be zero at steady state conditions.

Provision for detailed inspection of steady state data is provided in the computer program, including constituent gas flow rates, concentrations and relative density.

An interesting feature of the results is that relative density of the gas within the system is inversely proportional to load. A nominal density of 1.0 is assumed in the no fuel situation and the trend is summarised by the results presented in Table 5.

TABLE 5 Steady State Computer Results

Load Torque	Fuel %	Rel. Density
Motoring	0	1.0
Zero	15	.997
75%	80	.965
100%	100	.956

The effect is accounted for by the decreased oxygen concentration in the scrubber and associated pipework.

Transient Simulated Results

A limited number of transient response traces are included in this paper to provide an example of the way in which the model is used to assess the dynamic behaviour of the system when subject to step changes of between 0-75% of full load. The salient features of these figures are discussed below. Recorded variables appear on Channels 1 to 6 as illustrated:-

Relative density DR (0 - 200%).
Oxygen concentration at engine inlet $\%O_2$ (16%-32%).
Oxygen concentration in exhaust O_2 E (0-40%).
Oxygen replenishment flowrate Q_l (0- .008 kg/sec).
Engine speed N_e (1000 -1800 rpm).
Fuel rate X_f (0 - 100%).

High gain integral controller. Figure 3 is an example of the totally unacceptable result obtained when an unmatched controller is incorporated into the system. Following load rejection the increased flow of oxygen in the scrubbed exhaust drives the integrator into a shut off position. The oxygen regulator remains closed for 22 seconds and thereafter the system oscillates steadily with a periodic time of 40 seconds.

Low gain P + I controller with feed forward anticipation term. Initial computer runs with low gain P + I control resulted in a transient oxygen deficiency and unacceptable settling times. The trace which is reproduced in Figure 4 illustrates the response obtained when an open loop feed forward signal from the fuel flow metering device is used as an anticipatory control term, in an attempt to overcome these problems.

This particular technique has been shown to be effective in the absence of a transport delay. The transient response illustrated with a transport delay of 4.5 seconds, is unsatisfactory during load rejection due to the delayed arrival of concentration changes in the oxygen enriched scrubbed exhaust. After initially closing in due to anticipatory action, the oxygen control valve is seen to move transiently towards the open position before closing in once again under P + I control to eventually settle at the steady state position.

High gain P and P + I controllers. Simulated load rejection and load acceptance trials with a pure proportional controller set at the condition of limiting stability produced the traces illustrated in Figure 5a (Steady state offset = 5%, settling time = 10 seconds).

The fall in mixing chamber oxygen concentration due to the initial surge of nitrogen rich recycled exhaust gas produces corrective action by the oxygen controller which increases oxygen floweate, until the transport delay effect has passed. The steadily increasing concentration of oxygen then causes the oxygen regulator to close under proportional control towards the now steady state position. This is a feature common to all results obtained from systems which include significant transport delay effects.

The addition of an optimised integral term to the proportional controller identified with Figure 5a produces the traces illustrated in Figure 5b. Steady state offset is eliminated and transient response is smoothed. Total settling time is approximately 15 seconds, and fuel/oxygen ratios are maintained within acceptable limits.

Based on the data used in the preliminary simulation, these controller settings may therefore be regarded as being acceptable for the purposes of preliminary design specifications.

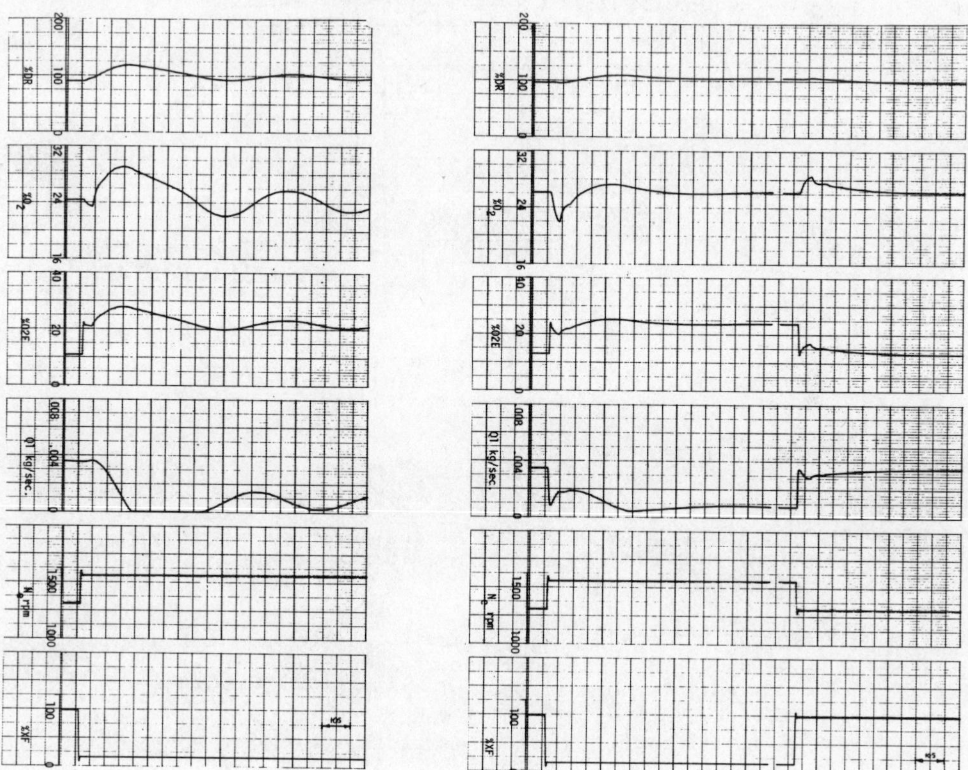

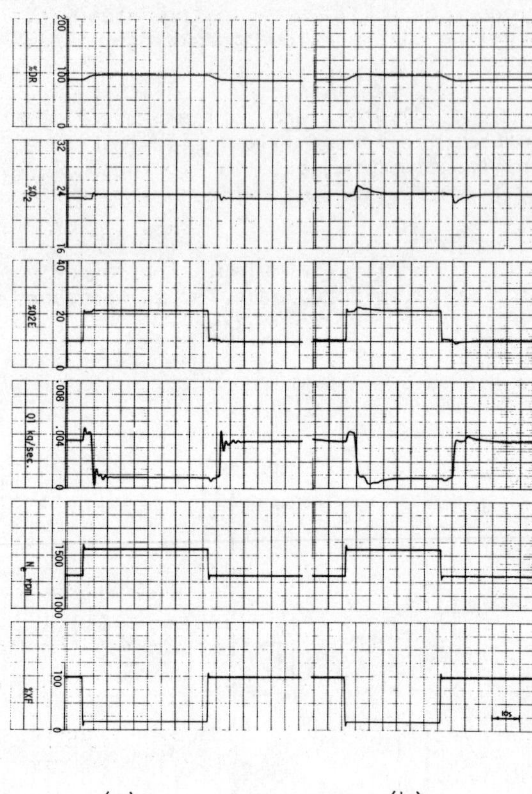

(a) (b)

Fig. 5 High gain contollers

CONCLUDING REMARKS

1. No single power system was found to be optimal with respect to all the desired characteristics. The selection of the power system is strongly influenced by the mission power profile and is mainly a question of assessing trade offs between performance data and cost.

2. The most appropriate power pack for medium length missions was found to comprise a combination of diesel oil and cryogenically stored oxygen as the energy source, whilst the most suitable energy converter was found to be the closed cycle diesel. The total unit was found to combine a low weight, volume and cost with well proven commercially available components.

3. A depth independent closed cycle diesel is proposed, employing nitrogen as diluent. It combines depth independence with simplicity, and employs well tried technologies.

4. Because of the potential offered by this system over the presently acceptable solutions, an experimental prototype unit is currently being constructed in the Department of Marine Engineering at Newcastle University to demonstrate the findings.

5. A fundamental prerequisite for the successful development of a practical nitro-diesel cycle, is the development of a cheap, safe and reliable control system, capable of ensuring satisfactory control under steady state and dynamic conditions. Hybrid computer simulation techniques are currently being used to design and evaluate a microprocessor based system.

6. Further power to weight advantages for the nitro-diesel will be investigated in a later phase of research, by raising the charge density in the closed cycle thereby effectively supercharging the engine without resort to turbo-blowers or other mechanical devices.

ACKNOWLEDGMENTS

The authors would like to thank the Science Research Council for their support and for permission to publish this work.

In addition they would also like to acknowledge the contribution made by Mr. Subbarao Bogarapu, formerly of the University of Newcastle upon Tyne.

REFERENCES

Economic Planning Department (1973). The Market for Underwater Power Sources. The Underwater Engineering Group, London, Report UR5.

Science Research Council (1976). The Report of a Task Force of the Engineering Board. Science Research Council, London.

Ayres, R.U. and McKenna, R.P. (1971). Alternatives to the Internal Combustion Engine. John Hopkins University Press.

Guthrie, J.A.S., Davies, J.G. and Moore, M.H. (1973). A Survey of Dynamic Power Systems for Underwater Applications. The Underwater Engineering Group, London Report UR4.

Macnair, E. (1975). Non-Nuclear Power Plants for Underwater Propulsion. Meeting of the Society for Underwater Technology, London, Parts I and II.

Lawson L.J. (1971). Design and Testing of High Energy Density Flywheels for Application to Flywheel/Heat Engine, Hybrid Drives. Intersociety Energy Conversion Engineering Conference. Paper 719150.

Rogers, G.T., Taylor, K.J., Moseley, P.T. and Turner, A.D. (1978). A Survey of the Future for Electrochemical Power Sources Undersea. A.E.R.E. Harwell, Report AERE-R8778.

Criddle, E.E., Gardner, C.L. and Wake, S.J. New Power Sources for a 14 ton Submersible. S.A.E. Paper 789042.

Overy, J. and Rayner, E.J. (1971). Power Pack for Underwater Applications. Systems Technology No. 12.

Puttick, J.R. (1971). Recycle Diesel Underwater Power Plants. S.A.E. Paper 710827.

Hoffman, L.C. Rudnicki, M.I. and Williams, H.W. Psychrodiesel. MTS Journal. Vol. 4 No. 6.

Rackley, A.R. (1969). Closed Brayton Power System for Deep Ocean Technology. S.A.E., Paper 690733.

Wimmer, R.E. (1970). Comparative Evaluation of Power System Types for Undersea Applications. Proceedings 6th Annual Conference on Marine Technology.

SUBMERSIBLE SYSTEMS FOR DRILLING SUPPORT

M. Hovland

Engineering Department, Statoil, Stavanger, Norway

ABSTRACT

Exploration drilling is moving into deeper waters and out of the practical and economical range of conventional saturation diving. This has forced the oil companies to seek alternative systems to assist the drilling operations on the sea bed. The various work operations associated with such assistance are summarized. The last few years has seen rapid developments within the field of manned and unmanned underwater work units. Some of the most promising of these diver alternative systems are reviewed. Their capabilities as drilling support systems are also discussed. Statoil has recently made inquiries for drilling support systems to be installed on a new drilling rig under construction. Some of the results from the inquiry are provided.

KEYWORDS

Diver alternative systems; underwater drilling support; work tasks; tethered remote controlled systems; tethered manned systems; free swimming systems; system assessment.

INTRODUCTION

In 1980 Statoil will start exploration drilling in water depths of more than 200 m, north of the 62nd parallel (depending on parliamentary approval). Although most of the drilling will be done in depths less than 300 m, Statoil is seriously considering and evaluating underwater drilling support systems capable of interactions in depths down to 500 m. Such a system is intended for installation on the drilling rig RR-12, of the "Ocean Ranger" type, currently being built in Finland.

There are at least three reasons why diverless work systems are favourable for drilling support in deep water. There will be an increase in safety, because personnel will not be directly exposed to ambient pressure. There is an expected reduction in drilling support expenditure, due to less personnel being involved in this work. And thirdly, there will be a weight reduction on the rig deck, mainly due to the mixed gas banks becoming redundant.

The past five years has seen rapid developments within the field of manned and unmanned diverless systems. This has resulted in a great number of different types of vehicles being offered on the market. Lloyd's register of 1979-80 (Lloyd's, 1979) thus lists over 250 systems. It is the objective of this paper to describe the work tasks incorporated in drilling support and review some of the most promising diver alternative systems available today.

DRILLING SUPPORT WORK TASKS

Before describing the work tasks it is desirable to take a brief look at the relevant components in offshore drilling. Figure 1 gives a general view of a semi submersible drilling rig.

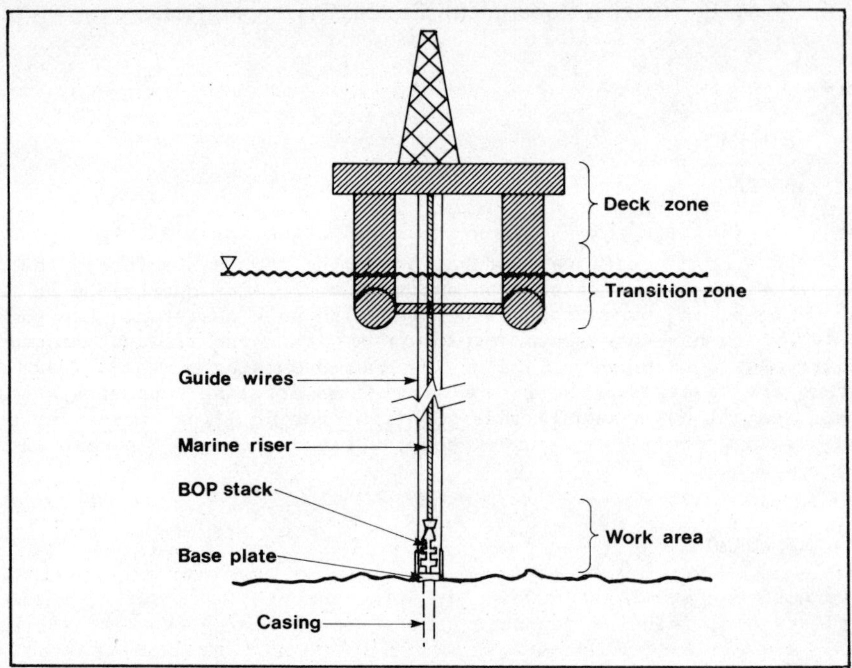

Fig. 1. A general view of a semi submersible drilling rig.

The underwater units are:

- Marine riser, housing the drilling string.
- Blowout preventer stack (BOP), which incorporates shut-off valves, and connects the marine riser to the casings.
- Base plate with guide posts (Fig. 2) upon which the BOP is aligned and resting.
- Four guide wires linking the drilling rig with the base plate.

The four guide wires provide guidance for installing and retrieving the BOP stack to and from the base plate during installation, servicing and well-abandonment. Various inspection and maintenance tools are also run up and down the guide wires.

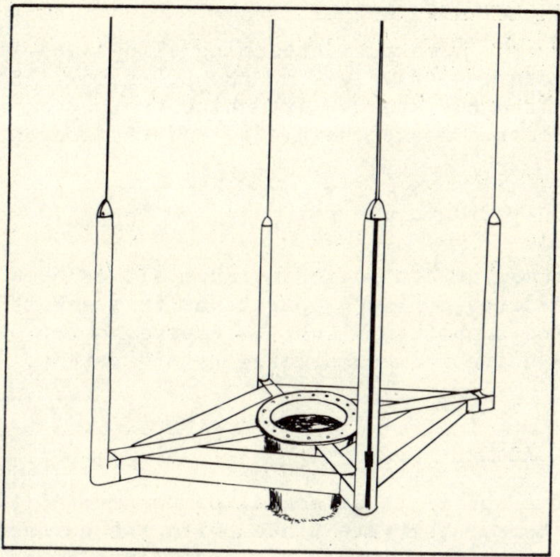

Fig. 2. The base plate with guide posts.

The work tasks to be performed by the drilling support system are related to these underwater units and may be divided in two main categories:

- Observation, inspection and surveying.
- Interactive manipulations.

The first category represents the simplest tasks and most often only involves a TV camera being directed to the point of interest. In the case of surveying, a bottom navigation system is required in addition. Many of the tasks require both observation and interaction. The main work tasks in underwater drilling support may be summarised as follows:

- Site inspection and surveying. This normally consists of visual- or TV-inspection of the sea floor in order to prevent obstacles.
- Aiding and observing the entry of casings. Physical guidance into the hole may be required in this operation.
- Aiding and observing the installation of the BOP stack on the base plate. This may be required as a precaution to prevent damage to guidewires and guide posts.
- Riser leak detection and observation, which is routinely done during descent and ascent.
- Asquisition of cement samples and monitoring of cement overflow. This is done when the 30" and 20" casings are cemented.
- Freeing and cutting of broken and entangled guide wires. This may be a complicated manipulative task if the wire clutters the BOP stack.
- Guide wire replacement. Various systems for guide wire connection to guide posts exist. The amount of manipulative work depends heavily on the system employed.

- Inspection, cleaning and replacement of the socalled AX-seal. This is a sealing-ring fitted to the 30" casing. It is accessible when the BOP is disconnected and taken off the base plate.
- Other manipulative work, such as: Attaching slings and lift lines to objects lost overboard. Operation of valves on the BOP stack. Stabbing hydraulic quick connectors to the BOP stack. Attaching explosives in the event of cutting off damaged guide posts. Beacon servicing and replacement.

DIVER ALTERNATIVE SYSTEMS FOR DRILLING SUPPORT

The systems to be reviewed in the following have either been employed for drilling support or have been offered for such operations in the North Sea. All submersibles suited for drilling support will not be covered here. It is only practical to review two or three units from each category of systems.

Large Tethered Manned Systems.

These units consist of a spherical personnel compartment fitted with external manipulators and thrusters. They are linked with the surface through a communications and power-cable (umbilical) which may also serve as a lifting line. The crew normally consists of one operator and one observer. Three different types are reviewed (Fig. 3), one of which is still in the prototype stage.

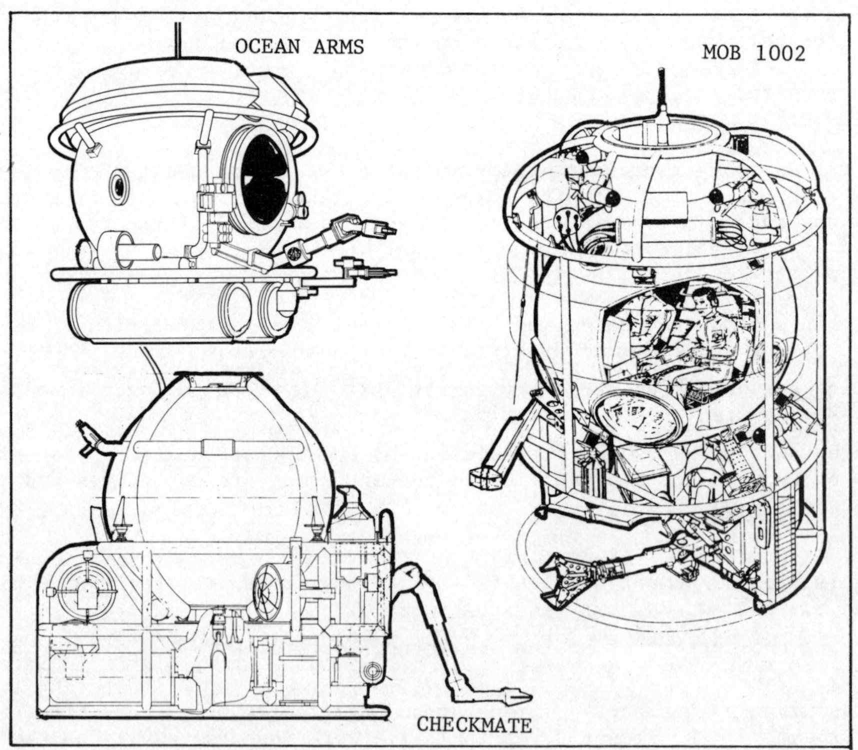

Fig. 3. Large tethered manned systems.

Ocean Arms III (Atmospheric Roving Manipulator System III), the hull built by A/S MØllerodden, Norway, is a standard diving bell modified for drilling support, underwater repairs and maintanance. The bell is fitted with a General Electric "force feedback", electro-hydraulic manipulator with 7 degrees of movement. This manipulator is a unique tool in that it is guided by a slave arm operated from inside the bell. The Arms III is also fitted with a hydraulic grabber arm used to lock the system onto the work area. This system has been in drilling support service since 1976 (Arms I). The Arms III has a depth capability of 500 m.

MOB 1002 (Manipulation & Observation Bell 1002), built by Comex Industries, France, is a specially designed bell for underwater work. It has one manipulator with 6 degrees and one with 2 degrees of movement. It is also fitted with a lifting winch capable of lifting loads with a weight in water of up to 300 kg. MOB 1002 has done drilling support work in 960 m water depths (Billingsley, Woods, 1978).

Checkmate, built by A/S MØllerodden, Norway, is still in the prototype phase. It consists of an acrylic, spherical personnel compartment and is fitted with six thrusters and one manipulator with 7 degrees of movement. It is presently fitted with four powerful suction cups which may be exchanged with a grabber arm. The prototype model has a depth capacity of 300 m.

One-man, Tethered Systems.

Perhaps the most daring and interesting developments within the field of manned underwater vehicles are the small, one-man systems (Fig. 4). The objective of developing such units lies in the desire to retain the agility of the diver without having to expose the operator to the ambient pressures. The development has gone from the armoured diving suit to the micro tethered submarine.

Jim, atmospheric diving suit, built by D.H.B. Construction Ltd., UK. This is an armoured one-atmosphere diving suit capable of operating at depths of 440 m. The suit was first tested in 1971 and was built over a patent application from 1922 (Trillo, 1978). The suit has no artificial propulsion, but Jim's legs are designed to walk along the seabed or a walkway adjacent to subsea installations.

Wasp, built by OSEL, Offshore Submersibles Ltd., U.K. The Wasp may be regarded as the next development step in that it maneuvers by the use of four hull-mounted thrusters. Essentially it is a one-atmosphere, armoured diving suit without legs. The Wasp has a depth capability of 610 m and has been used for drilling support since 1977.

Mantis, also built by OSEL, U.K., is the latest development in the tethered submarine field. It is fitted with eight or ten electric thrusters and has two seawater hydraulically operated manipulators. The Mantis was built in 1978 and has been used for rig inspection and debri-clearance operations.

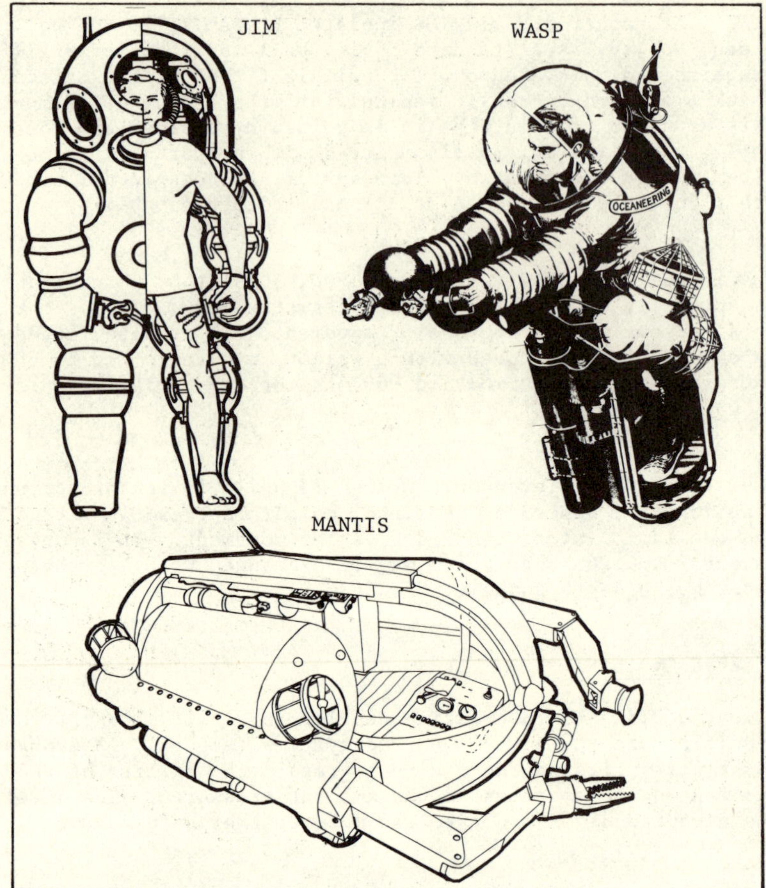

Fig. 4. One-man, tethered systems.

Free Swimming, Manned submersibles.

These systems have been used for a variety of tasks in the North Sea ranging from detailed sea-bed surveying to diver support and pipeline repair. They are normally fitted with two manipulators and two to six thrusters. They are battery powered and have a work endurance of 6 - 8 hours depending on type of work task. The types represented here (Fig. 5) may be called the underwater work horses of the North Sea.

Pisces VI, built by International Hydrodynamics Co. Ltd., Canada. This type of submarine is built up around a spherical, steel personnel compartment providing a depth capability of 2000 m. It has two electric thrusters and is normally fitted with a manipulator and a grabber arm. The Pisces VI has been used for drilling support from a drilling ship since 1977 (Hurd, Galerne, 1979) to depths of 1490 m.

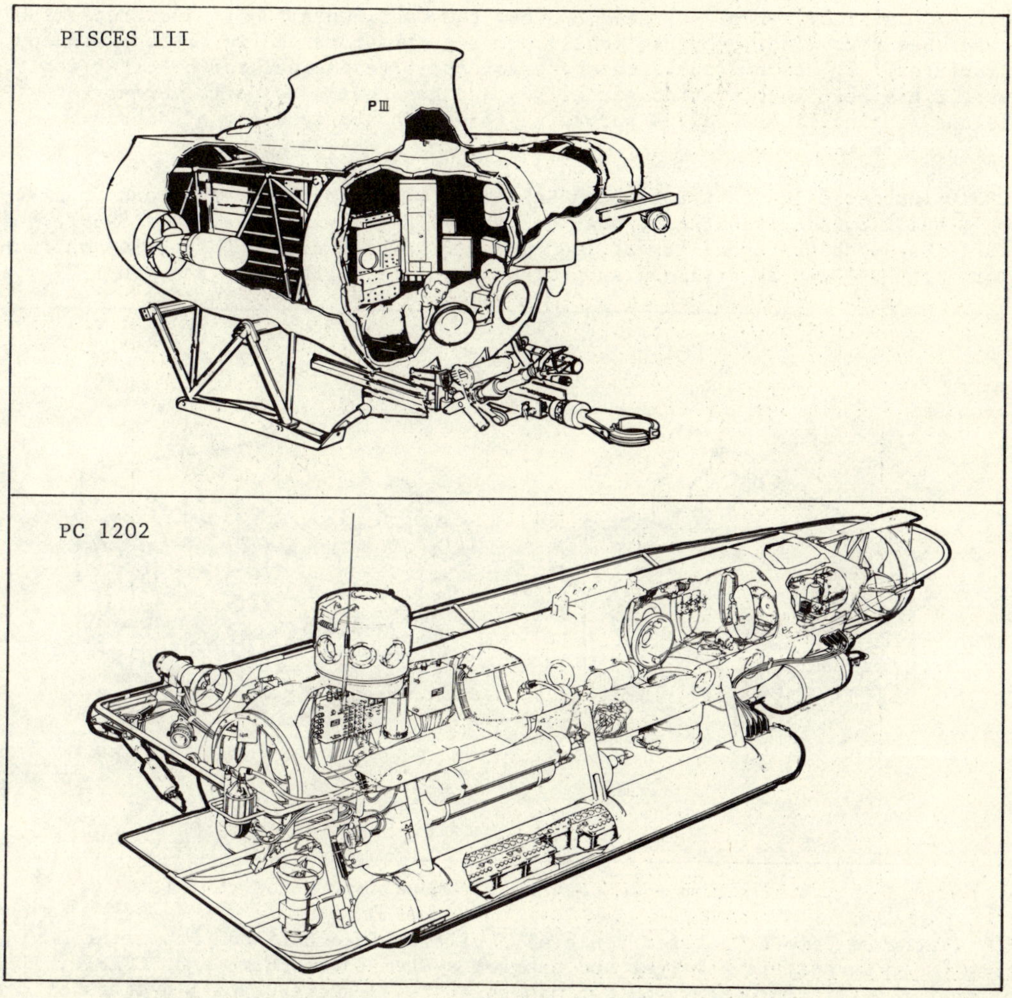

Fig. 5. Free swimming, manned submersibles.

PC 1202, built by Perry Oceanographics Inc., USA. In addition to the crew of three, it can carry three saturation divers to a depth of 330 m. It is fitted with one main propulsion motor, two auxiliary thrusters and two manipulators. The hemispherical plexiglass nose provides good visibility for work operations.

Tethered, Remote Controlled Vehicles

These systems are unique in that they normally do not require man to enter into water. They therefore provide the ultimate answer to the question of safety and diving. These craft (Fig. 6 and 7) are either powered by pure electric- or electro-hydraulic thrusters. They carry one or two manipulators and are fitted with one or more TV-cameras and lighting. The umbilical has normally a tripple function: high voltage power transmission, low voltage signal transmission and lifting of vehicle. Two medium sized and two large remote controlled systems are presented here.

Snurre I, built by Myrens Verksted A/S and the Continental Shelf Institute (IKU), Norway, has four electro-hydraulically powered thrusters and is fitted with one manipulator. It has hydraulic reserves for one more manipulator or other tools. Snurre I has been in operation since 1974 and has been used for a variety of missions ranging from detailed surveying to search and recovery of debris.

SCORPIO (Submersible Craft for Ocean Repair, Positioning, Inspection and Observation), built by Ametek, Straza, USA, has a similar size and design as Snurre. SCORPIO has a depth capability of 1000 m and is fitted with a 5 function manipulator. It has been in drilling support work in the North Sea since mid 1979.

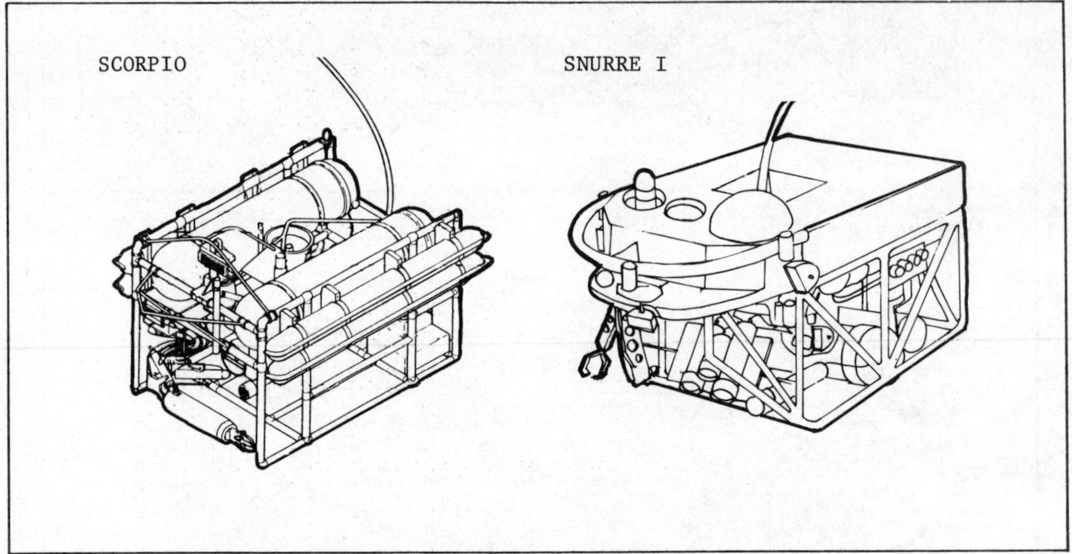

Fig. 6. Medium size tethered, remote controlled vehicles.

TROV (Tethered Remote Operated Vehicle), built by International Submarine Engineering Ltd., Canada, is a larger and heavier system than Snurre and SCORPIO. It has four electric thrusters and two manipulators (7 degrees and 4 degrees of movement). TROV has recently been used for platform cleaning and inspection in the North Sea.

Spider, built by Myrens Verksted, Norway, was designed as a remote controlled lifting hook, but has had a variety of other assignments. It was first put into operation in 1977, and has probably logged the greatest total number of operative hours of all remote controlled submersible systems, due mainly to the long lasting pipeline burial mission on the Ekofisk - Emden gas pipeline. The Spider has a dry weight of 3 tonnes. It has a depth capacity of 500 m and is fitted with four powerful electrohydraulic thrusters. It has not been used for drilling support yet.

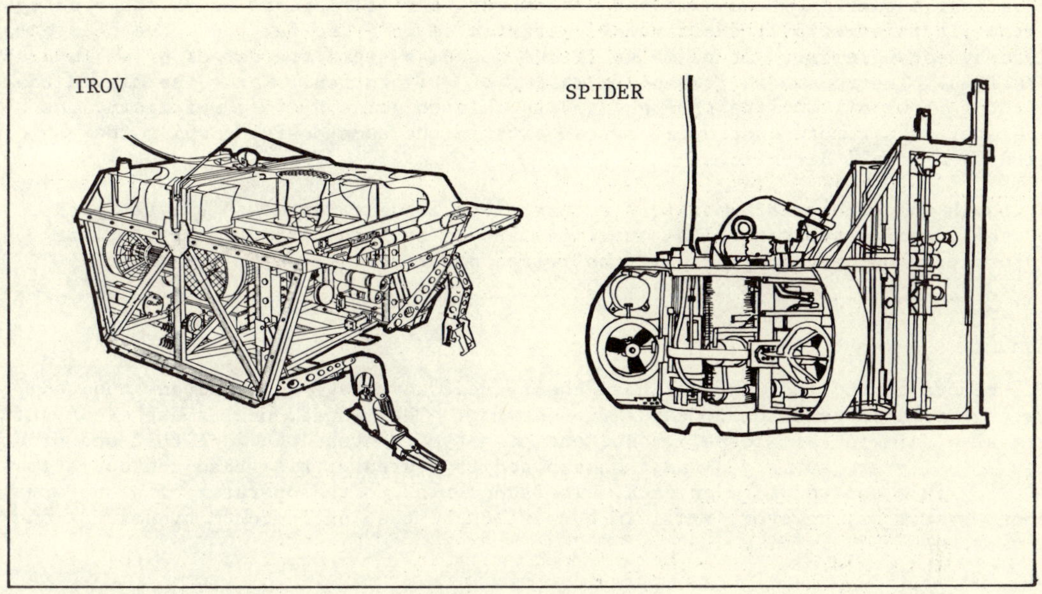

Fig. 7. Large tethered, remote controlled vehicles.

SYSTEM ASSESSMENT FOR DRILLING SUPPORT

There are four main factors to consider when selecting the system for a certain drilling support operation:

- Operational reliability and work ability.
- Safety.
- Weight on platform base.
- Cost of installation and operation.

Operational reliability and work capability

The work capability is dependent on three main factors: Size and maneuverability of the vehicle, dexterity of the manipulator and the overall view of the work area.

The physical size of the smaller vehicles, often combined with a high degree of maneuverability provides good access to the work site. A very sophisticated manipulator may thus not be of any great use if the vehicle itself is large and unable to move sufficiently close for all of the manipulator's functions to be utilized. Manipulative work is often time consuming, laborious and cumbersome. This is especially so when both the vehicle and the manipulations are remote controlled. In this respect the manned systems have the advantage of providing a man on the work site. With the human stereo and colour vision it is possible to achieve a full description of the whole work situation. A similar site description is virtually impossible with current TV systems.

The large, manned systems have an extra advantage in that they enable the client, or an expert, to view the work site. This is, however, not normally necessary in drilling support work, but may be of importance in special cases.

Umbilical entanglement is one of the greatest fears for operators of tethered systems. It is especially the unmanned tethered vehicles,so far,that have been hampered by entanglement. It seems as if the manned systems are not as prone to such problems. The reason is probably a matter of orientation. While the man in the craft follows all motions through, and is able to guide his way out again, the operator of a remote controlled system sits in the same position while the vehicle is driven in all directions.

Although the problem of umbilical entanglement is detrimental to the tethered systems, the advantages of the umbilical should be remembered, namely, online information on deck and virtually unlimited power.

Safety

The remote controlled vehicles have their great advantage in being one hundred percent safe in relation to saturation diving. The manned, one-atmosphere systems are also safe in this respect. But one cannot here evade the fact that one or two men actually are submerged and thus exposed to a greater risk than the operator sitting in a control room on deck. The added risk to the operator of a one-atmosphere system is, however, small in comparison to that encountered by saturation divers.

Weight on Platform Base

Deepwater drilling in remote areas calls for the drilling platform to carry as much stores as possible. Every tonne saved on utility systems, such as drilling support, is a precious weight release. The weight of a fully operational saturation diving-spread in the North Sea, south of 62°N, is thus about 80 t. The weight of the smallest one-man system, in comparison is only a tenth of this and it occupies only a fraction of the deck space (Bovie, 1979). If the weight factor is critical for system selection, then the larger systems have a great disadvantage as opposed to the smaller ones.

Cost of Installation and Operation

The all overriding factor in any assessment of system is most often the cost. For a long-duration assignment the equipment expenses become less important than personnel expenses,since material costs are written off over a longer period while personnel expenses are constant.

A deep diving, saturation system has a large number of divers and operators at hand, a diver alternative system only employs a maximum of 5 - 6 persons. It will therefore be the smaller remote controlled systems that are cheapest in operation over a longer period.

The most cost effective system to operate is maybe the smaller remote controlled vehicles. However, in an overall assessment of cost, the work efficiency and reliability must also be considered. What is, for example, the loss in the event of a major entanglement of a tethered craft, whereby the whole drilling operation is shut off for some period of time? This must also be treated as a relevant factor in system assessment.

AVAILABLE SYSTEMS FOR A NEW DEEPWATER RIG.

Statoil invited ten diving companies and submersible operators to bid their services for periods of two and four years for drilling support on the new RR-12 rig.

This inquiry gave the companies an opportunity to offer various categories of drilling support systems, ranging from conventional diving with a 300 m depth capability, to one-atmosphere, manned systems and remote controlled systems capable of working at 500 m depths.

The final bid evaluation has not been completed at the time of writing. Some of the results may, however, be disclosed here.

Out of the nine bids entered, five bidders offered conventional saturation diving, two of these in combination with a modified diving bell, and one in combination with a TROV system. Six of the bidders offered large, tethered, manned systems. Only one bid included free swimming, manned submersibles.

Six companies offered remote controlled vehicles. Three of these were with the use of Snurre II and the other three with SCORPIO. Spider and TROV were also offered as additional alternatives. Five out of the nine companies offered one-man, tethered systems for drilling support. Four of these bids were with the Mantis system.

When incorporating a diver alternative system on an exploration drilling rig it is important that the operators are familiar with drilling support. Careful training of pilots and system-operators is therefore necessary prior to commencement of offshore drilling.

CONCLUSIONS

The Statoil inquiry has shown that the majority of conventional diving companies are willing to disregard saturation diving for drilling support to 300 and 500 m in favour of alternative diver systems, manned or remote controlled.

In the North Sea the drilling support scene is currently governed by conventional diving to depths of about 200 m. In deeper waters some drilling operators work without any back-up system. They rely upon independent remote controlled components and automatic drilling equipment, more or less standard on deepwater drilling rigs. Others prefer to employ diver alternative systems of which the Wasp and Ocean Arms bell probably are the most widely used manned systems, and SCORPIO the most used remote controlled vehicle.

The fact that four companies out of nine proposed to use the Mantis for drilling support shows that there is great faith in this newly developed system. The Mantis probably represents the most exciting development within the field of one-man systems. The main advantages are low weight, low cost and versatility. However, the unit still introduces a man in the water. This is by some held as an unnecessary risk factor. On the other hand, current technology is perhaps not at the stage yet where a remote controlled system may be developed which can substitute man totally. Even replacing the man in the tethered manned vehicles will take time. The full takeover is dependent on developments within videosystems (colour, stereo), remote manipulation (force feed back) and a memory system for orientation. Even if elaborate electronic systems are incorporated in the vehicles making them compatible with manned systems, the price tag may be inhibitive. The present situation where there is room for all types of systems, will remain, but probably not forever. The ultimate situation will undoubtably have been reached when the drilling supervisor himself is provided full control over the underwater drilling support system from his office on shore.

REFERENCES

Billingsley, L. E., and J. W. Woods (1978). Drilling assistance at 3300 foot depths. Offshore-Technology Conference, Vol. II, 3196, 1257-1259.

Bovie, B. A. (1979). Water depths create diving challenges. Offshore, Vol. 39, 140-148.

Galerne, A. (1979). Pisces VI - Deep Eyes and Arms for Discoverer Seven Seas. Ocean Industry, No 11, Vol. 14, 64-68.

Harris, L. M. (1976). An Introduction to Deepwater Floating Drilling Operations. Petroleum publishing Comp., 82-160.

Hurd, D. R. (1979). Deep water drilling support using manned Submersible Pisces VI. Offshore Technology Conference, Vol. III, 3575, 1899-1904.

Lloyd's (1979). Lloyd's Register of Shipping. Lloyd's Register of Shipping, London, 112-183.

Trillo, R. L. (1978). Jane's Ocean Technology. Jane's yearbook, London, 1-225.

ADDENDUM

At the time of the conference, Statoil have completed the bid evaluation mentioned in the paper. Two companies have been requested to provide their services for drilling support on two exploration drilling rigs, the RR-12 and the Nordraug. Each rig will thus be equipped with one Mantis system together with one Mantis system on standby, either on board the rig (remote areas) or within 24 hours of mobilization on board.

The Mantis system was chosen mainly due to the following factors: low cost and weight, a high degree of mobility, versatility and an assumed degree of work ability and reliability.

SUCCESSFUL USE OF UNMANNED SUBMERSIBLE "SCORPIO" ON DEEPWATER EXPLORATION DRILLING PROVES FEASIBILITY OF REMOVING DIVERS

K. Lenning and E. Archer

Norsk Hydro, Norway

ABSTRACT

For the well 33/5-1 drilled by "Treasure Seeker" in 340 m of water the on board diving system was replaced with the remote controlled vehicle "Scorpio".

The decision was taken after an investigation into available systems, after establishing relevant work tasks and after the capability of the rig's own guide line running tools had been established.

"Treasure Seeker" with its present outfit is basically capable of providing its own underwater services. "Scorpio" is included as a back-up, to do trouble shooting and to be able to do more comprehensive subsea surveys.

To prove the capability of "Scorpio" an extensive test program was set up, that were successfully completed while the rig was on location. The test program included work on a two ton test structure, set on bottom, about 20 m from the wellhead.

The test structure consisted of a concrete clump weight, pad eyes for launch/recovery, a full size test post, quick connectors, valves and its own guide wire to surface. The work done by "Scorpio" on this test structure and the assistance provided to the drilling operations proved the feasibility of eliminating divers on drilling rigs.

In November 1979 the "Scorpio" was employed as the primary system for locationing and reentering of the well 15/5-2 in 110 m of water. The vehicle detected the wellhead target after a 6 minute sonar search and latched all four guide lines within 4 hours 30 minutes.

To say that divers will never be required is, of course, impossible. Salvage type operations may occur, but then it can be justified to call in a suitable diving vessel.

As the usage rate per rig is low, and the equipment is highly mobile, several rigs could shear one system. Only then the vehicle can be utilized effeciently providing the operators with a very cost effective tool.

KEYWORDS

"Scorpio", remote control vehicle, exploration drilling, replacement of divers.

INTRODUCTION

When Norsk Hydro was awarded the deep water license covering the blocks 33/2 and 33/5 in the Norwegian North Sea, a study was initiated to find alternative methods for underwater support. As the water depths ranged from 315 m down to 356 m, the option of a convential saturation diving system was immediately ruled out.

Also after having studied the possible work tasks, it was found that a system capable of doing more than pure inspection work was highly desireable.

The subsea service market was investigated for availability, work capability and cost effectiveness of manned and unmanned submersibles, atmospheric suits and diving bells with manipulator arms.

The most attractive option in all respects was found to be Ametek Straza's "Scorpio".

At this point the decision was made to replace the on board diving system with the "Scorpio" RCV and let the tool prove itself during an extensive test program on location while drilling in 340 m of water.

DEFINITION OF WORK TASKS

In order to find the most suitable system, a careful evaluation of possible work tasks was performed.

The most complex underwater tasks were related to repair of the BOP stack. However, even with divers available, this work would never or very seldom be performed as it nearly always will prove more cost effective to secure the well and pull the BOP for repair on surface. The complexity of our subsea BOP stacks equipped with emergency retrieval system and acoustic back-up control system makes it virtually impossible for divers to get access to shuttle valves, hydraulic connection, etc. that might need repair. At the same time improved BOP handling systems makes it possible to roundtrip a BOP stack in less than 24 hours in 350 m of water with gale conditions (Beaufort 8) while shorebased divers will require 48 to 72 hours mobilization time.

The required work tasks were defined as follows:

1. Inspection of areas which were not covered by the rig's TV camera.

2. Capability to handle wire and hook for picking up of various objects.

3. Cutting of wire. This may be a broken guide wire which has entangeled with the riser and stack preventing releasement of the lower riser connector or the stack connector. Cutting of wire may also be required for cleaning guide post previous to latching of a new guide wire.

4. Connection of guide line latch. When a guide wire has broken, this will normally be reestablished using a conventional guide line latch utilizing remaining three guide lines as guidance. However, when more than one wire is broken or when reentering previously abandoned wells, subsea assistance is welcome, but not absolutely necessary with todays techniques.

5. Stabbing of hydraulic quick connect and turning of valve. There has been instances when a BOP stack or a hydraulic latch (pin connector) cannot be disconnected from the wellhead. For this purpose we fitted our hydraulic connectors with an additional shuttle valve, a 90 degree ball valve as isolation valve and a hydraulic stabbing port. Hence, a hydraulic line with a quick connect can be run directly to the connector's open port and stabbed and thereby by-pass any fault in the primary hydraulic circuit.

6. Placement of shape charge. A special charge has been developed for severing of a bent guide post which may prevent the landing of a BOP stack.

The system to be selected should be as uncomplicated as possible to execute the above tasks. The time requirement to execute each required task was considered of less importance. The manning of such a system should be kept to a minimum such as 3 operators for intermittent work tasks and 5 - 6 operators for continuous 24 hours diving.

SELECTION OF SYSTEM

The Ametek Straza RCV "Scorpio" was selected for the following reasons:

1. Availability. The vehicle could be delivered within the required time.

2. Cost effectiveness. The vehicle was found to be the most attractive and would be cost saving compared to conventional diving systems and most other viable alternatives.

3. Safety. As the system does not involve any personnel in the water, in any manner, it was found to be the ultimate answer to safety for underwater support.

4. Personnel requirement. The system is operated by two operators and one supervisor.

5. Task requirements. The vehicle being equipped with a five function Perry manipulator and a set of necessary tools, it was deemed satisfactory for the fulfilment of the requirements. However, this would be necessary to prove in an extensive test program.

6. Mobility. The vehicle including umbilical and control station weighs less than five tons. The system can easily be shared between several rigs as it is highly mobile. The limited weight is also a definite advantage over more heavy systems on floating drilling rigs as every kilo saved adds to the available deck load.

7. Launching system. Considerable effort was made to find a satisfactory "all weather" launching system which may be employed either in transit draft or in drilling draft. The latest version utilizing the "basket" principle with a guide line to the pontoon has proved effective in this manner.

8. Guidance system (ref. Fig. 1). A primary design requirement was a solution to the entanglement problem. By the utilization of a separate umbilical guide line, the vehicle is only provided with enough free umbilical to reach the target. In this manner the possibility of entanglement is greatly reduced.

TECHNICAL DESCRIPTION

General

The "Scorpio" is an unmanned, undersea vehicle system that has been developed to perform a variety of offshore work tasks. It is configured for underwater observation, sonar search, and bottom survey as well as mechanical work tasks.

System configuration as installed on board D/V "Treasure Seeker"

(ref. Fig. 2 and Fig. 3)

The major components of the standard "Scorpio" system consists of the vehicle, tether cable, cable reel and winch, power distribution unit and operator station. In addition, a handling system is required. On board "Treasure Seeker" the handling system consist of a pivoting davit for launching/recovering the vehicle, an umbilical wheel and a winch for handling the vehicle guide wire clump weight. Also, a separate winch for handling the basket is installed. The system configuration as installed is shown in Fig. 1.

Vehicle

The vehicle, shown in Fig. 2, is designed to be slightly positively boyant in the water. Four hydraulically powered thrusters give the vehicle the required mobility. An on board hydraulic unit supplies power to the four thruster motors and to the actuators for the manipulator arm, TV pan and tilt unit, sonar tilt mechanism and any optional tool. A search and navigation sonar provides critical guidance information and acoustic beacon homing for navigation through hazardous areas. A five motion manipulator arm with a claw provides guiding, turning and gripping functions. All electronic assemblies aboard the vehicle are located in the starboard float tank. The vehicle is equipped with two TV cameras, one pan and tilt unit and one fixed unit.

Control hut description

The major components of the control room is shown in Fig. 3. The components consist of the power distribution unit, operator control unit, two TV monitors, sonar display and remote control unit.

Power distribution unit

The power distribution unit provides a central point for all the power required by the "Scorpio" system. It contains all of the power switching, circuit protection and transformers necessary for power control.

Operator control unit

The operator control unit is a collection of the displays and controls used for the operation of the "Scorpio" system. It consists of the vehicle propulsion controls, TV camera controls, manipulator controls and status indicators.

Remote control unit

The remote control unit contains the basic propulsion controls. The operator can carry this small portable unit out on deck during launching and recovery where there is full visibility of the vehicle handling equipment.

TV-monitor

Two TV monitors provide the display for the vehicle TV cameras. These monitors are again connected to the central TV monitoring system on the rig as well as to a video tape recorder.

Sonar display

The sonar display provides range and bearing indications to sonar targets and bearing indications to acoustic marker beacons.

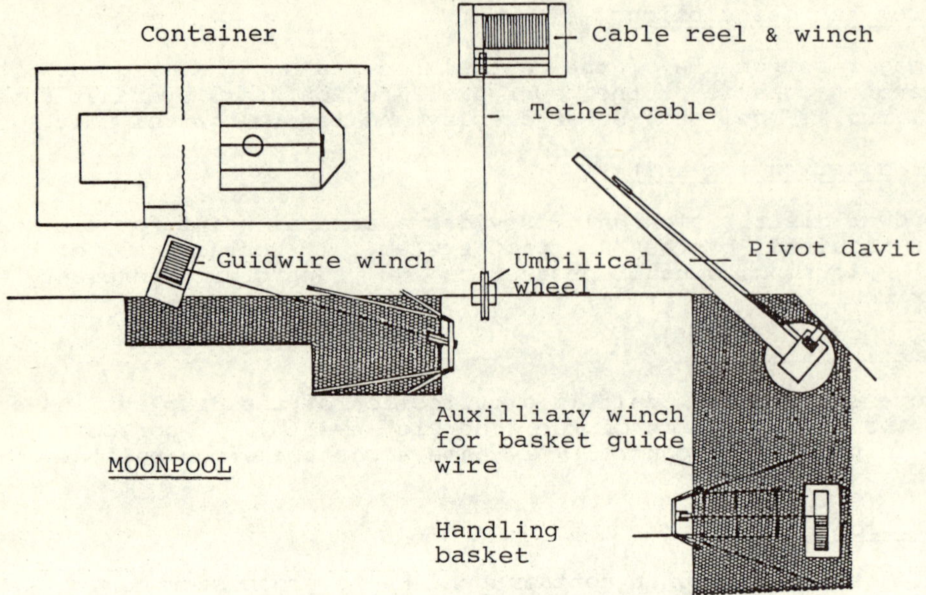

Fig. 1 System configuration

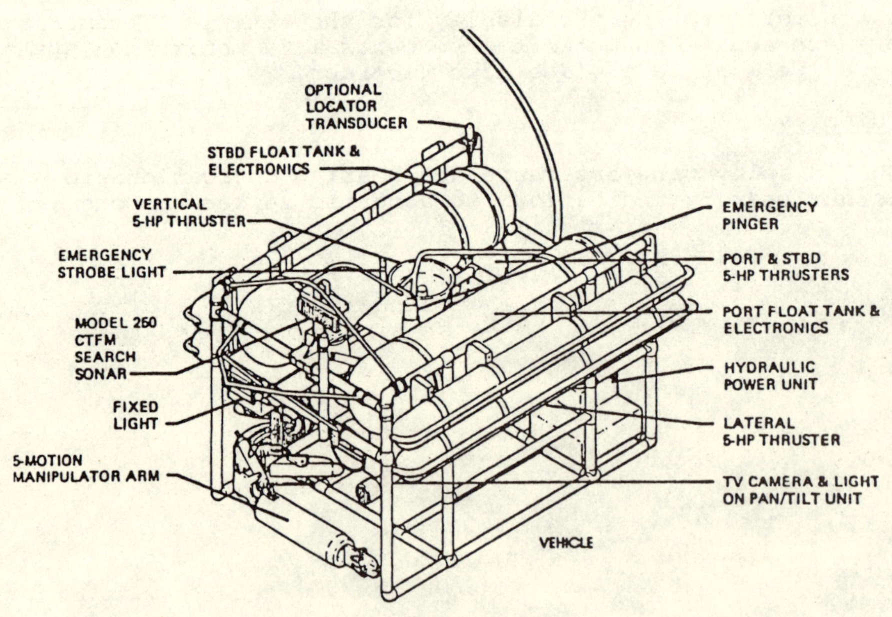

Fig. 2 Scorpio

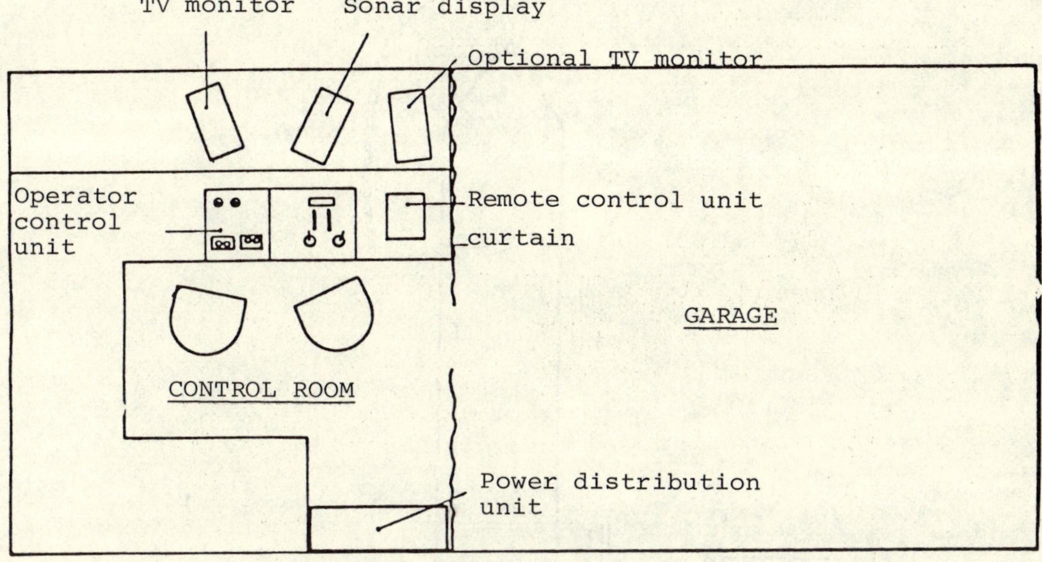

Fig. 3 Control hut lay-out

TEST RESULTS

The basic test post configuration is shown in Fig. 4

Inspection

Inspection was carried out on:

- Temporary guide base
- stab in of 30" casing
- Permanent guide base
- Wellhead

"Scorpio" is well suited for inspection tasks with its high degree of manouverability. For tasks close to the wellhead, the rig TV is generally better due to its ease of intervension and its better stability.

During the stab in of the 30" casing "Scorpio" was used to show how the casing shoe was positioned relatively to the temporary guide base. Based on this information the rig could be moved into the correct position.

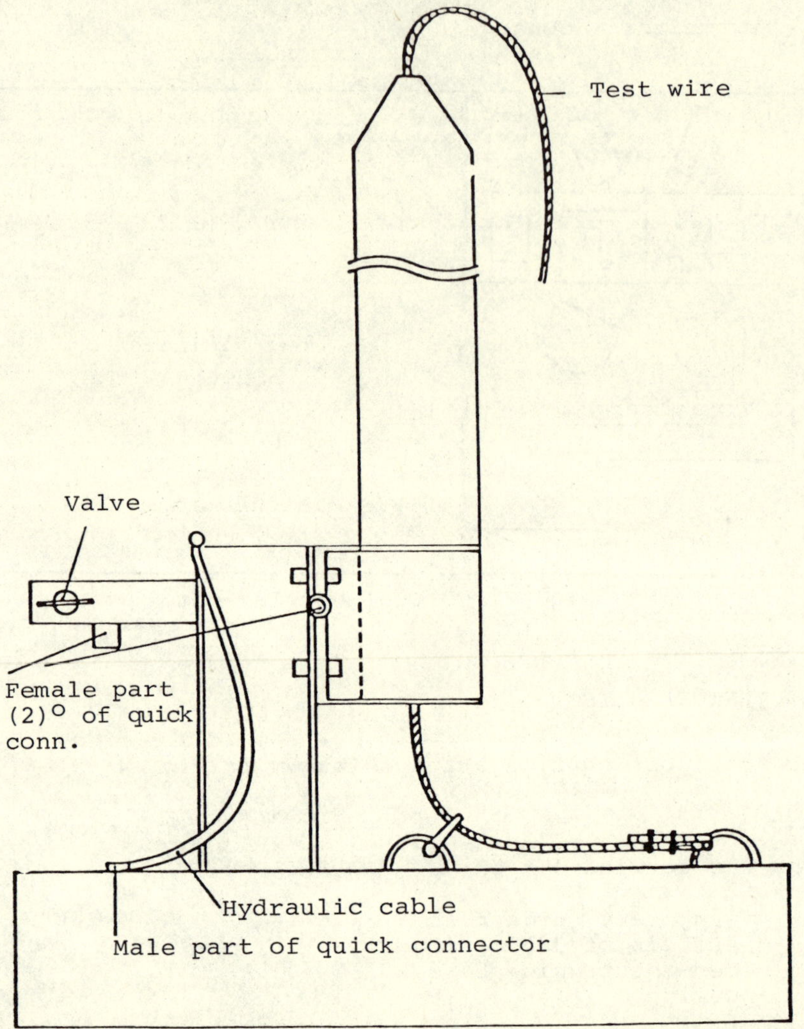

Fig. 4 Test post configuration before installation on seabed

Handling of wire and hook

"Scorpio" proved to be well suited for hooking up to wire slings and to release hooks. Typically such an operation would be done in 2 - 5 minutes, once the hook had been brought to the sling. When a hook from surface was to be hooked up to a sling on bottom such as for equipment recovery, the hook should be lowered all the way onto the bottom before "Scorpio" was launched. This was done to prevent entanglement between the hook and the "Scorpio" umbilical.

Once at bottom "Scorpio" would search with the sonar for the hook wire and the object to be lifted. Once the hook wire was found and "Scorpio" had manouvered into position the hook would be raised slowly to a suitable height above bottom where "Scorpio" would grab it. "Scorpio" could now move the hook to the sling and latch on.

When releasing the wire sling from the hook it was easiest to lift the wire out from the hook rather than turning the hook to release the wire.

Turning the valve

The valve used was a ball valve with a 90° open/close action. The turning handle was replaced with a flat plate, about 5 x 10 cm for the arm to grab.

After the arm and the TV had been put in a suitable position, "Scorpio" would move to position the claw and it could be closed, locking onto the actuator plate. The valve could then be opened/closed.

"Scorpio" managed this task fairly well, although several attempts were required before the actuator plate was properly grabbed. The plate was also rather shallow so that at one time the arm grabbed the valve body, inside the actuator plate.

Stab-in of quick - connect

Initially on the test post the male part of the connecting mechanism was installed. No guidance on either the male or female part were used. Such a stab-in proved to be the most difficult for "Scorpio". The vehicle could not obtain sufficient stability to move the connector into the correct position. It was jerking around, repeteatedly trying to hit the target. A stab-in was achieved on the first dive, but when the exercise was repeated on the subsequent dive no success was the result. This would have been caused by stronger current, or that the first connection made was pure luck. A guidance have now been fabricated which have facilitated this operation.

Wire cutting

The wire cutter was installed on the manipulator arm of "Scorpio". On bottom "Scorpio" headed for the test post and about 2 m above the top of the post it manouvered the wire into place. On the first attempts the cutter did not work and considerable manouvering took place to ensure the wire was laying properly in the tool. As this did not help, the cutting tool was brought close to the TV for inspection. On first trial the cutter blade was not activated, but when repeated, the cutter blade was activated properly. The cutter was then positioned on the guide wire and the wire was cut, leaving a clean out.

The two next cuts were made with the guide wire hanging beside the post. An attempt was made to cut the wire just above the post, but the wire end just moved away when touched by the tool. A 1 m wire stump was therefore left sticking out the post.

On the next dive the wire stump was easely removed by pulling at the wire sticking out of the post. Alternatively, the manipulator arm could grab the wire end and push it down through the top of the post.

Connecting Regan guide line latch to test post

(Ref. Fig. 5)

A Regan guide line latch was lowered down to approximately 3 meters above sea bottom. It was detected by "Scorpio" TV camera and on inspection, was found to be located nearly straight over the top of the guide post. "Scorpio" flew up to the tool and grabbed the handling wire near the top of the Regan guide line latch. "Scorpio" manouvered the Regan latch into position straight above the guide post. By lowering the Regan latch down onto the guide post while kept in position by "Scorpio", the Regan guide line latch was successfully connected to the guide post.

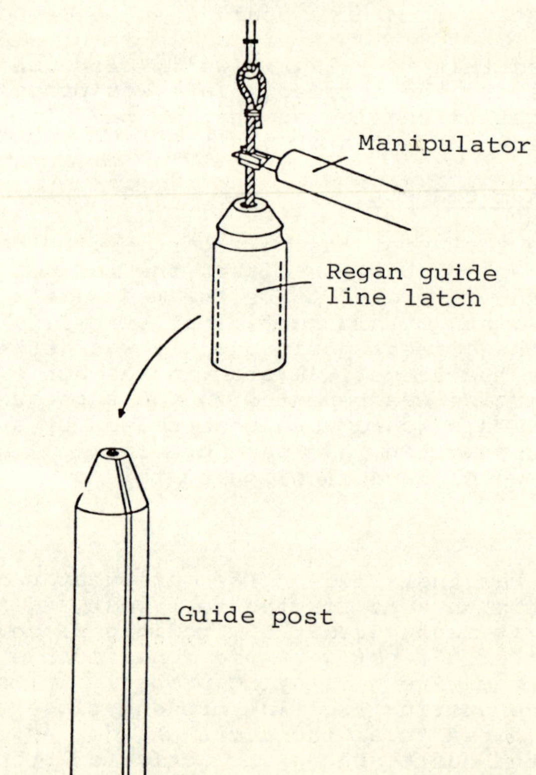

Fig. 5 Connection of Regan guide guide line latch

Connect Regan latch retrieving tool onto Regan latch

(Ref. Fig. 6 and Fig. 7)

The Regan latch retrieving tool, fitted with a dead weight to make it heavier, was lowered down to the test site. The vehicle went up and grabbed the tool just above the cone of the tool. With the assistance of "Scorpio", the retrieving tool was put into position right above the Regan latch. It was then lowered slowly down onto the Regan latch. The weight was not sufficient to release the tool.

The wire of the Regan latch on tool was then inspected to see if it had caught the release tool in any way. The manipulator arm grabbed the wire and moved it freely inside the cone of the retrieving tool. It was then decided to lift the tool slowly and adjust it down. While paying out wire fast the dead weight fell off the unlatching tool and the hook was released from the unlatching tool handling wire.

The handling hook drifted out of sight, and had to be found again. After some searching with the sonar, the hook was found together with the umbilical near the vehicle guide wire. "Scorpio" grabbed the hook and went to the test post. The first try to hook in was unsuccessful, as the hook slid out of the sling when the wire was tightened up. However, on the second try the hook up was successful and the handling wire could be tightend up. In so doing the Regan latch came off the guide post. Evidently the extra adjusting of the retrieving tool was sufficient to connect the Regan latch retriever.

Attach explosive to guide post

A special shaped charge has been prepared in a clamp to be fitted around the 8" guide post. The explosive charge was lowered to bottom and picked up by the vehicle. Within two minutes the charge was attached to the guide post and the job completed. The charge, however, did not severe the guide post when fired so that additional work is being done to improve the explosive charge.

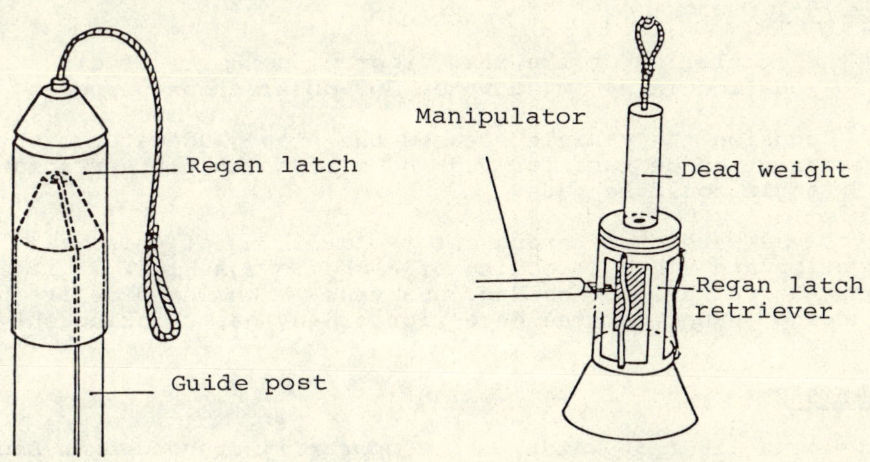

Fig. 6 Connection of Regan latch retriever

Fig. 7 Releasing Regan latch from guide post

WORK EXPERIENCE

Well 33/5-1

After the completion of the test program, a number of dives were made on this location in 340 m of water for observation purposes.

On one occasion the vehicle cleaned the slope indicator mounted on the permanent guide base for mud and silt so that observations of the BOP angle could be made.

Weekly dives were made throughout the drilling of this well for the observation and video recording of a shallow gas leak migrating out between the 20" and 30" casing. Surveys were also made of the surrounding seabed for the detection of any gas breaking out through the seabed.

Well 15/5-2

This well in 110 m of water, was temporarily abandoned in December 1978 to be reentered in November 1979 for production testing.

The rig "Treasure Seeker", was moved into position using a Pulse 8 positioning system. A side scan sonar fish mounted on drill pipe was planned to be the primary tool for exact determination of the wellhead position, but this tool was found not to be working. The "Scorpio" was immediately prepared for diving and was on bottom two hours later. The wellhead was found after a 6 minute sonar search. The "Scorpio" was moved over the wellhead and the rig's anchor chains were manipulated using the "Scorpio" position plottings as reference.

The four guide lines with guide line latches were lowered utilizing a four arm guide frame on drill pipe (see Fig. 8). Within 4 hours and 30 minutes the four guide lines were latched.

A total diving time of 9 hours were spent on the complete job. Strong bottom currents and entaglement problems hampered the operation somewhat.

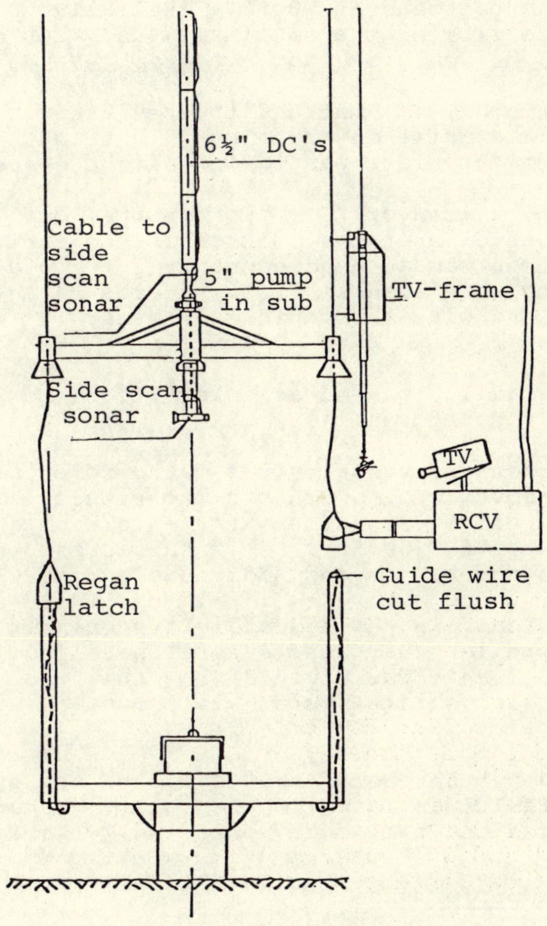

Fig. 8 15/5-2 Reentry

OPERATIONAL EVALUATION

Initial and technical problems

Not untypical, "Scorpio" also had its problems in the early stages. In the first dive the "Scorpio" was released before the operator had got it under control, and he was unable to regain the control. It was quickly established that the "Auto heading" (gyro controlled heading to ensure correct heading of vehicle) was not working properly. Then the panel alarm indicated loss in hydraulic pressure, resulting in the decision to abort the dive and pull "Scorpio" in by the umbilical. When checking the power supply it turned out that the rig power was 440 V instead of 480 V, resulting in "Scorpio" being starved for power.

When going down to bottom it was difficult to control the vehicle until it got below the pontoons at 70 ft. Initially the operator would follow the guide wire of the umbilical, and when he lost the guide wire on the TV, he would be lost.

Amongst the pontoons, columns and trusses the sonar was of little use. Some confusion did arise on the first dives in this situation. On later dives when the riser was followed, the descent was controlled.

The auto heading system of "Scorpio" dropped out intermittently resulting in loss of operator control in the early dives. Later, the vehicle was run on manual heading control. Two attempts to repair the auto heading were unsuccessful. As the failure was only occurring intermittently and with the vehicle in water, it was difficult to track the fault.

The fault was finally located as intermittent power supply to the gyro and was properly corrected.

The vehicle launch and recovery was found to be somewhat cumbersome, although it improved with operators experience. The main problem was to get the vehicle from the water interface and down to sonar operating depth (70 - 80 ft).. This has later been improved with the introduction of the basket and additional guide wire.

Salt water was found in the hydraulic system. It had to be flushed out and new propeller shaft seals were installed. First only the propeller control unit was flushed, but this was not sufficient. Flushing the whole vehicle system was a complex undertaking requiring most of a day.

The sonar faded out on one of the dives and the system was examined in detail. A fault was found in the rotary mechanism of the sonar. This was repaired and the sonar worked well thereafter.

General vehicle performance

"Scorpio" has a high degree of manouverability and responds quickly to its commands. During stab-in tasks, where the target was less than 2 square cm, the vehicle could not be moved with a sufficient degree of accuracy. The vehicle should, however, be able to meet a fixed target of 5 - 10 square cm. Current, and particulary side current affected the vehicle performance adversely and resulted in a more "jerky" vehicle. "Scorpio" handled loads, e.g. wire slings well, but its stability decreased rapidly with increasing loads.

To ease vehicle operation "Scorpio" is equipped with "Auto heading" and "Auto depth". When entering Auto heading mode (by flipping a switch) the vehicle keeps itself at that heading by aid of the on board gyro. Auto depth is included to automatically keep the vehicle at its depth when entering the auto depth mode.

The 5 motion Perry arm moved slowly and appeared a bit clumpsy. The arm was therefore first put in a suitable position, letting "Scorpio" do the final manouvering.

To work with one TV only was not easy and requires a trained eye. The lack of depth perception was one effect, another was the common lack of perfect sharpness associated with underwater TV. For this reason a second fixed TV camera has been installed after the initial testing.

Field organization

During the evaluation period on board "Treasure Seeker" a four man crew was available:

1 Supervisor
2 Operators/Technicans
1 Engineer

The supervisor had his background as diving supervisor on drilling rigs. The two operators had extensive experience from other vehicles and adapted well to "Scorpio". They were able to first assist and later assume responsibility for the in field repair tasks. Repair tasks were:

- fix the auto heading
- change TV light
- reterminate the umbilical when this was badly kinked during entanglement
- replace propeller shaft seals
- flush out the hydraulic system, removing intruded sea water
- sonar repair

It is apparent that a three man crew should be considered a minimum and is only likely to give max. 8 hrs. in the water per day, and not 7 days a week. To operate the vehicle to perform tasks two operators should be considered a minimum. One would operate the vehicle and one navigate and operate manipulators. Thus also the supervisor should be able to handle the vehicle.

Should longer dives be required, caused by an urgency to complete a longer job, then another two operators should be made available.

Launch and recovery

The original system took "Scorpio" down into the water when the rig was at operating draught only. The operator would receive instruction to manouver "Scorpio" to guide it to a line of reference for its descent. This could either be the umbilical guide line or the riser. Once the line of reference was shown on the TV monitor then the operator could start the descent. At the early stage the sonar was of little use due to the interference from pontoons, trusses and columns. When the operator lost visual contact with his line of reference, he would be lost. With current and wave action working on the vehicle, loss of control happened. Following the riser was fairly reliable, but particulary ascent along the riser could not be recommended due to the congestion of the area.

Below, say 80 ft, the sonar could be used to locate the reference line when outside the TV range.

The launching system is now improved by the use of a basket system (ref. Fig. 8). The "Scorpio" vehicle is locked onto the trolly with two hydraulic swing cylinders. When preparing to run the vehicle, the trolley is rolled out of the garage, lifted around to the lifting frame by the davit and rolled into position in the framework. The trolly is bolted to the frame work and ready to be launched by using the main rig crane. When in drilling draft the basket is guided by a permanent guide wire attached to the pontoon. When in gransit draft, a separate guide wire with a clump weight attached is lowered by use of the auxilliary air winch and the basket is attached to this wire.

In addition, the guide wire for the umbilical will always be employed. A two function hydraulic line is run from the basket to the deck for the control of the swing cylinders.

When the basket is below the splash zone, the locking cylinders are opened and the vehicle is free to swim out. When recovering the vehicle, the reverse procedure is employed. The "Scorpio" swims into the basket and is guided into its excact position by fenders so that the cylinders can lock the vehicle in place and the whole package retrieved.

Underwater navigation

The following equip ent for underwater navigation were available:

- compass with digital readout
- depth recorder
- sonar
- TV

Further on a drilling rig the guide lines, drill pipe or riser were available as reference lines for orientation.

In principle this should suffice, but it must be recognized that underwater navigation is difficult and the "Scorpio" operators did loose orientation on occasions. To navigate between the pontoons, columns ar trusses were particulary difficult as the sonar could not identify any one object and as the vehicle was affected by wave action.

During descent and ascent "Scorpio" followed a line of reference on TV and sonar. This could be the umbilical guide line, a wellhead guide line or the riser. Approaching bottom the sonar could be switched to "Bottom scanning mode" telling the operator the distance from bottom. In this mode the operator would have to rely nn his TV to keep contact with his line of reference.

On bottom "Scorpio" would head for its work site. Loss of orientation while on bottom did occur and was caused by:

- Misinterpretation of sonar
- Wrong definition of object on TV e.g. guide post 3 instead of guide post 2
- Loss of visual contact with specific object
- Pulling at object with the vehicle to achieve specified task and then the grabber would loose its grip

The loss of orientation incidents could be minimized with experienced operators and more stringent navigation procedures. Also a minibeacon will be installed so that the vehicle can be continuously tracked on the rig's positioning system.

Umbilical entanglement

Umbilical entanglement will always be a concern with tethered vehicles. Entanglement can be caused by the vehicle itself as may happen when operator looses orientation and the vehicle goes "astray". Entanglement may also be caused by foreign objects on bottom such as debris, or a lifting wire from surface drifting to the umbilical guide wire/vehicle system. Some basis precaution can be made to minimize the risk of entanglement:

- Survey bottom first for debris
- Work with minimum length of umbilical out. Always use the umbilical guide wire
- Minimize the number of lines from the surface and keep the ones required well tensioned to minimize drift off.
- Ensure good control of vehicle by keen navigation and seek to minimize free swimming time of the vehicle.

The risk of entanglement should, however, not be exaggerated, and there should be no need to place undue limitations of the vehicle operation.

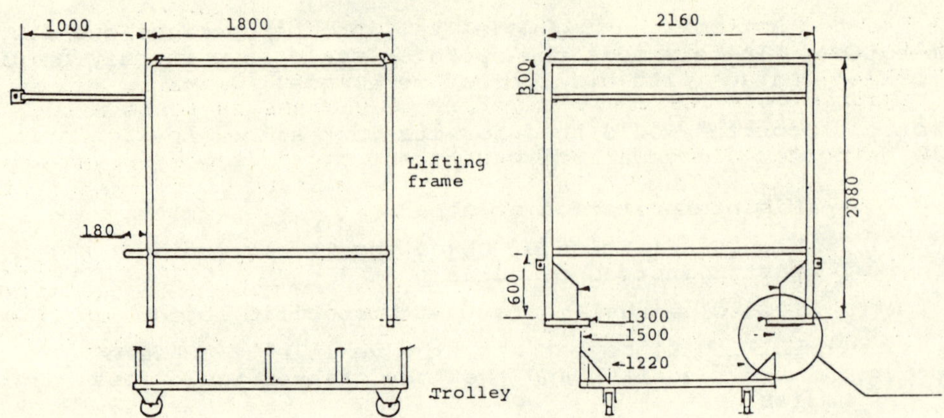

Fig. 9 Scorpio handling basket

SUMMARY AND CONCLUSIONS

During the test and evaluation period the "Scorpio" made a total of 13 dives. The dives were carried out in the period July 19 - 29, 1979 and the accumulated in water time were 40 hours. In addition a number of dives have been made since that time as outlined in the chapter on work experience.

The "Scorpio" proved to be a highly manouverable vehicle with an excellent sonar that proved mandatory in underwater navigation. The 5-motion Perry arm fitted on the "Scorpio" is rather sluggish and all the positioning was made by the "Scorpio" itself. This procedure worked well when working with wires, hooks and other larger objects. For finer tasks, such as stab-in of the quick connect, sufficient precision in the vehicle movement could not be obtained to accomplish the task with a degree of certainty. For this reason special guides were installed for the stabbing of hydraulic connections.

Technically "Scorpio" had its problems, ranging from sonar failure to salt water intrusion into the hydraulic systems.

The faults were addressed as they evolved and repaired by the crew itself. No shortage of spares did occur.

The modified launch and recovery system should enable operation in most any sea state as long as the rig cranes can be operated.

Underwater navigation proved to be difficult and requires well trained and experienced operators. The separate umbilical guide line have proven to be mandatory.

All required work tasks have been performed with satisfactory results, which proves that the system is a viable alternative to conventional diving service. However, careful planning and analysis of required work tasks are required if success is to be anticipated.

There may, of course, be requirements for a suitable diving vessel to perform salvage type operations, but this type of operation will be extremely infrequent and unpredictable.

The RCV system described here is highly cost effective compared to conventional diving. Further, the system is mobile so that several rigs may share one system. Only then the vehicle can be utilized efficiently providing the operators with a very cost effective tool.

THE USE OF DIVERS AND SUBMERSIBLES IN UNDERWATER INSPECTION—A COMPARISON

S. Freeman

Engineering Department BP Petroleum Development Limited, Dyce, Aberdeen, UK

ABSTRACT

This paper reviews the operational performance of three different underwater structural inspection systems and attempts to establish their overall cost effectiveness. Using data gathered over the study period operational costs are established and combined with a factor for inspection efficiency to allow realistic comparison of the systems. The results of the study suggest that conventional diving systems still represent the most cost effective method for complete structural inspection programmes.

KEYWORDS

Underwater Inspection; Methods of Underwater Intervention; Diving; Underwater Vehicles; Offshore fixed steel installations; Non Destructure Testing.

INTRODUCTION

The object of this study is to examine three different underwater systems which were contracted to BP Petroleum Development to carry out structural inspections during the 1978 season in the Forties Field.

Using data gathered at the time the separate performances of the systems will be analysed and compared in an attempt to establish the relative operational and overall cost effectiveness.

By highlighting those aspects which offered the most scope for improvement, proposals for more cost effective inspections will be made.

1. Principles of Structural Inspection

Although this study is concerned with how well the three underwater systems perform structural inspections, it is necessary to first establish why these inspections are carried out and what possible defects are being looked for. This is to enable the inspection requirements and thus suitable techniques to be identified.

Fixed steel installations are facilities for producing hydrocarbons from below the sea bed. The production equipment is situated on the superstructure, the safe support of which is the main function of the immersed jacket beneath. This support in Forties takes the form of four main legs carrying the compressive deck load which is transferred to the sea bed by a system of skirt piles. The rest of the jacket structure is designed to maintain the load bearing legs in the correct configuration and to offer support to other items such as conductors and risers. The legs are not only designed to resist compressive deck loads but also those bending and shear loadings resulting from environmental forces.

It is the need for assurance as to the capability of the structure to safely withstand all loading that gives rise to the requirement for periodic underwater inspections.

These inspections seek to establish two positions:

The suitability for service of the structure at the time of the inspection and the capability of the structure to remain in service, principally until the next such inspection.

The aspect of present suitability involves two considerations:-

that the structure is physically sound - free from distortions and discontinuities -and that the load bearing structural members have not suffered such a reduction in material thicknesses or an accumulation of marine growth that would cause the service stresses to exceed the design limits.

These considerations necessitate inspections for:

1. loss of structural straightness and local indentations - general visual survey.

2. complete or incipient failure of attachment welds - specific visual or a specialised crack detection technique.

3. metal loss due to corrosion or abrasion - specific visual supplemented by wall thickness measurements.

As regards future suitability it can be seen that this concerns, primarily, the performance of the structural protection systems and their remaining effective life. With the Forties structures, these protection systems include a sacrificial zinc anode cathodic protection system, an overall protective paint coating and a splash zone fendering system. These considerations necessitate inspections for:-

1. the effectiveness of the cathodic protection system - potential measurements at selected places, estimation of anode wastage by measurements on or from photographs.

2. the loss or deterioration of the structural protective coating -specific visual and colour still photographs with reference scales.

3. other protective features - specific visuals.

It must be noted that the use of specific visual usually implies some cleaning content.

A table can now be drawn up which will give the requirements for an inspection programme to provide assurance about the suitability for service of a structure.

TABLE 1

Present Suitability	Future Suitability
General visual survey	Anode measurements
Specific visuals	Specific visuals
Specialised crack detection	-
Wall thickness measurements	-
-	Potential measurements

Combining all these gives the following list of established inspection requirements:

TABLE 2

General visual of structure
Specific visual inspections
Potential measurements
Wall thickness measurements
Specialised crack detection

In a later section, the study will examine how well the different underwater systems carry out these various inspection requirements.

Underwater Inspection Organisation

Several principles determine the organisation of BP's underwater inspection operations.

The first is that divers are not inspection engineers, having almost invariably, not the qualifications, background and training necessary to take real responsibility for either overall or detailed inspection of a complex structure. For these and other reasons as much interpretive work as is possible at this time is transferred from the man in the water to a BP inspection engineer on the surface. To do this the diver is equipped with a high quality head mounted wide angle videcon camera and a suitable light which allows visual coverage on a suitable monitor in the dive control station. Aural communications are maintained at the highest possible efficiency in order to allow free and easy interrogation and direction of the diver.

The second is that the results of the inspections are recorded in such a form that data can eventually be manipulated, analysed and compared in order that long term trends and patterns can be more easily identified. To do this the results of the inspection are first added to separate uniquely numbered inspection sketches representing each element of the structure -brace member, leg member, node, bracket etc. each completed sketch being the actual inspection report. Because of the mass of information recorded BP find it generally advantageous to employ specific data collectors on site for this task - allowing the BP inspection engineer to monitor the operation in a suitable manner. Generalised data is then extracted from the inspection sketch and processed so that it can be entered into a computerised data storage and retrieval system. Thus the operational organisation of BP inspection work whatever system is used provides for monitoring by BP engineers and specific data collectors.

Operational Programme

The three underwater systems forming the subjects of this comparison study were: a conventional surface and saturation diving system, a lock out submersible facility and a Remote Controlled Submersible (RCS) fitted with various inspection packages. The inspection package range was designed to provide the same capability as a diver based system.

The study was set up to take advantage of the employment of the different systems. It had been intended, originally, to undertake all the required structural inspection work using the usual ship based diving system. It was only because two platforms were rendered inaccessible by accommodation barges that manned lockout submersibles were used. It was felt that these could be used to gain access, despite anchor wires, by being launched away from the platform.

The fourth platform not being the subject of a routine inspection was used at a site for field trials of the RCS.

The underwater inspection programme was therefore:

TABLE 3

Platform	System	Work Scope for 1978
FA	RCS	Trial programme
FB	Manned submersible	Remaining 60% of base line inspection
FC	Diving Spread	Up to 100% of base line inspection
FD	Manned Subersible/ Diving System	up to 100% base line inspection

UNDERWATER SYSTEM DESCRIPTION

Bell Diving System

The conventional bell system was based on a converted fish factory vessel, fitted not only with an adequate four point mooring system but also with a computer controlled dynamic positioning system. The built in saturation system, capable of holding eight men, operated in an orthodox manner through an aerated moonpool amidships and was fitted with a cross haul device. A purpose built surface diving station was situated amidships on the starboard side. This system operated on a 24 hour basis with 21 diving operators, 5 inspection staff and 2 BP inspection engineers. Total ships complement was limited because of accommodation and safety craft places to a maximum of 60.
Diving operations were controlled from a bell van which was also equipped with the necessary facilities for monitoring and data recording

Manned Submersible System

Two separate submersible facilities were employed during the season. One, a single medium sized lockout submersible had been employed on a pipeline inspection and was used for structural inspection over only a limited period. This was not only to provide early experience of operational and technical problems, as this was a relatively new concept, but also to allow any lessons learnt to be incorporated into the working up of the main facility. This latter facility consisted of two submersibles based on a large support vessel designed to be operated back to back on a 24 hour basis. The two submersibles, one large and one small covered for each other and were never both deployed in the water at the same time. The built in saturation system was fitted with two transfer locks and trunking to the separate submersible positions. Both submersibles were launched over the stern using the same 'A' frame. Diving and submersible operating staff including BP inspection engineers totalled 38 with the balance of the total complement of 70 being made up of ships crew.

RCS (Remote Controlled Submersible)

The RCS based on the FA platform was one developed specifically for platform based structural inspection work. The development work had been directed towards achieving three objectives.

1. Ability to operate from within platform designated hazardous areas. This involved pressurised control packages with sophisticated gas detection and shutdown systems.

2. A launching system based on a composite umbilical embodying both electrical and strength members which would allow uninterrupted and recovery operations

3. A range of inspection packages offering all the capabilities of a diver based system. It was intended that this would be a step on the diver replacement road.

The vehicle was a large space frame machine built in Canada by McElhanney, weighing about one ton with the principle addition of an on-board electro/hydraulic HP water jetting system.
The field trial was intended to allow the vehicle and its inspection systems to be worked up and then by following a routine inspection programme to be evaluated in comparison with the diver based system.
The RCS system operated on a 12 hour basis and the seven contractors crew and one BP Inspection Engineer were accommodated on the platform.

STUDY ORGANISATION

Collecting the data necessary to satisfy the study objectives took the form of identifying the relevant operational activities and logging them in absolute time sequence. The BP Inspection Engineers were required to monitor activities in two ways. The first was to compile a daily log of all main events, together with a description and comments. This provided a method of collecting useful general data but during underwater activities it was felt that a more detailed and precise recording method was necessary. The second duty of the engineer was thus to monitor underwater activities by completing an 'event and function' log for each dive. This event and function log was devised in the following manner -Each separate underwater inspection system was divided into separate identifiable events such as bell or submersible off transfer hatch, bell at inspection depth etc., selected on the basis of productive/non-productive time. The events were also selected so that the elapsed time between related events comprised a significant activity/function. These events and related functions were then entered onto an event and function key, Appendix 1 and working logs were also drawn up Appendix 2.
The BP inspection engineer during the dive was required to identify the occurrence of each event and record its absolute time on the working log. The engineer then classified the elapsed time between sequential events using the function key as a basis, e.g., bell transit time, diver dressing time etc. the resultant data being used to isolate and totalise the different productive/non-productive times. Although some interpretation was often required on the part of the engineer, especially where the sequence of events was unusual and despite it not being generally found possible to detail times spent on each separate inspection function the system proved to be successful in producing the required data.

Analysis Periods and Assumptions

The original intention was of course to compare the three systems over exactly the same period of time whilst they were all undertaking identical duties. In practice, however, this did not prove possible mainly because of the initial unreliability of the manned lockout submersibles and the RCS. Appendix 3 shows the operational periods of the three systems during the season and it can be seen that the diving system was demobilised before the start of the second periods of the subs. and RCS. It was only during these second periods that the operational reliability of these two systems improved sufficiently to allow representative data to be collected.

Various assumptions were made in order to facilitate the comparison of the different systems.

It was assumed that all dive time during the contract period analysed was spent exclusively on routine inspection work. This was to allow the functional (inspection) analysis results to be combined with the operational analysis results in order to produce a figure for effectiveness/inspection hour. Certainly the RCS was only engaged on inspection during the operational period but the diving system and to a lesser extent, the submersible system performed other duties such as debris removal of the structures. Allowance has also been made for surface diving from the diving system and it must be borne in mind that the three systems are essentially compared only from -80' depth to the mud line. These assumptions permit the selection of periods for functional analysis which represent the inspection performance to be expected during future contracts to enable the results to be useful for planning purposes. The analysis periods were also selected to avoid including the 'learning curve' periods for the less well established lockout submersible and RCS systems.

ANALYSIS OF RESULTS

Operational Utilisation of Contract Time

Conventional Diving System

The period analysed was 25th May to 21st June during which time the system was exclusively engaged on structural inspection work. The following is a breakdown of operational time in terms of hours, as based on data extracted from daily logs.

TABLE 4

		Hours Involved	Percentage of total
1.	Total Contract Time	1392	100.0
2.	Diving Time	552	39.6
3.	Diver Changeout	109.7	7.9
4.	Equipment Maintenance	16.4	1.2
5.	Mooring Operations	66.6	4.8
6.	Waiting on Platform	137.2	9.8
7.	Ships Passage Time	70.5	5.1
8.	Weather Downtime	439.6	31.6

So on average a contract day consisted of:

9 hours 31 mins diving (2)
4 hours 34 operational delays and maintenance (3 + 4 + 5 + 7)
9 hours 55 mins unavoidable downtime. (6 + 8)

Manned Submersible System

The period analysed was from 28th July to 19th August inclusive, during which time both submersibles carried were used although not in the manner originally envisaged.

Much of the time it was only found possible to use the larger of the two submersibles. The performance of this submersible, however, was dramatically improved during the contract period particularly with regard to diver heating capacity and dive duration.

The following table is a breakdown of the operational time in terms of a 24 hour day.

Total number of days were 23.

TABLE 5

		Hours	Percentage
1.	Total Contract time	552	100%
2.	Dive time	154.7	28.0
3.	Launch & Recovery	67.0	12.2
4.	Dive change out	13.1	2.4
5.	Battery Charging	243.0	44.0
6.	Maintenance time	5.5	1.0
7.	Weather downtime	52.7	9.5
8.	Platform operations	16.0	2.9

It must be noted that maintenance has only been isolated where it extended beyond the battery charging period thus delaying operations.

The period considered does not cover the complete contract but is one during which the performance was reasonable, and representative of that normally to be expected.

On average therefore a contract day consisted of:-

6 hours 43 mins submersible in the water (2)
13 hours 18 mins operational delays and maintenance (3 + 4 + 5+ 6)
2 hours 59 minutes unavoidable downtime (7 + 8)

RCS System

The period analysed was from 22nd July to the 26th August inclusive.

The following table is a breakdown of operational time in terms of hours based on a 12 hour contract day. Total number of dives involved were 42.

TABLE 6

		Hours	Percentage
1.	Total contract time	432.00	100%
2.	Total dive time	111.64	25.9
3.	Routine work permit and pre dive checks	35.41	8.2
4.	Platform operations	14.24	3.3
5.	Weather downtime	16.00	3.7
6.	Permit withdrawals	18.3	4.2
7.	Breakdown and maintenance	235.66	54.7

So on average over the period of 12 hours a contract day consisted of:

3 hours 6 mins RCS in the water (2)
7 hours 32 mins operation delays, breakdown and maintenance (3 + 7)
1 hour 22 mins unavoidable downtime. (4 + 5 + 6)

Dive Time

This is defined as the total time the bell/submersible/RCS spent underwater during the contract period analysed.

Weather Downtime

It must be noted, at this point, that for both the lockout submersible and the remote controlled submersible much weather downtime was included within 'battery charging' for the lockout submersible and 'breakdown and maintenance for the RCS.

Downtime Generally

Downtime is only included under a heading when operations were prevented specifically by it.

Summary of Operational Utilisation

The separate results are compared on a percentage basis of total contract time:-

TABLE 7

	Bell	Sub	R.C.S
Underwater time	39.6	28.0	25.9
System Maintenance and operational delays	15.4	59.6	62.9
Unavoidable downtime	45.0	12.4	11.2

Productive Utilisation of Contract Time

Productive utilisation is defined as the proportion of underwater time actually spent carrying out inspection related tasks.

Data has been extracted from the time analysis over that period for each system which produced a performance considered to be reasonably representative of that to be expected during future inspection contracts.

Bell Diving System

A period was chosen covering 45 bell runs on typical structural inspection work (26th May to 19th July).

Average Dive time (9 hours 12 mins/bell run)	100%
Inspection time 56.35%	
Non productive dive time	21.72%
Bell handling time	16.31%

Manned Submersible System

A sequence of 18 submersible divers using one large submersible on structural inspection was chosen (28th July to 17th August).

Average submersible time (11 hours 54 mins/dive)	100%
Inspection time 53%	
Non productive time	14.3%
Sub handling time 32.7%	

R.C.S. System

With this system it was considered best to analyse 14 dives which occurred in a sequence at the completion of the trial (10th August to 26th August).

Average RCS time (3 hours 29 mins/dive)	100%
Inspection time	71.5%
Non Productive time	6.8%
RCS handling time	21.7%

In all the above, handling time covers launch and recovery operations and positional movements. Non productive dive time covers all in water waiting time, rigging etc.

Overall Utilisation

Combining the results of the analysis into both operational and productive (inspection) utilisation gives the overall utilisation of the system.

This can be expressed as the percentage of total contract time devoted to useful underwater inspection activities.

TABLE 8

	Bell	Sub	RCS
Inspection time at % total contract time	22.32%	14.84%	18.52%

Operational Cost Effectiveness

This is defined as the cost/inspection hour, which is the hourly cost of the system spent performing useful inspection at the structural metal surface.

Costs

1. Hire rate - Cost of personnel and equipment including vessel.

2. Consumables - Those items consumed on demand. Food, bunkers, gas, video tape and photographic film etc.

3. Platform charges - Direct costs only of basing contractors on a production platform.

TABLE 9

	Bell	Sub	RCS
Av. daily rate (1 + 2)	£20,000	£19,500	£4,000
Direct costs (3)	-	-	£2,000
Total	£20,000	£19,500	£6,000

$$\text{Operational Cost Effectiveness} = \frac{\text{Average Daily Hire Rate}}{\text{Average Daily Inspection Hours}}$$

$$= \frac{£20{,}000}{24 \times 0.2232} = £3750 \text{ approx. for bell system.}$$

$$= \frac{£19500}{24 \times 0.1484} = £5500 \text{ approx. for submersibles}$$

$$= \frac{£6000}{12 \times 0.1852} = £2700 \text{ approx. for RCS}$$

FUNCTIONAL (INSPECTION) EFFICIENCY

I would like to review how capably the underwater systems undertake inspection tasks. The previous section highlighted the proportion of operational time actually spent on inspection activities and the operational cost effectiveness of the systems. It will be obvious that the ultimate cost effectiveness of the system must involve combining operational efficiency and inspection efficiency, i.e., there is no advantage being able to inspect for longer unit dive time if the system requires an even longer time on inspection to produce results of an equivalent standard. Planning for future inspection operations also requires that some assessment of inspection efficiency be made.

Both operational and cost effectiveness were calculated objectively from data based on the time analysis, but for a number of reasons it is necessary to be more subjective when dealing with inspection efficiency.

The comparison in this section is between the systems using divers and the system using an RCS to carry out inspections. Where a similar inspection function is performed the comparison can be on a time analysis basis as in the previous section. In practice, however, the RCS was required to follow a programme calling for the use of inspection methods idfferent enought to make the efficiency assessment rather subjective. Another difficulty arose because of the very nature of the RCS programme which was primarily a field trial undertaken as part of an R & D programme.

Now inspection efficiency can be reviewed by considering the following aspects:

Access

Diving System

With vessel mounted bell systems the first problem is to get alongside the platform. The present position is that BP will not allow diver to penetrate to the interior parts of the structure unless the support vessel is secured on suitable moorings. If however, a semi-submersible is already anchored alongside, the diving vessel is denied access.

Given that the vessel is in a suitable position alongside the platform it has been found that the available umbilical length permits diving access to all parts of a Forties structure.

Manned Submersible - Although access to the structure is not denied this system by anchored semi-submersibles, the location of the manned sub. during lock out diving could be a problem. The submersibles were positioned for lockout diving by being set down across suitable conjunctions of structural members. Once in position offering safe level support, the buoyancy of the submersibles was made sufficiently negative to prevent any inadvertent submersible movement. Additional means of securing were also employed as thought necessary. Because of the limited number of set down positions over a horizontal level and the available diving umbilical length some parts could only be reached from the largest sub. The smaller submersibles used for this work with their shorter diving umbilicals only enjoyed limited access.

Even with a different securing system, these subs would have to reposition more frequently than those with longer umbilicals thus reducing this system's effectiveness.

The submersibles were affected by tidal direction in that the selected set down position could only be enjoyed whilst the sub could float clear of the structure following a power failure. Although it was possible to minimise the effect of this total constraint on inspection site selection during the analysis period there is no doubt that it remains a problem. If the programme position is that the remaining inspection is confined to one general area, say the north west gradiant, then the operational utilisation might be reduced by 50%. (See Appendix 4).

The submersibles were also affected by current speed, particularly during manoeuvring.

RCS - Access to the structure was severely limited in three ways.

From any one launch position on the platform deck the most that could be covered was two sides of the structure. Therefore a general structural inspection implied at least two launch positions.

For operational reasons it was not possible to inspect the interior parts of the structure. The RCS was also not manoeuvrable enough to cover outside ones. It also suffered the same problems with the tide direction as the manned lockout submersibles (See appendix 4)..

Inspection Speed

As divers operating from bell and manned submersible perform at a similar speed this comparison can be reduced to one between divers and the RCS.

Divers

Over a contract period of 1392 hours bell divers completed inspection on 278 structural elements. As inspection hours for bell diving comprise 0.2232 x contract hours the inspection speed =

$^{278}/1392 \times 0.2232$ = 0.894 elements/ Inspection hours

RCS

Because of the trial nature of the RCS operation and because of limitations outlined earlier, e.g. access, it is not easy to compare or even establish performance. However, by breaking down an acceptable element inspection into its component inspection functions and using the total number of separate functions attempted by the RCS during a known period, an indication of performance can be.

Now a typical structural element inspection on a Forties jacket is made up of 5 inspection functions.

1. General visual

   Involving the complete length and circumference of the element (member, node etc.) the identification of all features, the condition of the coating and whether any corrosion can be seen.

2. Saddle Weld Inspection

   The location, cleaning and visual inspection for defects.

3. Butt weld inspection

   The location, cleaning and inspection.

4. Cathodic Potential Measurements.

5. Wall thickness measurements.

These functions are not listed in any order nor are they weighted in importance but it is considered that for the purposes of the calculation they can be regarded as equal and separate.

Over a contract period of 19 days the RCS attempted 209 of the above listed inspection functions on separate structural elements. Therefore, the assumed number of element inspections becomes $^{209}/5 = 42$.

As inspection hours for the RCS operation comprises 0.1852 x contract period the inspection speed = $\frac{42}{19 \times 12 \times 0.1852}$

= approx. 1 element/inspection hour.

It must be strongly emphasised, however, that this figure for performance disregards access and quality because the separate RCS functions used in the calculation were mainly general visual, and because of the many assumptions involved, it is felt that the speed of inspection of the two systems is sufficiently close for them to be regarded as equal.

Capability

Briefly the functional capability of the three systems was as follows:-

TABLE 10

Function	Bell	Sub	RCS
Visual	YES	YES	NO
Video	YES	YES	YES
Photography	YES	YES	YES
Stereophotography	NO	NO	YES
Potential Measurements	YES	YES	YES
Wall Thickness	YES	YES	YES
Cleaning Ability	YES	YES	YES
Sampling	YES	YES	NO
Crack detection (NDT)	YES	YES	NO

Given the basic organisation and system capabilities how are the inspection requirements summarised earlier, fulfilled?

Dividing into diver based and RCS based systems we have:-

TABLE 11

Diver Based

Inspection Requirement	Inspection Procedure
1. Structural Survey -	Divers eyes supplemented by CCTV, and high resolution colour stills where doubt exists and records necessary.
2. Member attachment -	Selected spot marine growth removal and then as for 1.

3. Crack Detection -	Cleaning weld over the whole length down to bare metal then:- Divers eyes, close up high resolution still photo mosaic and applicable NDT technique.
4. Metal Surface -	Selected wire brushing to remove marine growth, then divers eyes supplemented by high resolution colour still photography.
5. CP Survey -	Either contact or dip technique with display either at surface station or visible using divers TV.
6. Wall thickness -	Suitable instrument with display available to monitoring enegineer as in 5. Operated by diver following a selected procedure depending on the nature of thinning encountered.

TABLE 12

RCS Based

Inspection Requirement	Inspection Procedure
1. Structural Survey -	High quality low light television system for general work backed up by still photography.
2. Member attachement -	Where access is possible and marine growth is minimal as video TV, otherwise water jetting at selected places to remove covering as access permits
3. Crack detection -	Cleaning selected weld lengths to bare metal by water jetting, as access permits and high resolution still photograpy.
4. Metal surface survey -	At welds where 2 and 3 undertaken otherwise only where evidence requires confirmation.
5. Cp Survey -	Dip technique equipment with spring stand off half cell in manipulator. Potential measurements displayed on control consul. Avoiding necessity for cleaning.
6. Wall thickness -	Similar equipment system to 5 but using a developed stand off ultrasonic immersion probe calibrated on 2nd and 3rd back wall echo to avoid necessity to prepare metal surface.

Quality

In an attempt to reduce the subjective nature of the comparison between diver and RCS based systems regarding quality of completed inspection the following method was observed:

The inspection procedures devised to satisfy the various inspection requirements, as listed in Table 6 for the two types of systems were reviewed in the light of four relevant attributes. These attributes were identified as those most necessary in an inspection system and are defined:-

Perception	-	Awareness of the unlooked for and the unexpected.
Precision	-	Accuracy Definition - Degree of distinctness of outline Resolution - Resolving power
Flexibility	-	Facility of operation Manoeuvrability Coverage
Spontanaity	-	Immediacy of acceptable interpretation.

Table 13 was then drawn up which gives a rating of 0 to 5 for each system depending on how well it is considered the procedure satisfies the demands for the attribute.

TABLE 13

Inspection Requirements

Inspection Procedure	Perception	Precision	Flexibility	Spontanaity	Diver	RCS
(See Table 11 & 12) 1	5	4	4	4	17	
	2	2	2	2		8
2	4	3	4	4	15	
	1	2	1	1		
3	4	4	4	5	17	
	1	1	1	1		4
4	4	3	4	4	15	
	1	2	1	1		5
5	-	4	4	4	12	
	-	3	2	4		9
6	-	5	4	4	13	
	-	3	2	4		9
TOTALS Diver	17	23	24	25	89	40
RCS	5	13	9	13		

This allows an inspection factor to be produced using the totals for each system.

$$\frac{\text{RCS}}{\text{Diver}} = \frac{40}{89}$$

$$\therefore \text{RCS Quality} = \frac{40}{89} \text{ Diver Quality}$$

RCS Quality = 0.45 Diver Quality

So Inspection Factor = 0.45

Ultimate Cost Effectiveness

Accepting the possible inaccuracies of the previous subjective assess ment of quality we can still get some indication of the overall inspection effectiveness of the three systems.

From a previous section the overall utilisation of the three systems was calculated, and expressed as the percentage of total contract time actually spent inspecting. The figures were:

Diving bell system - 22.32%
Lockout submersible - 14.84%
RCS System - 18.52%

From the previous section on operational cost effectiveness the overall daily hire rate was used in conjunction with the above results to produce figures for the cost of each operational inspection hour.

By introducing the inspection performance factor obtained in the last section the ultimate cost effectiveness of an inspection system can be produced in terms of the cost of an hour's actual inspection to a comparitive adequate standard.

TABLE 14

System	Operational Cost Effectiveness	Inspection Performance Factor	Ultimate Cost Effectiveness
Bell Diving	£3750	1.0	£3750
Lockout Sub	£5500	1.0	£5500
R.C.S.	£2700	0.45	£6000

REVIEW OF STUDY RESULTS

In a number of ways, as is often the case with offshore projects, the comparison study in practice did not achieve all its expected targets. Because of operational difficulties with both the manned submersible and RCS systems it was not found possible to select exactly the same operational period for study purposes with both the submersibles and RCS systems. A reasonably representative performance was only produced towards the end of their contracts, after bell diving had been completed. These same operational difficulties had another effect on the submersible system which was that the operation virtually concentrated on one submersible.

It is still however, considered that the study results are valid and represent the comparitive performances of the systems, at that time. Certainly at the time of the study the RCS system could not realistically be considered either an economic or a technical alternative to a conventional diving system for a complete structural inspection programme in the Forties Field.

FUTURE DEVELOPMENTS AND PROSPECTS

How can a system improve its comparitive position?

From this study it is clear that, in a strictly competitive situation, there are three factors contributing to system ultimate cost effectiveness:-

1. Absolute daily cost

2. Operational Efficiency

3. Inspection Efficiency

Daily Cost

The direct costs of any system are of course strictly subject to market forces but some influences can be identified. It is difficult to envisage any significant reduction in saturation diving costs except where gas reclaim techniques reduce cost of helium. In fact, as the support vessel and associated equipment become more sophisticated it is reasonable to expect the costs to rise.

There is also no reason to expect a dramatic lowering of manned submersible costs either, especially as the number of service companies has been reduced since the time of the study. Considering RCS's however, in order to predict the trend of costs the machine type has first to be defined. Similar machines to that involved in the study are likely to be subject to higher capital investment and so will increase in cost. There is however a trend towards smaller, more limited machines which are offered at a much lower daily hire rate.

In summary, the relative daily contract costs will remain much the same if independent systems are employed.

Operational Efficiency

Advances in utilisation of bell diving systems are being achieved because of more stable support vessels and forms of heave compensation. There is scope for up to a 30% improvement (using study figures) by reducing weather downtime though any such improvement will almost certainly be matched by higher hire rates.

Submersibles suffered from poor weather capabilities and battery charging time. During the study it was intended to operate two submersibles back to back but this was never successfully achieved. It is a matter of judgement whether those companies still left in this field can supply the support facilities necessary to operate in this fashion.

With RCS the main problem was and remains one of reliability. Although the study of the RCS system only involved a 12-hour working day, on many occasions, maintenance needs would have prevented any extension of the day.
Environmental problems also reduce RCS utilisation. Sea state restrictions for both operational and insurance reasons, especially for ship based systems, affect RCS more than some modern bell systems. One of the advantages of the type of RCS deployment arrangement involved in the study, should have been one of better weather capability but this has not been realised. Tidal currents too play a restricting role, as RCS operators are reluctant to risk their machines during periods when the current tends to carry it into the structure. The high loss rate of these machines because of the understandable commercial effect on insurance costs and limitations is of course responsible for this attitude. This is perhaps a problem that might lessen with increased reliability.

Inspection Efficiency

Apart from the use of new inspection equipment it is difficult to see how increases in speed and quality of diver inspection can be obtained. Certainly the biggest scope for improvement lies with RCS. It seems however, that several problems demand solution:-

Access

Routine access is not only required to all parts of outside structural elements but also to those internal parts remote from the outside.

The principle development concerning access has been submersible garages. By restricting RCS umbilicals to short horizontal runs it is hoped that snagging will be eliminated as an operational problem.

Access to interior of members is expected by increasing RCS manoeuvrability.

The problem of RCS position fixing inside steel jackets, however, remains.

Cleaning

If an RCS is required to undertake any inspection task beyond general survey work some cleaning capability is usually required. Any cleaning tasks dramatically reduce the inspection speed of an RCS system.

Hopes have existed that inspection techniques could be produced which might eliminate cleaning requirements but these hopes have not so far been realised.

CONCLUSIONS

The study has produced results in three main areas.

It has confirmed that at the time of the study the diving system was ultimately more cost effective than the others.

It has produced various figures such as inspection speed which will be useful for planning purposes.

It has identified parts of the systems performance where there exists scope for improvement. This particularly applies to the RCS system.

Given a commitment to removing or reducing the necessity for diving what are the prospects for RCS regarding inspection in the Forties field type of situation?

One aspect of the RCS performance in the study was how much faster it was at performing general survey work. It was only the requirement for cleaning for detailed inspection which slowed performance so much. By taking advantage of the good parts of the RCS performance some reduction in diving effort could be obtained. It is envisaged that after a detailed base line inspection has been made further inspection programmes could perhaps be split between the general, undertaken by small RCS and the detailed undertaken by diver based system. In this way costs could be reduced but quality maintained where essential.

ACKNOWLEDGEMENTS

I would like to acknowledge with thanks the advice, assistance and encouragement received from my colleagues P.D. Reeves and G.F. Cammack

APPENDIX I

EVENT & FUNCTION KEY - BELL DIVING

E. No.	EVENT	F. No.	FUNCTION
1.	Start of pre-dive checks		
2.	Bell off transit hatch	A	Pre-dive check time
3.	At inspection depth	B	Bell transit time
4.	Down line established	C	Time to establish/re-establish down line
5.	Diver at the inspection site	D	Diver transit time
6.	Diver stops inspection work	E	Inspection time
7.	Diver returns to the down line	G	Return time to the down line
8.	Request for NDT (other inspection)		
9.	NDT completed	F	NDT (other) inspection time
10.	Diver returns to the bell	H	Diver return transit time
11.	Change out diver leaves bell	J	Diver change out time
12.	Bell back on transfer lock	K	Bell return transit time
13.	New diving team in bell	L	Surface change out time
14.	Bell in new position	M	Bell repositioning time
15.	Diver leaves bell	N	Diver dressing time
16.	Diver leaves bell stage	P	Equipment fitting and checking time
17.	Debris removal completed	Q	Debris removal time
18.	Ship in new position	R	Ship repositioning time

APPENDIX II

WORKING LOG OF EVENTS AND FUNCTIONS

DIVE NO. SAT 90
DATE 12.7.78
LOCATION F.D. - 330'

DIVER BARBER/MAAR
BP Eng. HART
DATA RECORDER BRAUR

Event No.	Absolute Time	Function	Elapsed Time	Description
13	0950	L	0136	BELL READY PLATFORM OPS STOP DIVE.
2	1108	A	0014	PRE-DIVE CHECKS
3	1112	B	0014	BELL LAUNCHED.
15	1126	N	0019	DIVERS DRESSING
18	1128	R	0021	REPOSITION SHIP
5	1150	D	0003	TRANSIT TIME
6	1249	E	00.59	INSPECTION TIME - 330 HORIZ.
10	1251	H	0.02	DIVER RETURNS TO BELL
18	1301	R	0.10	REPOSITION SHIP
5	1304	D	0.03	DIVER RETURNS TO INSPECTION
6	1430	E	1.26	INSPECTION TIME - 330 HORIZ.
10	1432	H	0.02	DIVER RETURNS TO BELL FOR CP METER
16	1437	P	0.05	EQUIPMENT CHECKS ON BELL STAGE
4	1445	C	0.08	REPOSITION DOWN LINE & RETURN TO BELL
18	1449	R	0.04	REPOSITION SHIP CLOSER TO STRUCTURE
4	1505	C	0.06	REPOSITION DOWN LINE (FIND STRUCTURE)
18	1537	R	0.32	REPOSITION SHIP ADJACENT N.W. LEG
5	1539	D	0.02	TRANSIT TIME TO N.W. LEG - 330

APPENDIX III

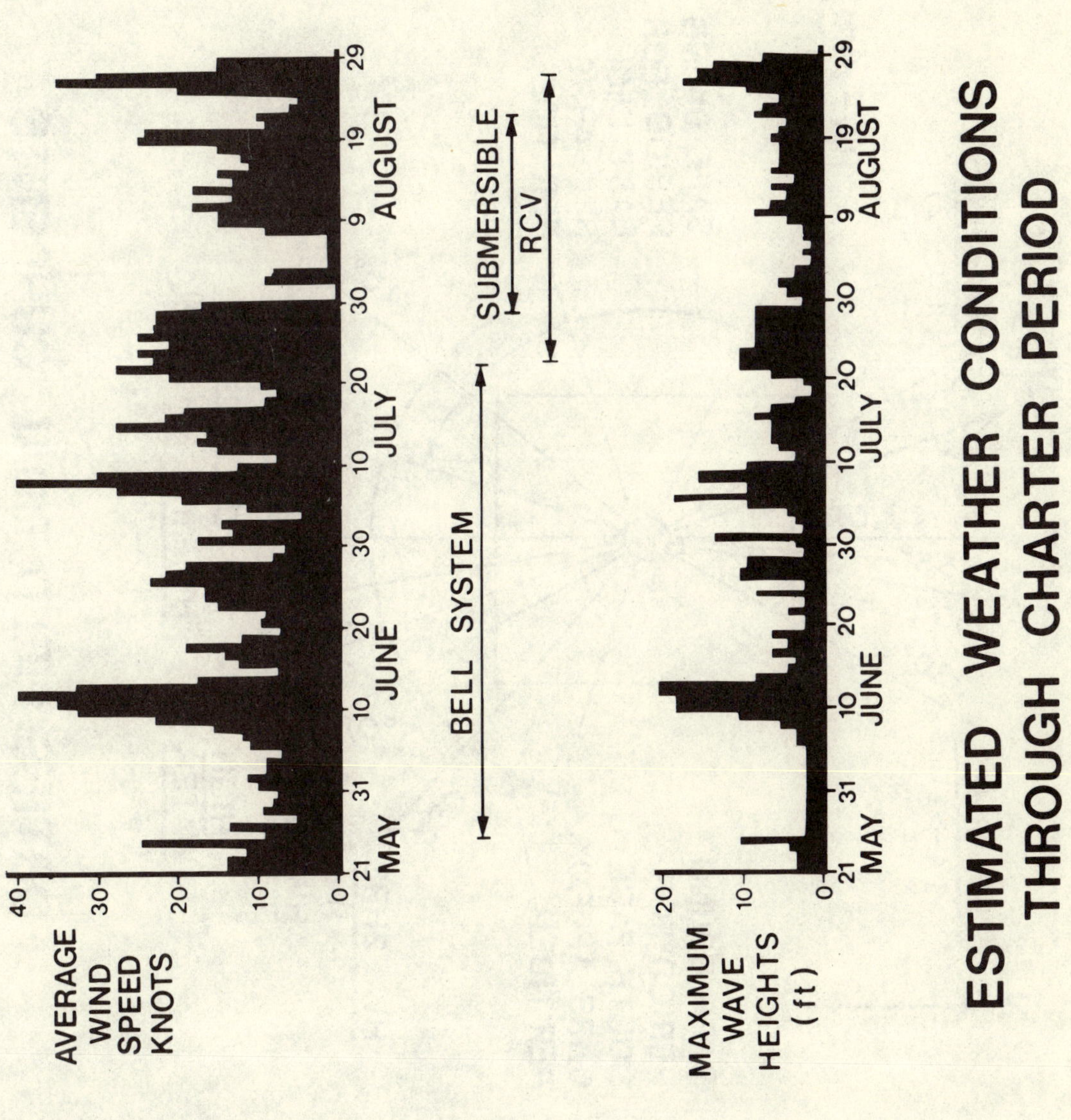

APPENDIX IV

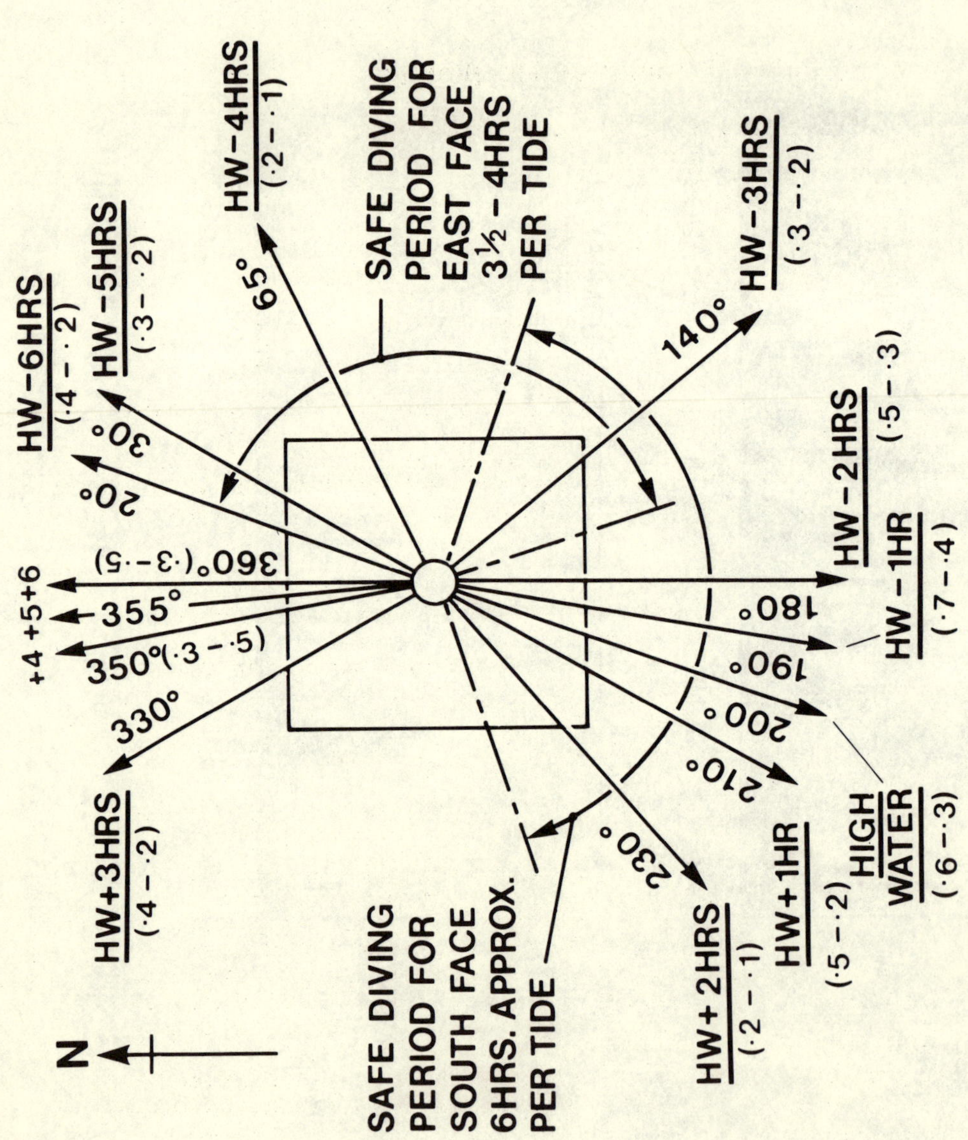

FORTIES FIELD TIDAL CURRENTS

THE DEVELOPMENT OF AN UNDERWATER TEST AREA IN BERGEN

G. Evensen and H. Ringnes

Det norske Veritas, Høvik, Norway

INTRODUCTION

This paper will give the Status of development of what we have named "An Underwater Trials Area", UWTA, which is facilitating research activities using the ocean as a laboratory. The paper will outline why we need such a laboratory facility. Status as regard present and immediate future activities will also be given.

Up to recently, the work with UWTA has solely been carried out by Det norske VERITAS, but as the long term strategy of UWTA is a co-ordinated R&D-effort in underwater technology in Norway, VERITAS is now co-operating with the Norwegian Council for Industrial Research in order to establish the optimal solutions.

Background

The conceptional idea of an underwater trials area for full scale testing and trials of underwater systems utilizing the deep fjords in protected waters of Western Norway is fairly old. The idea was first discussed in 1972 at VERITAS during the feasibility study of a Norwegian Ocean Simulator. At the time of the studies it turned out that it was not very realistic to endavour into the large investments of the facilities necessary for full scale testing in the ocean environment. However, an amputated part of the total concept was carried out, and the result of this was eventually the foundation of the Norwegian Underwater Institute in 1976. The Norwegian Underwater Institute is owned jointly by the Norwegian Government and VERITAS, each owning 50 %.

Since 1976 the situation has changed significantly. The energy situation in the world has made it realistic to prepare for the exploration of the deep oil and gas fields off the west-coast of Norway along with the other deep fields areas of the world. However, the optimum technology for this type of exploration is not fully demonstrated. A full scale trials area will be an important tool for demonstrating reliability and technology, economy feasibility as well as development of new technology.
It seems that the time for deep water technology is emerging. Developing a full scale trials area is our way of being prepared.

Justification of UWTA

Deep water field development is one key word justifying the development of an underwater trials facility.
Figure 1 shows the Norwegian Continental Shelf. The white area shows the depth between 0 and 200 metres. The blue area shows the depth between 200 and 500 metres, and the deep blue area shows the depth between 500 and 1000 metres. The figure demonstrates that quite a bit of the shelf is in what we could term deep water. The development so far has been mainly in the shallow part area south of the 62. parallel.

Figure 2 shows the depth distributed south of the 62. parallel, between the 62. and the 72. parallel and north of the 72. parallel. In the areas between 62. and 72. parallel which will be developed next, 72% lies in the range between 200 and 500 metres. North of the 72. parallel we are down to 60% in this depth range. However, at these latitudes we talk about areas where there will be iceproblems. Going sub sea would be a natural thing to do in these areas provided the technology is available.

The conclusion of these two slides would be that deep sea technology is fundamental for full utilization of the potential oil fields off the Norwegian coast.

Figure 3 shows the depth distribution from the fourth round of concession. The fourth round of concession is recently completed. The figures demonstrate that we are not talking about tomorrow's needs when we are talking about sub sea, but rather today's situation.

Figure 4 illustrates 3 different uses of subsea production schemes:

- Marginal fields
- Additional production/injection
- Deep water production

An additional reason for using subsea production systems is the early cash flow advantage.

Figure 5 shows the oil fields in the Norwegian sector off the west-coast of Bergen. Far south in the Ekofisk area we have the shallower fields. Sub sea may represent a more efficient usage of these areas and possibly development of marginal fields. The northern part of the map shows where Statfjord is located. Around Statfjord water depth is 150 metres. As we move towards land, water depth increases sharply. This is the main areas of the fourth round of concession.

Figure 6 shows schematically tomorrow's oil field. Tomorrow's oil field may contain components like sub sea satellites, multiwell system, manifold, risers to floating structures, buoy loading as well as elaborate service systems for installation, inspection and maintenance. An underwater trials area has to focus upon the problems relating to this.

Inspection, needs for testing

My attention this far has been focused upon the deep water developments. That does not mean that we have solved all the problems for the existing fields.

Figure 7 shows the estimated value of the inspection effort under water in the North Sea. The figure is taken from CIRIA UA13. As shown here the value of the necessary inspection in the North Sea will increase sharply in the years to come. The value in 1976 being between 450 mill. NOK and 600 mill. NOK.

Figure 8 shows the estimated need for qualified diving inspectors in the North Sea. Utilizing present technology, the need in 1987 will be somewhere between 250 and 300 men. Is it reasonable to expect that there will be sufficiently many well qualified diving inspectors available for work in the North Sea in the future? VERITAS being an organization working with the product of quality assurance does not think that the correct attitude is to passively sit back and hope for this to happen. We want to activily stimulate the industry in such a way as to facilitate training of qualified diving inspectors.

Part of our trials facilities philosophy is centered around this.

Description of UWTA

General

Figure 9 shows schematically our concept of the trials area which should be a useful tool for the solution of the problems described so far. The components shown in green here, are phase 1 of the trials area. Phase 1 comprises a full scale steel structure which is a model of a typical north sea jacket, and a 1000' long 36" pipeline. Phase 2 is shown in red, and comprises a barge with a large center moon pool for handling equipment. Phase 2 also contains a large bottom fundament with standard templates for simulation of deep wells. It also contains some service equipment for service and inspection of tests at sea bottom. The bottom fundament can be handled by the barge and moved to any position in the fjords where we find a suitable place for specific tests. Phase 3 is shown in yellow. It is composed of a pump station on land and a pipe going to specific predetermined positions where the bottom fundament can be hooked up to the pipe. Flow lines back to land concludes the loop. A trials center containing phase 3 will permit testing and simulating live wells. Phase 3 finalizes the existing plans for the test area facilities.

Phase 1

Figure 10 shows an isometric view of the steel test rig. The test rig measures 20 x 20 x 15 metres, and weighs 150 tons. It is located in 30 metres of water. The purpose of the structure is to provide training facilities for our underwater inspectors, to provide training facilities for the diving companies/diving inspectors, to certify diving inspectors and to develop and verify new methods and equipment for underwater inspection. The structure is thoroughly mapped. We know exactly what sort of welding failures we have in each specific joint. We also have a couple of other defects built into the structure such as geometrical deformations, dents and so on.

The structure has a loose corner. This corner can be replaced by a set of similar corners where we have built in various solutions to the joint structures with a number of different failures, including internal failures. This corner is basicly intended for non-destructive testing. An inspector inspecting this will have to handle his inspection gear, handle the situation of limited visibility, report his findings in a similar way as he would in a real situation offshore. Having a number of loose corners we will be able to use this as a certification test for underwater inspectors. Last November the steel structure was put under water in 30 metres depth outside Bergen.

Figure 11 shows a slide taken during the installation.

Current activities:

Figure 12 shows a list over current & past activities.

Figure 13 shows one of our diving inspectors working in the steel structure assisting a research project on cathodic protection on steel structures which is pressently being carried out. The goal of this study is to supply field data for an assortment of cathodic systems and a significant part of the work is carried out through in situ measurements on the steel jacket.

Figure 14 shows the reference sensors used in this project.

Figure 15 shows a remote controlled vehicle doing underwater inspection in the steel structure during a recent R & D programme aimed at remote controlled methods for inspection of offshore structures.

Figure 16 shows the vehicle on dock side.

Figure 17 shows the control console during ultrasonic thickness measurement with the vehicle. In the middle of the picture one can see the scope of the thickness-measuring device, demonstrating clearly that one can read thickness with this technique.

Figure 18 shows the particular ultrasonic probe used in this test.

The project demonstrated that it is fully possible to perform all three levels of underwater inspection, visual overview, visual close-up, NDT-measurements, with existing remote controlled vehicle techniques provided some fairly simple adaptions are made.

However, the time efficiency of the inspection is not good. A large development programme will be initiated in 1980 where the steel structure and the pipe will be basic tools.

Phase II

Figure 19 shows an isometric drawing of the barge of phase 2, presently in the projecting phase. The barge has principal dimensions, 58 x 22 x 3.8 metres. The displacement is 220 tons. It can handle equipment as heavy as 150 tons. The crane shown in the slide lifts 40 tons. The barge has a deck area of 450 square metres.

Figure 20 shows a bottom fundament, an artist's conception of a full scale test of sub sea equipment, handled through the center well. The figure illustrates an operation on the bottom fundament, simulating maintenance on a subsea well by divers. The figure is highly schematic.

Planned activities

Figure 21 shows planned activities. Presently two projects have been planned. The first project is called "Reliable Operation of Subsea Production Systems" and has started in 1980 and will last until the end of 1982. The second project is a full scale testing of a marine riser.

The project called "Reliable Operation of Subsea Production Systems" has the main objective of improving the confidence in subsea production systems. We believe that the only way to meet this objective is to include full scale testing into the scope of work of the project. However, full scale tests is an expensive research activitiy and the project will first carry out a Definition Phase in order to pin point the most critical areas of subsea production systems and thereafter continue to plan a test phase of the project which will be carried out through 1981 and 1982. However, focus have already now been put on the following topics which most probably will be included in the test programme:

1. Diverless Installation/Retrieval/Re-entry
2. Inspection/Maintenance methods and philosophy
3. Control and Signal transmission systems
4. Underwater connectors

Several tests are in the planning stage. This includes a flowline tie in test where parameters influencing the operations are carefully monitored and documented. Another test which is visualized in this research programme is installation/re-entry in deep waters where vessel motion is simulated and limitation on operation caused by weather conditions/vessel motion might be determined.

The aim in this project is basicly - to establish the limits of equipment/operations and thereby making reliable data available which might be used in connection with decicion on field development schemes utilizing subsea production equipment. Such limits can only be established in a safe and economically feasible way by simulated operation where the different parameters can be varied.

Figure 22 shows an artist's conception of full scale testing of a marine riser at 300 metres depth. Figure 23 shows the same concept from another angle of view.

Figure 24 shows this test in more detail. The riser is fixed to a bottom structure and goes all the way to the surface barge where it is hooked up to a trolley on the barge. This trolley can move the top of the riser back and forth along a slide in the barge, thus simulating horizontal movements of a surface vessel. The riser will be instrumented to yield information for the excitated riser like displacement, accelerations, stresses etc. The project will be extended to a second phase, where the effect of ocean waves will be investigated.

Again, external parameters affecting the riser are varied, and this gives information on the riser behaviour under different conditions. This test yields good understanding of the phenomena under investigation.

The final result of the project will be to verify/adjust an analytical model for dynamic non linear analysis of deep water marine risers.

The project will also yield some practical results on performance/reliability of key elements in a riser concept.

Other proposed activities

In addition to the described activities, we also have continous proposals for new tests. The most realistic ones are:

- Testing of a dry sub sea completion system. Proposed by Kongsberg Våpenfabrikk and Cameron Iron Works.
- Testing of a submersible system for oil/gas collection after a subsea blow-out. The project is proposed by Bergens Mekaniske Verksteder.
- Simulation of a submarine rescue operation including sub to sub docking.

Time schedule

The time schedule for the trials facilities is shown in figure 25.

Phase I:

The planning of phase I was finished early 1979. The structure was installed in November 1979. The pipe will be put in the sea in April 1980.

Phase II:

The planning of phase II will go on untill 1980. We expect the barge to be constructed during this year and be operational in April 1981. At the same time we will have the bottom fundaments operational. Auxillary equipment will be bought continously between now and through 1981.

Phase III:

We have started the feasibility of phase III and expect to start on the detail planning towards the time where phase II planning ends. The planning stage of phase III will run into

1982, resulting in a pump station between 1982 and 1983. Loops and simulated wells will be finished at the same time.

Cost estimates

Phase I:

Value of the steel structure	2,5 mill. NOK
Value of the pipe line	4,0 " "
	6,5 mill. NOK

Phase II:

Value of barge and equipment	16,0 mill. NOK
Value of observation bell	5,0 " "
Value of bottom fundament	0,6 " "
Value of auxillary equipment	5,0 " "
	26,6 mill. NOK

Sum total phase I and phase II will be approx. 33,0 mill. NOK

Phase III:

Oil/gas well simulator
TFL-loops
Shallow-water simulation well

Value pump station	10 mill NOK
Value TLF loop and wells	15 " "
	25 mill. NOK
Total	60 mill. NOK

Concluding remarks

As time passes we expect new ideas to develop as a result of the feed-back from the market. This feed-back will result in additional equipment and installations in all phases of the facilities, most so in phase III, however.

For this reason we expect some changes in presented plans and also in the finished facilities.

At the time of writing the concept of a phase IV is being discussed. The idea is too new to present at this time, however. It needs additional user feed back before being realistic.The reason for mentioning it here is to indicate that the presented concept of an underwater trials area is not a static one but rather a dynamic one with great emphasis on meeting the needs of the marked at any time.

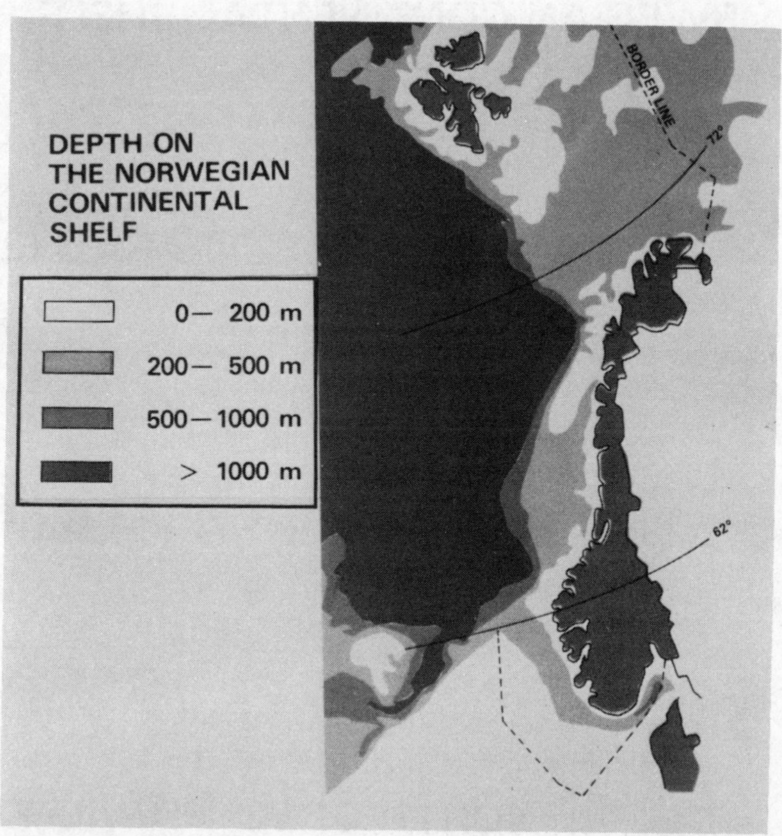

Fig. 1.
Depth distribution on the Norwegian Continental Shelf.

Fig. 2.

COMPARISON OF DEPTH DISTRIBUTION ALONG THE NORWEGIAN CONTINENTAL SHELF
3%
55%
42%
SOUTH OF 62° PARALLEL
13%
15%
72%
BETWEEN 62° & 72° PARALLEL
8%
32%
60%
NORTH OF 72° PARALLEL
0 — 200 m
200 — 500 m
500 — 1000 m

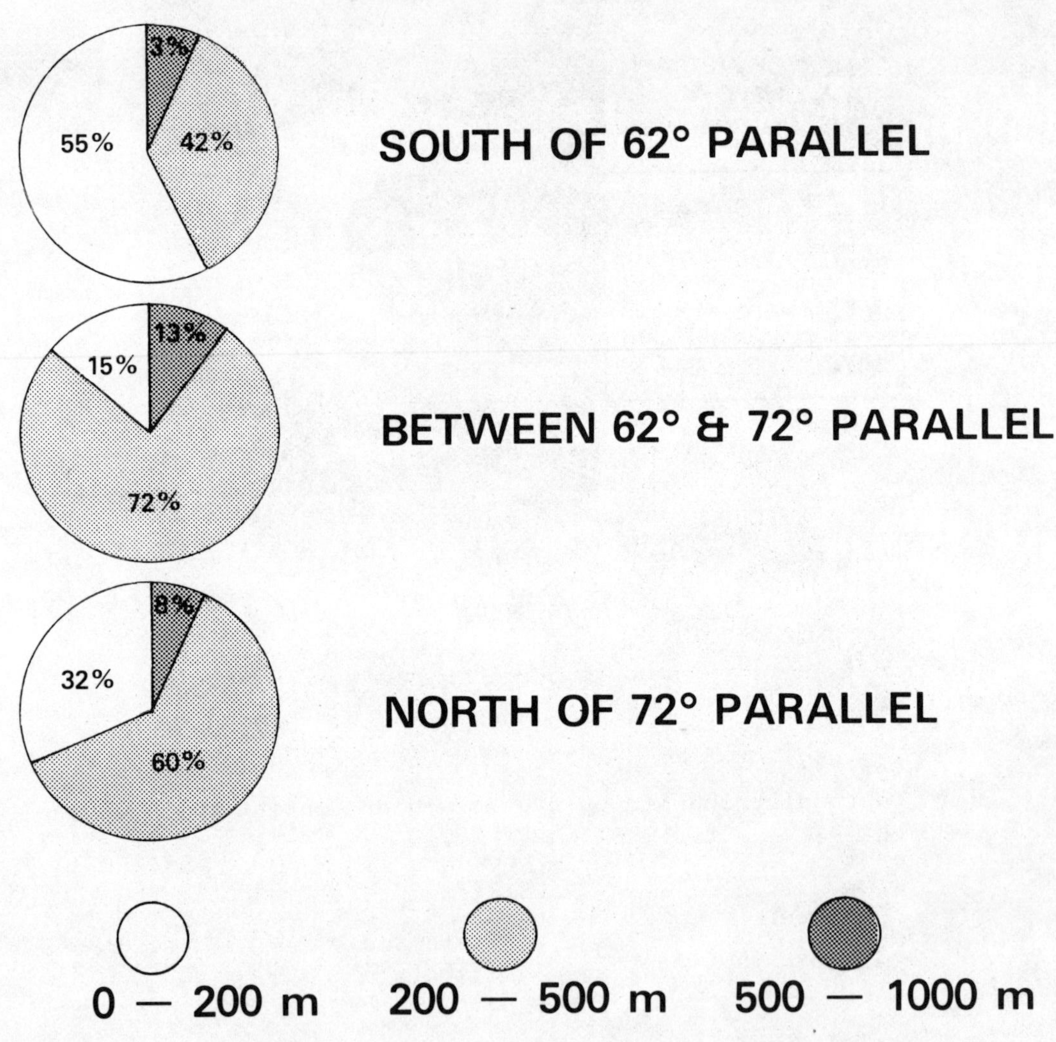

Fig. 3.

DEPTH DISTRIBUTION NORTH SEA
100 m
200 m
300 m
400 m
500 m
Ekofisk
Frigg
Statfjord
Depths at 4th round of concession

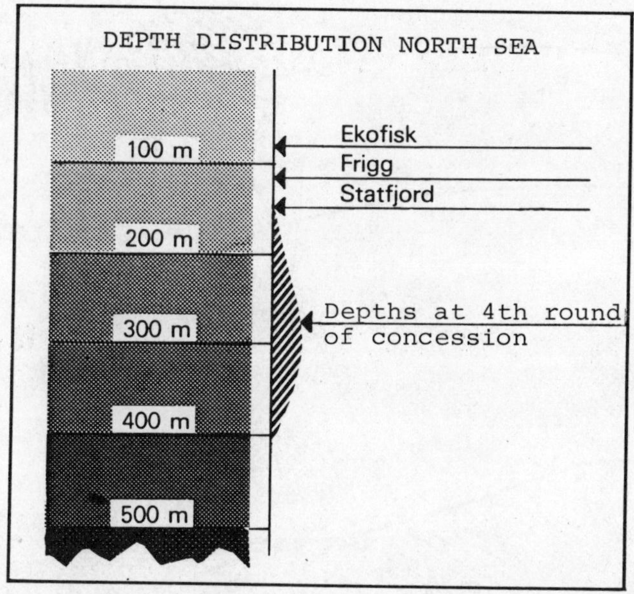

Fig. 4.

PRESENT AND FUTURE APPLICATION OF SUBSEA PRODUCTION SYSTEMS

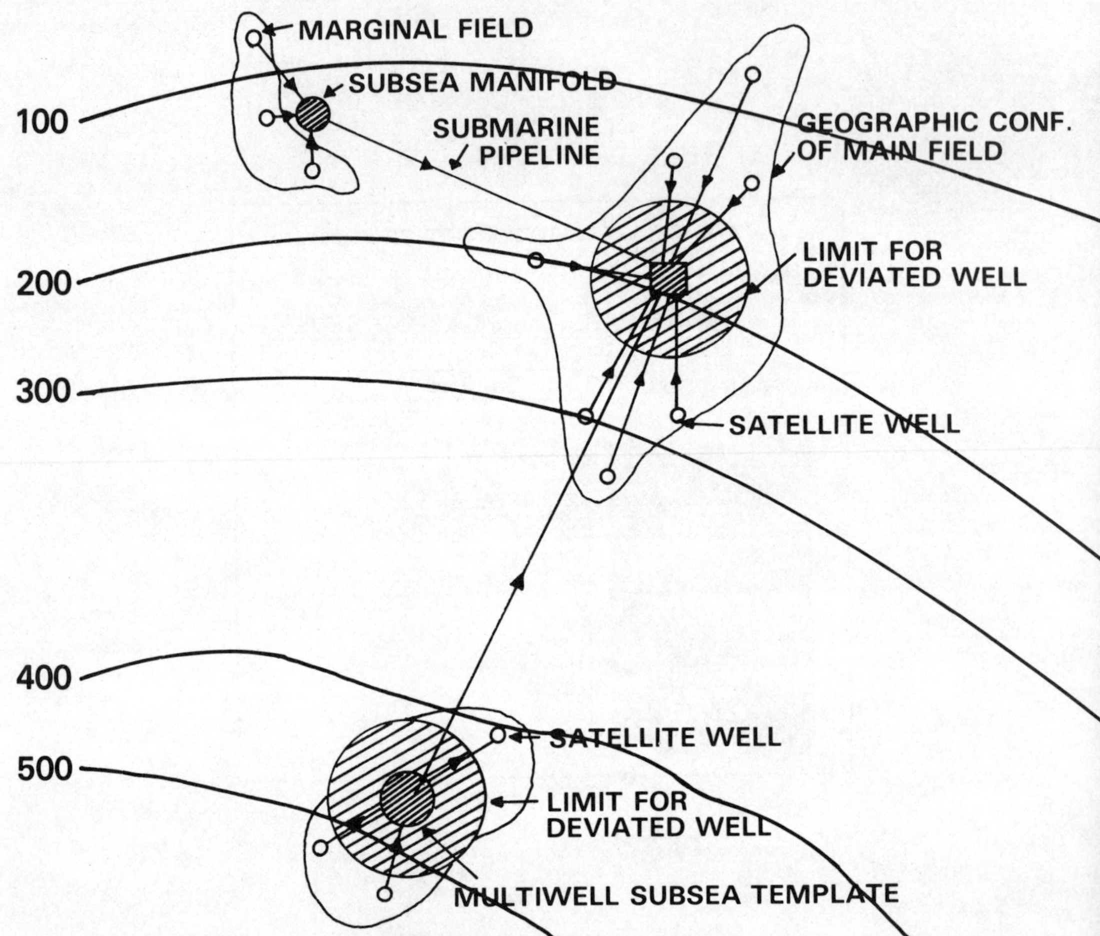

- TO OBTAIN A HIGHER EFFICIENCY RESERVOIR DRAINAGE PATTERN
- EARLY CASH FLOW
- MARGINAL FIELDS
- WATER DEPTH IN EXCESS OF FEASIBILITY OF FIXED PLATFORMS

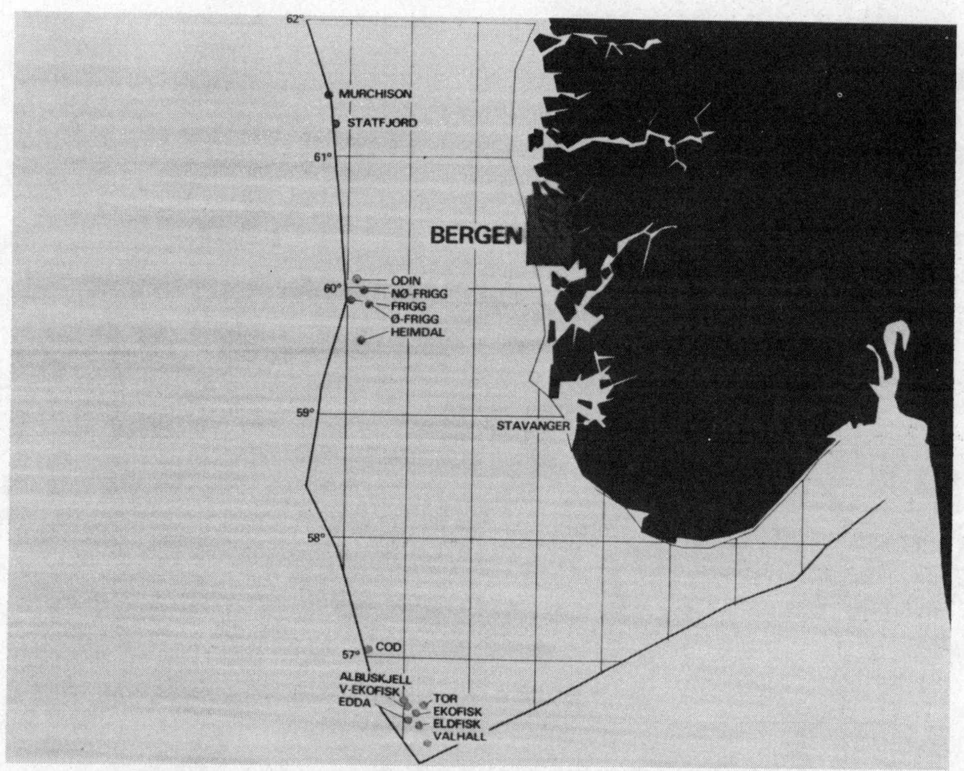

Fig. 5.

Position of the trials area and present oil fields on the Norwegian Shelf.

SURFACE REPAIR
AND MAINTENANCE
FACILITY
OFFSHORE
FLOATING
PRODUCTION
LOADING
BODY
UNDERWATER
SERVICE SYSTEM
SUBMERSIBLE
SERVICE AND
INSPECTION
VEHICLE
RISERS
SATELITE
X-MAS TREE
MULTIWELL
SYSTEM
PIPELINES
MANIFOLD
CONFIGURATION OF NEAR-FUTURE UNDERWATER PRODUCTION SCHEME

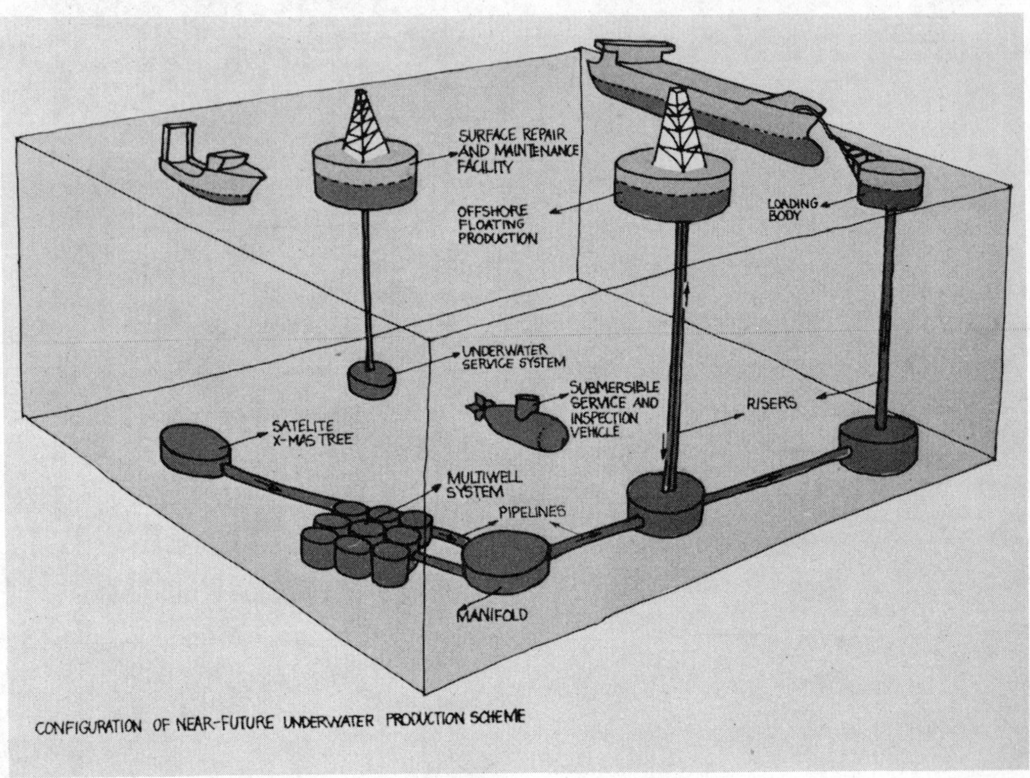

Fig. 6

100 mill.
NOK
Estimated value inspection effort underwater
North sea
high
low
6
5
4
3
2
1
0
80
year
85
From CIRIA UR 13

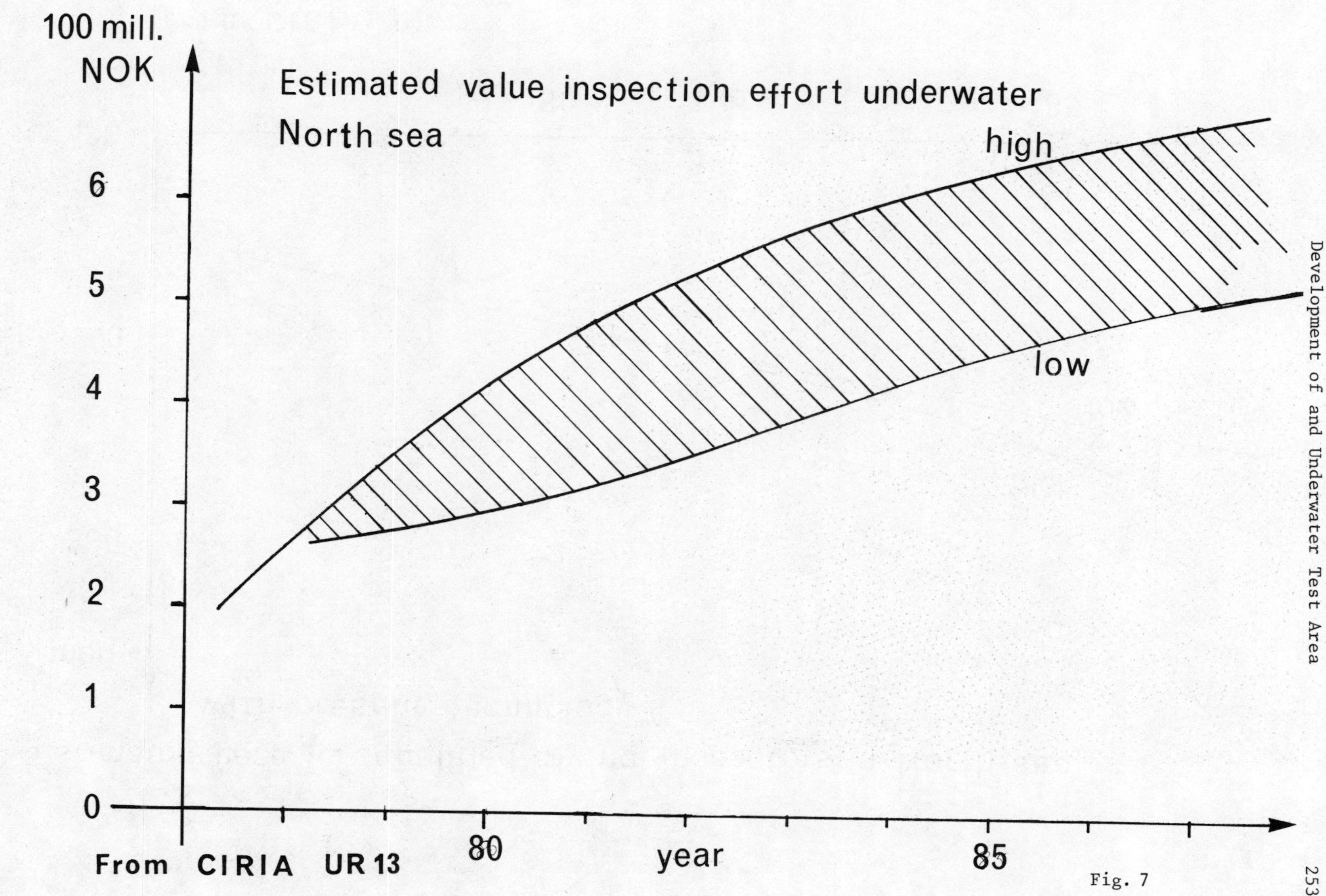

Fig. 7

Fig. 8

Estimated need for qualified diving inspectors in North sea
with present technology
men
300
200
100
high
low
80
year
85
From CIRIA UR 13

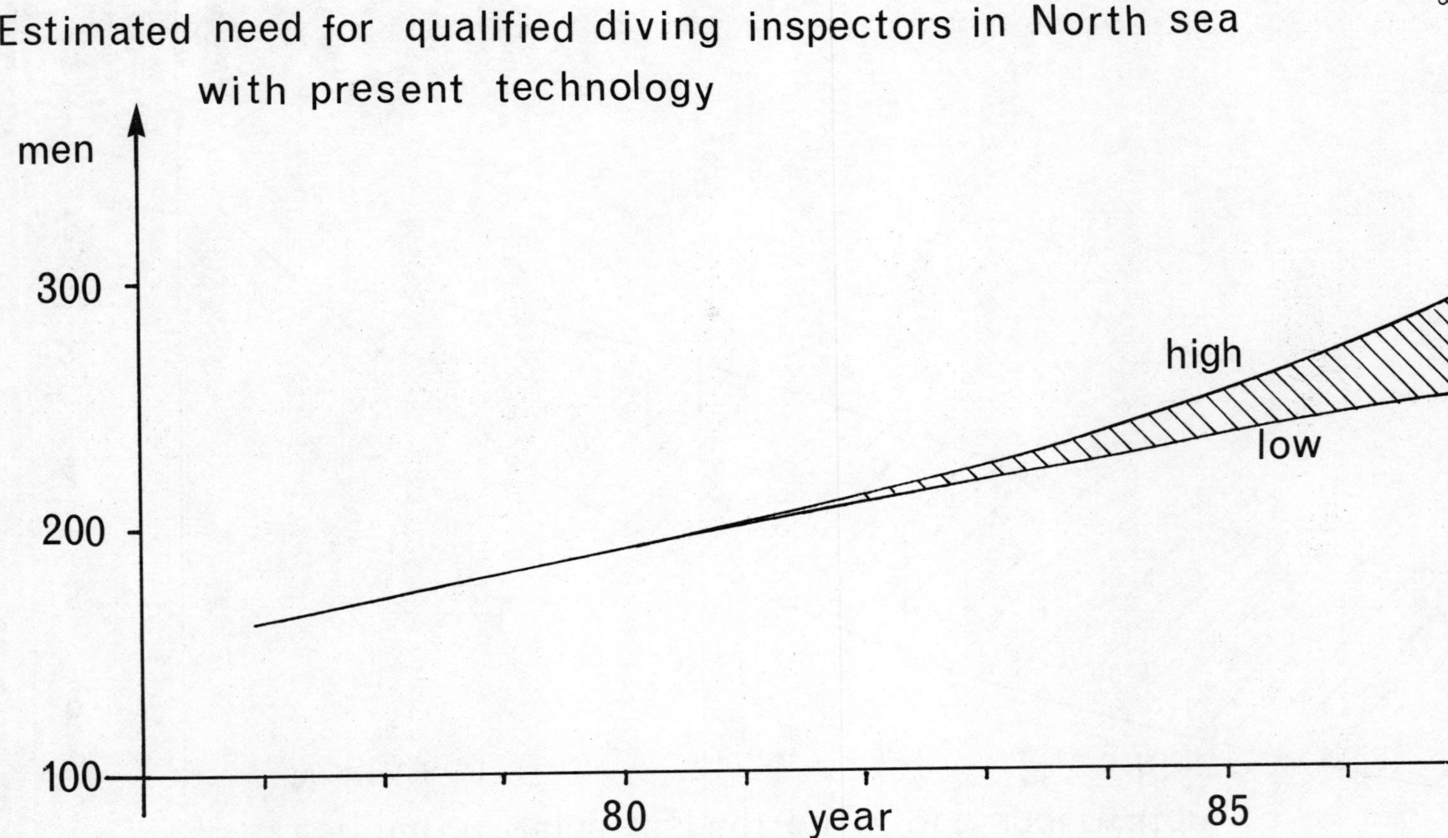

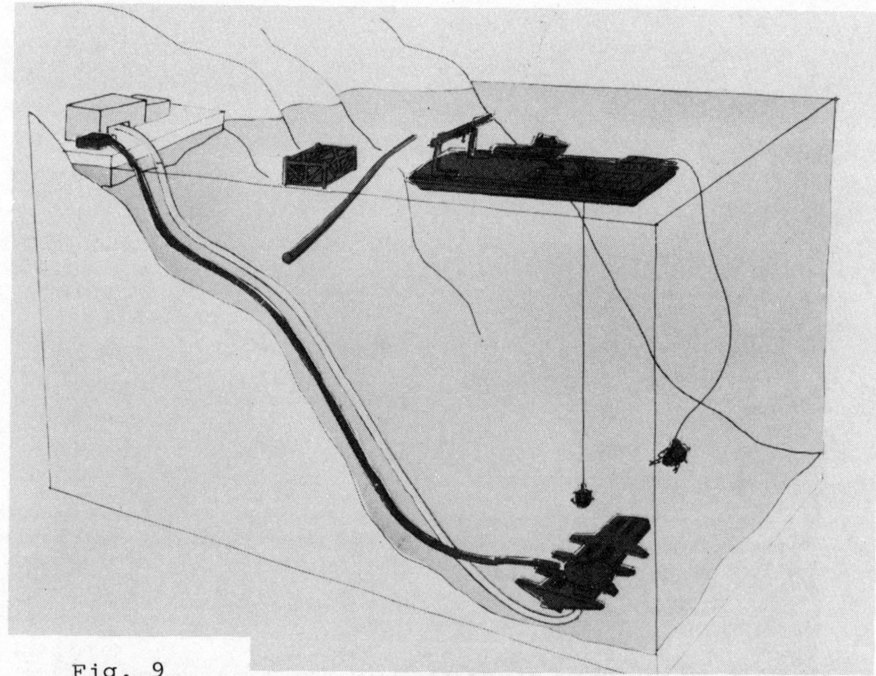

Fig. 9

The concept of the underwater trials facilities.

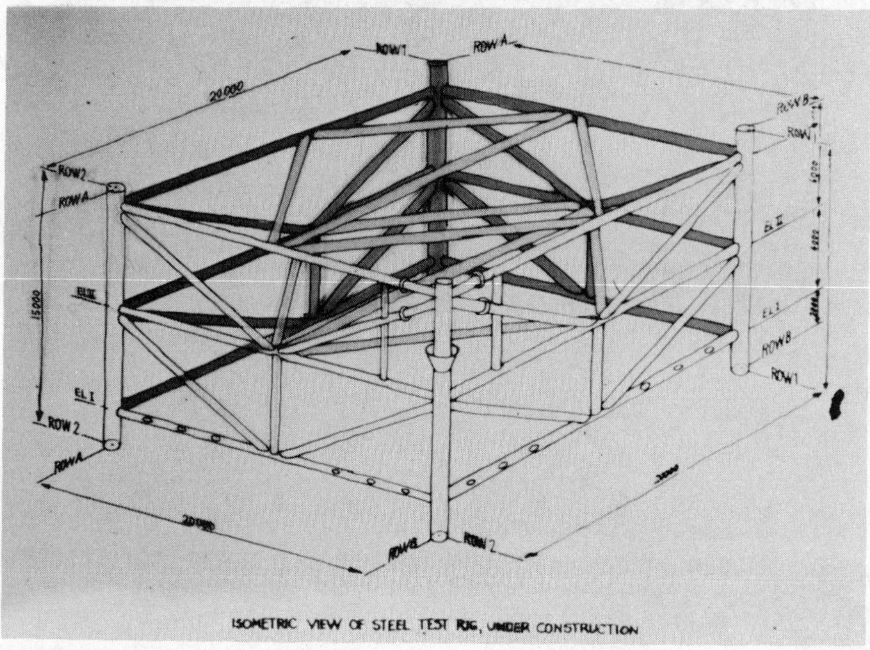

ISOMETRIC VIEW OF STEEL TEST RIG, UNDER CONSTRUCTION

Fig. 10

Fig. 11

The steel structure during location

Fig. 12.

PAST AND CURRENT ACTIVITIES:

	Time
• Corrosion testing	1979 —
• Concrete reinforcement testing	1976 —
• Remote underwater inspection performance trials	1979
• Stereo photography in underwater inspection	1980
• Acoustic methods in underwater inspection	1980 —

Fig. 13

A Veritas surveyor working on a corrosion test project in the structure

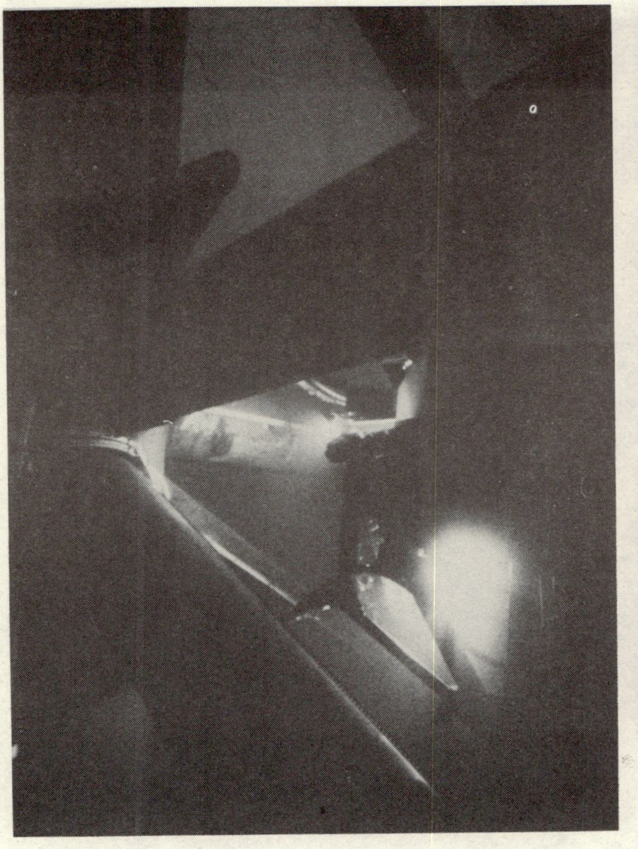

Fig. 15

Picture taken of a remote controlled vehicle during a remote controlled vehicle inspection program in the steel structure

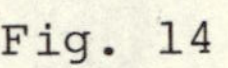

Fig. 14

Sensors used in the cathodic protection field measurement project in the steel structure

Fig. 16. A picture showing RCV assessed in the RCV inspection program in the steel structure (SNURRE)

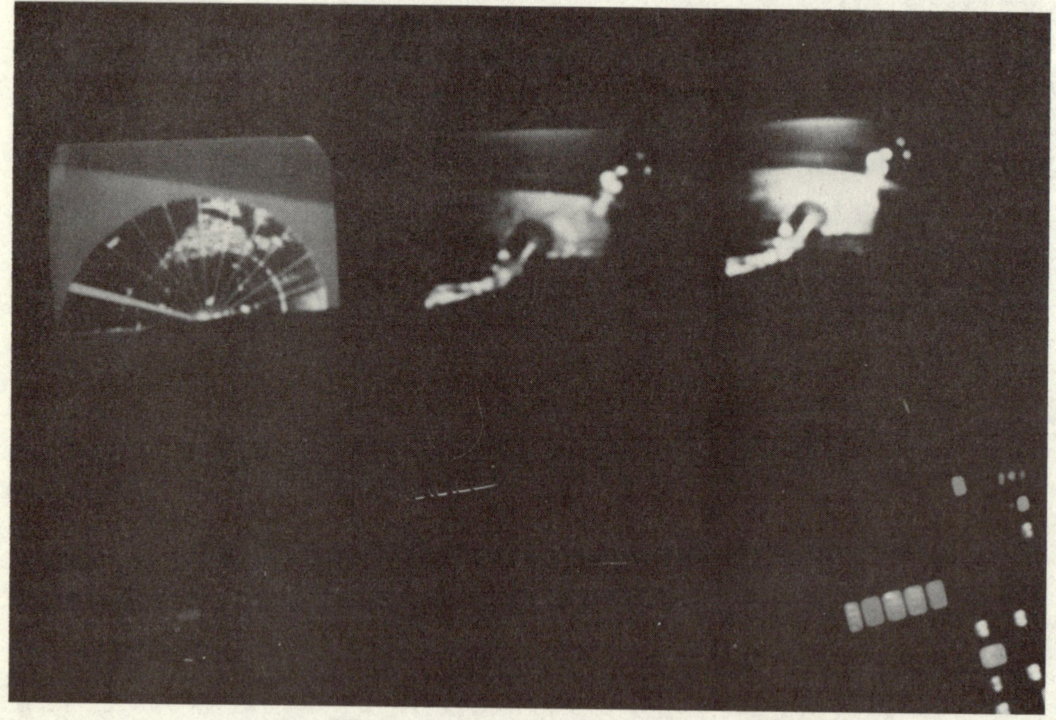

Fig. 17. A picture of the control console of SNURRE during thick-nessmeasurements with SNURRE in the steel structure.

Fig. 18

Details of ultrasonic probe, manipulator and suction pads used in thickness-measurements in the steel structure

Fig. 19. An artists impression on the Central Surface Unit (Test Barge)

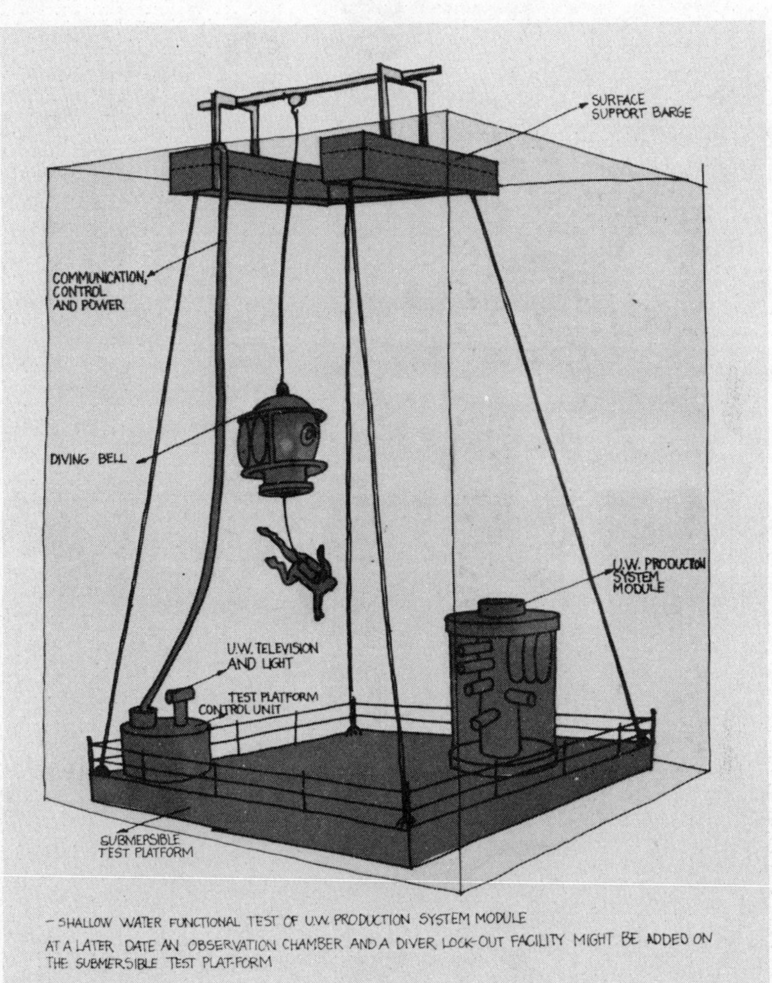

- SHALLOW WATER FUNCTIONAL TEST OF U.W. PRODUCTION SYSTEM MODULE
AT A LATER DATE AN OBSERVATION CHAMBER AND A DIVER LOCK-OUT FACILITY MIGHT BE ADDED ON THE SUBMERSIBLE TEST PLAT-FORM

Fig. 20

PLANNED ACTIVITIES:

- **Reliable sub sea production 1980 - 82**
- **Full scale testing of marine riser 1981**

Fig. 21.

Fig. 22 Deep Sea Riser Test.
Simulation of horisontal movements.

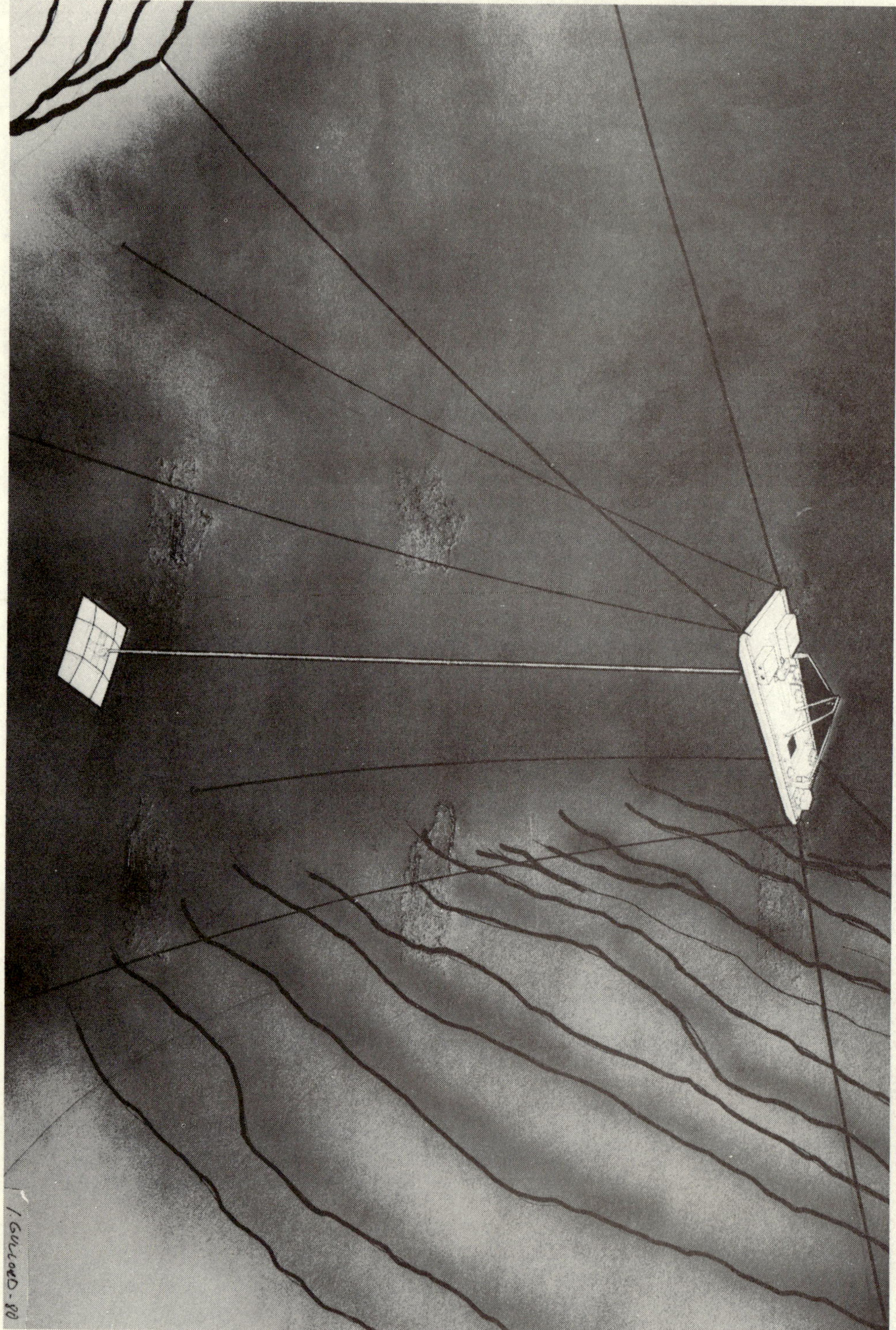

Fig. 23 Deep Sea Riser Test. Anchor configuration.

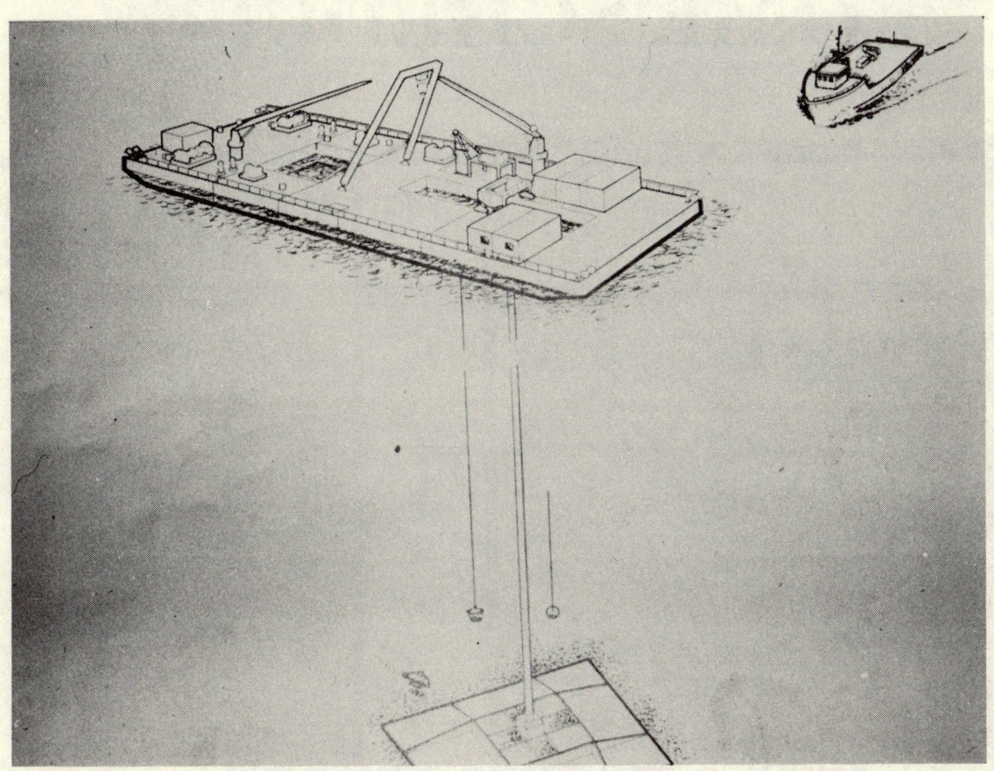

Fig. 24 Surface and subsea services during Deep Sea Riser Test.

Fig. 25

TIME SCHEDULE
79
80
81
82
Phase I
Planning
Structure
Pipe
Phase II
Planning
Barge
Fundament
Aux. equipment
Phase III
Planning
Pump station
Loop & wells

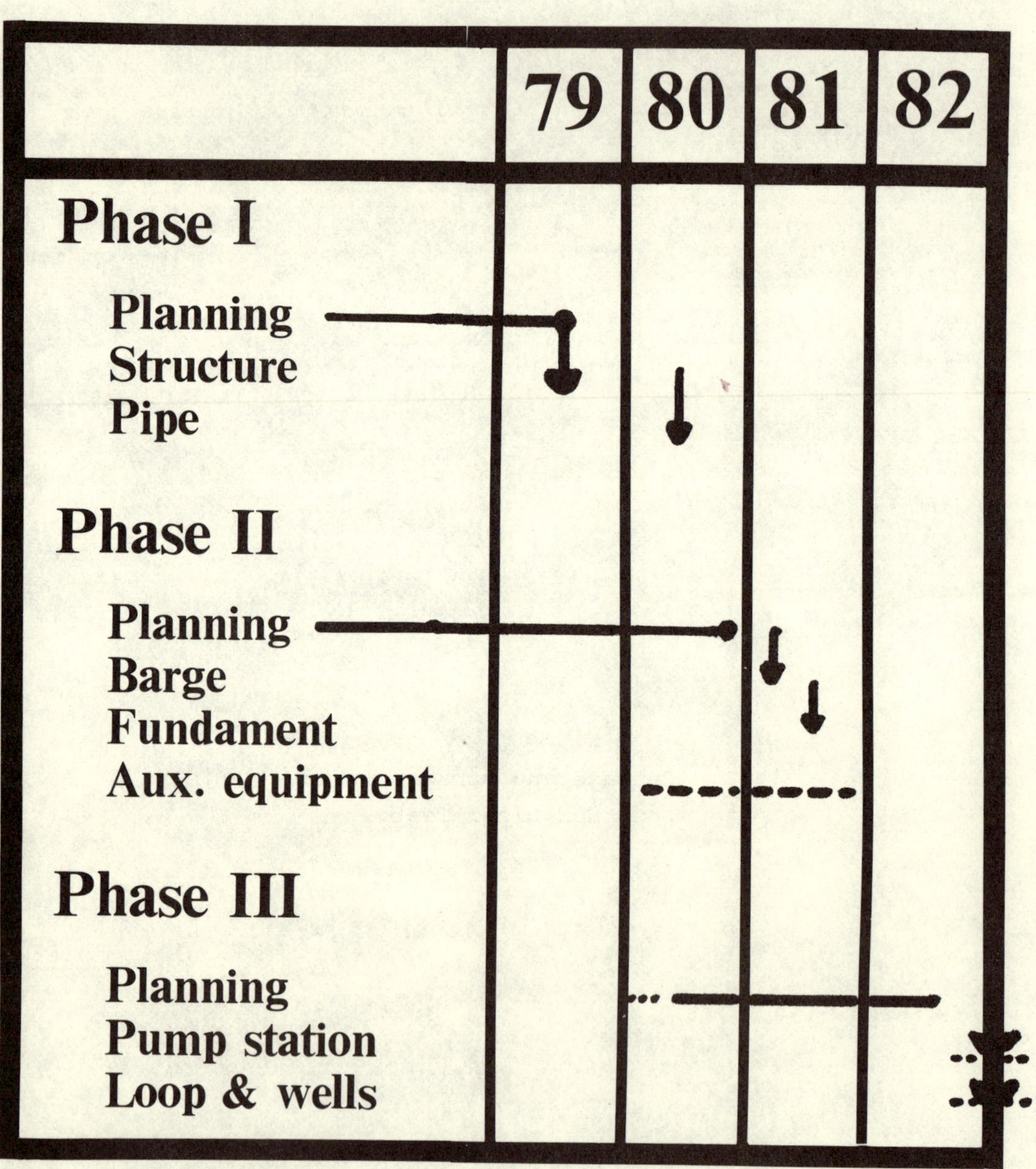

INSPECTION-RELIABILITY-CONTROL

QUALITY REQUIREMENTS FOR UNDERWATER INSPECTION AND REPAIR METHODS

R. Sletten and B. S. Carlin

Veritas, Høvik, Norway

ABSTRACT

The paper summarizes the experience gained from inspection and repair of offshore structures in service. In addition it will highlight the research and development work that is performed by VERITAS.

The different inspection methods will be reviewed with reference to procedure, quality requirements and relative cost-effectiveness. Based on this, the choice of both inspection programme and methods will be discussed. Also the suitability of documentation and reporting procedures will be commented and advice be given on expected development in this field.

In the North Sea relatively few platforms have suffered damages that have required repair, but obviously there is a need for a contingency that includes suitable repair methods. The traditional methods used for repair above water may in many cases suffer a certain decrease in quality and most certainly increases in cost and efforts when applied below water. This means that a number of problems have to be looked into and solved before the same techniques readily can be applied below water. Possible methods will be compared with the quality requirements that have to be fulfilled in order to satisfy the regulations. From this material some thoughts on the choice of method will be expressed. In addition, the problems related to site inspection and testing in connection with underwater repair will be discussed.

KEYWORDS

Underwater inspection, underwater repair, offshore structures, damages, inspection methods, quality requirements.

INSERVICE INSPECTION

General Inspection Principles

The principle followed in connection with inservice inspection varies from country to country and from owner to owner. In the North Sea the national authorities having jurisdiction over the different areas have laid down certain miminum requirements that must be met.

However, the general practice followed in most areas is the following.

- An inspection program is prepared annually by the owner. In the North Sea this program is to be approved by a certifying authority.
- Inspection is carried out under surveillance by representatives of the owner and/or the certifying authority or directly by the owner's inspectors.
- Inspection reports are evaluated to determine the need for remedial actions or other changes. This is often done by special review panels or experts within the owner's organizations. In the North Sea the results of the evaluation will also be submitted to the certifying authority.

A more thorough description of the principles followed in the Noth Sea is given by Sletten, Fjeld and Røland (1977).

The preparation of inspection programs is based on:

- the general experience of the operator regarding the type of structure in question
- documentation related to the design, fabrication and installation of the structure
- operating experience with the structure

In order assure that programs effectively meet the objectives and requirements set forth by the operator and the authories, the documentation must be reviewed in a systematic fashion by competent personnel. The aim is to identify the areas or items of the structure that has the highest potential of failure in a relative sense. These items/areas are referred to as significant areas. The reasons for significance in this respect may be e.g. high stress, low fatigue life, repair during fabrication, exposed location, extreme consequence of failure. When assessing the relative importance of the significant areas, one should consider both the probability and the consequence of failure. By assigning numerical values based on the degree of probability and consequence of failure for each area and regarding the product as a relative criticality number it is possible to grade the significant areas into cathegories according to this number. The review of design, fabrication and installation documentation is therefore a most important part of the preparation of the inspection program and should extract and preserve in a readily accessible way the information that will be required for assessment of the seriousness of damages and determination of remedial measures. This information, including the list of significant areas, may be collected in a suitable form by utilization of timely data technology, and referred to as the Design, Fabrication and Installation resumé.

To obtain a surveyable record of the condition of a structure, significant areas and the inspection results may be plotted on specially prepared diagramatic representations of the structure. These diagrams together with additional condensed tabular or computer stored information is referred to as the structure condition record, that will be used as input for preparation of the inspection programme.

The sheer size of most offshore structures makes it necessary to distinguish between general visual inspection covering large parts after structure and close visual inspection and testing that will have to be performed for significant areas.

The annual inspection programme will then in addition to the general visual inspection of the entire structure consist of significant areas selected for close visual inspection or testing. What methods and means are available for detection and examination of defects will be discussed in the following sections.

General Visual Inspection Methods

The different inservice inspection methods will in each particular case have procedures to follow that depend on the equipment and personnel used.

The aim of the general visual inspection methods is to detect major defects and other changes in the overall condition of the structure such as marine growth, debris and scour.

Structural defects that will be detected by this type is inspection are:

- out-of straightness
- major cracks, dents, gouges and holes in tubular members
- cracking, spalling or disintegration of concrete
- missing members/items
- coating damage on risers or pipelines
- corrosion attack

Most of these defects will normally develop due to external factor such as installation damage, collisions, dropped items, weather damage, explosions, overloads, earthquake or material deficiencies.

Some defects will however develop more slowly and will then normally have a chance to be detected by close visual inspection at a size much smaller than the ones looked for by general visual inspection.

General visual inspection may be performed using inspection divers, inspection submersibles or remote controlled vehicles (RCV).

Inspection divers may operate from a diving system onboard the platform or a separate diving support ship or lock-out chambers in an inspection submarine. The quality of results obtained by general visual inspections performed by inspection divers will depend on the experience of the diver and the preparations performed in briefing the diver about conditions to be observed. In this connection it is most important to define a reporting guideline so as to be able to compare reports on the same conditions made by independent observers. Among the items to be covered in this guideline is a standard terminology of definded terms.

The main advantage of using inspection submersibles compared with divers especially for general visual inspection lies in the possibility of using trained engineers with thorough knowledge of the structure as inspectors and thus reducing the need for thorough briefings. This will give better quality as the evaluation can be based on first-hand information from skilled engineers trained for this purpose.

The use of RCV has the same advantage except for the loss in detail of the view on a tv-screen compared with the direct view by the human eye. This is especially pronounced when using black and white recordings and may to a certain extent be overcome by use of colour-tv and colour photographs.

Close Visual Inspection Methods

The aim of the close visual inspection is to detect cracks, pittings or spalling of concrete before it develops so far that it is detected by the general visual inspection.

The same means as for general visual inspection are available. The main difference is that close visual inspection requires advance cleaning of the surface of the area to be inspected. In some cases this may be achieved by use of diverheld brushes or waterjetting equipment. The quality requirements on cleaning will in most cases differ between concrete structures and steel structures. For concrete structures where normally larger cleaned surfaces are required some remaining hard deposits may be left without obscuring the view too much. For steel structures the cleaning normally has to be performed down to bare steel to make it possible to observe surface cracks or pittings.

The possibility of detecting cracks by the bare eye of an underwater inspector is depending on factors such as lighting, visibility, cleaning and crack dimensions.

In our opinion it should be possible after cleaning of a concrete surface and observation at a distance of about 1 metre under normal visibility conditions to observe cracks with nearly the same accuracy as in air. For steel structures we expect rather small cracks to be visible provided that the cleaning and the ligthing is adequate. The difficulties here are more related to the possibility of detecting cracks in the transition zone between the weld and the parent material due to eventual undercutting that may hide the actual crack.

For close visual inspection the recording methods are of greater importance and all observations should be documented by colour photographs in addition to eventual video recordings. Already today methods using fixed frames, to give constant focusing distance and controlled lighting conditions have been used for documenting the conditions of welds at nodes in jacket structures. This combination of photographing techniques with diving skill and inspection experience is one of the more important developments in underwater inspecion over the past years. The use of 3-dimensional photos and photogrammetric methods will in certain cases make it possible to substitute direct measurements.

Non-destructive Testing (NDT)

Non-destructive testing techniques currently available for underwater use are in most cases direct applications or modifications of techniques used onshore. Most such techniques properly applied under water will give quite acceptable results.

The following instrumented methods are commonly used for underwater inspection:

- Magnetic particle testing for fatigue cracks in tubular joint welds
- Ultrasonic testing for revealing corrosion
- Potential measurements for revealing corrosion

Ultrasonic testing of welds is considered too complicated to perform under water on a larger scale and is only used for diagnozing purposes and in connection with repairs. Further, radiography is also used for controlling repair welds.

Potential drop measurements may be performed in connection with magnetic particle testing for determining crack depth as input for fracture mechanics evaluations.

Radiography may be used for corrosion mapping, ultrasonics however being the preferred method for this purpose requiring access to one side of the object to be examined only and giving more reliable results on remaining wall thickness. In table 1 the different methods are compared with regard to advantages and limitations.

METHOD	DEFECTS	ADVANTAGES	LIMITATIONS
Visual	Surface cracks, Impact damage.	Easy to interpret. Findings can be photographed or transmitted to topside by television and video rec.	Limited to surface defects. Surface must be cleaned for detailed observation.
Magnetic Particle	Surface cracks, laps, seams, and some near-surface flaws.	Indications can be photographed or televised topside for evaluation and video recording.	Requires thorough cleaning. Weather dependent in splash zone. Limited to surface and near surface defects. Present equipment limited to diver use. Cumbersome to perform underwater.
Ultrasonic	Cracks, Inclusions, Lack of fusion and incomplete penetrameter in welds	Especially sensitive to cracks. Can be used to evaluate subsurface integrity. Equipment is lightweight. Results can be transmitted topside and video recorded.	Thorough cleaning required. Operator skill is required. Surface roughness can affect test. Present equipment limit-limited to diver use.
Radiography	Internal defects such as shrinks, inclusions and incomplete penetration in welds.	Provides a permanent record. Standards have been established and are accepted by codes and industry.	Potential health hazard. Water should be displaced between source and subject. Requires access to both sides.

TABLE 1 Present NDT Techniques: Advantages and limitations in underwater Applications.

To be able to benefit from the advantages of ultrasonic testing for as well weld inspection as corrosion mapping, and reduce inspection costs and hazards to diving inspectors, automatic equipment has to be used. Such equipment is under development for use by divers, RCV's or manned vehicles, and will be available for operation soon. Especially for corrosion mapping the advantages are predominant to present methods as full description of corrosion state by a large number of measurements and 100% coverage of area examined allows calculation of residual strength and corrosion rate.

The dominating underwater ndt-method used so far in the North-Sea is magnetic particle inspection (mpi). The normal procedure after having observed a crack indication is to grind the weld a few millimetres. If the indication still persists the crack is considered confirmed. These remaining crack indications in the joints vary quite much in appearance and the sizes of 25 crack indications observed in totally 18 joints out of totally 382 tested has the distribution given in Table 2. The length varies from 6 to 940 mm and only the most severe cases have been repaired. The remaining ones are kept under observation.

Crack length in mm	Number of cracks observed	Number repaired
-25	10	-
26-50	4	-
51-100	1	-
101-200	7	-
200-	3	3

TABLE 2. Cracks observed by mpi related to crack length at discovery.

Special Equipment for Internal Pipeline and Riser Inspection

Pipelines and risers may be inspected using self-contained pigs or umbilical devices. Umbilical devices based on ultrasonics have been designed for corrosion control of offshore risers. These devices give an efficient mapping of the corrosion state, but are cumbersome to use from an operational and safety point of view. Shut down is required and gas risers are difficult to inspect due to lack of ultrasonic coupling medium.

There are pigs on the market, which partly fulfil inspection needs such as:

- Mechanical calliper pigs to reveal gross defects such as dents and ovalization.
- Magnetic pigs for corrosion detection.
- Eddy current pigs for detection of longitudinal cracks.
- Acoustic pigs for leak detection (not proven under water).

Magnetic pigs for corrosion detection are suitable for detection of pitting on the inside of the pipe, but less sensitive to pitting near welds and on the outside. General corrosion is not detected and the quantitative information gained poor.

Pigs based on ultrasonics are under development. The first available of these will probably be aimed at corrosion detecion in risers and flowlines as corrosion has proven to be the predominant cause of deterioration and the consequences of failure are more substantial for the pipeline parts close to other offshore installations. Equipment for crack detection on trunk lines will be developed later.

Quality requirements

Because of the great costs involved in underwater operations the quality requirements for inservice inspections should be based on the best professional results obtained within the different disciplines involved, e.g. engineering, diving and photography.

Skilled engineers should be used for the planning and evaluation of the inspection work and emphasis should be given to continuity and reproducibility of operations.

An experienced diving contractor should be used for the under-water work. The recording equipment should be handled by trained operators and photographers and supervised by people with experience in documentary film or tv.

This seems to sum up to a very expensive kind of operation, but we are convinced that by a professional approach there will be much time and effort saved in the evaluation phase. Only records with a certain minimum standard of quality should be presented to authorities and other bodies.

In order to achieve results that can be used for further evaluation and comparisons it is necessary to standardize the reporting by using a simple and uniqe structural identification system and a common terminology for description of defects. Measures should be confirmed by use of rules and not only estimated from a distance. When photographing under water a rule should be placed close to the motif in order to give a relative scale measure and a measure of the resolution achieved in the picture. If colour is critical, a standard colour spectrum may be attached as well.

In order to make planning and execution easier and more exact each inspection task should be as well defined as possible. The following procedure for close visual inspection of a concreting joint could serve as an example.

1. Find selected area and make a first examination of marine growth and general condition. Record status with colour still photographs covering the inspection area 1 m x 2 m.

2. Clean by waterjetting an area 0.5 m on both sides of concreting joint for the total length of 2 metres to be inspected.

3. Inspect area carefully for signs of cracking, spalling or corrosion. Record status after cleaning with additional colour still photographs.

4. Report only significant findings or state that inspected area is in a good condition.

The confidence that can be attached to an inspection system depends on detection probability, characterization/findings and reporting.

Available information on inspection efficiency is scarce or qualitative in nature. This prevents the full optimalization of inspection efforts at the present stage. Future developments to make the inspection more cost effective should therefore in addition to development of new methods and equipment include the establishing of inspection reliability data.

To illustrate the uncertainties involved during inspection and the type of information that has to be available about inspection performance, some preliminary results from a recently performed study of the efficiency of ultrasonic and radiographic weld testing are given in Fig. 1 and 2.

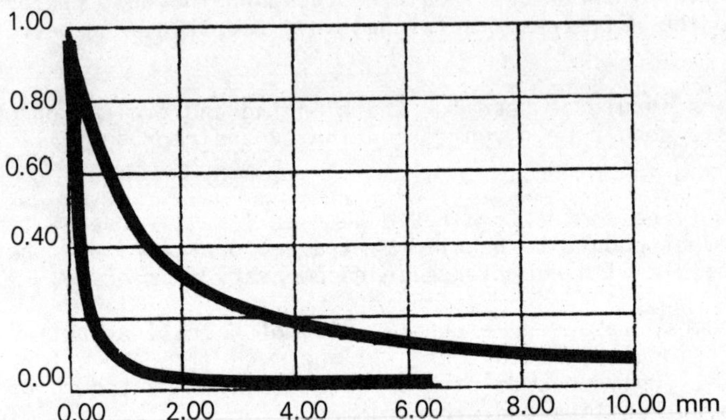

Fig. 1. Probability of Accepting a Lack of Root Penetration Defect when Examining According to ASME, Sect. VIII, as a Function of Defect Height (mm) (acceptance curves). Top curve is valid for ultrasonics and bottom curve for radiography.

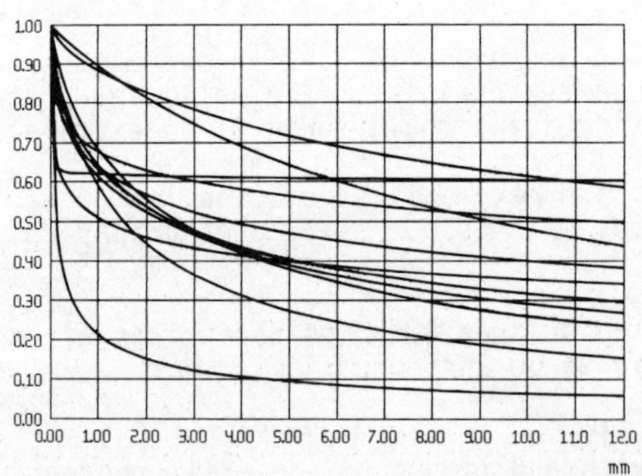

Fig. 2 Acceptance Curves for 12 Differient Ultrasonic Operators.

These results also clearly indicate that the defects may be present even after a rigorous production control and thus may represent a hazard to operation and require in-service inspection especially with respect to fatigue.

Choice of Inspection Method

The cost effectiveness of an underwater inspection system as such is related to keeping the inspection costs low while maintaining the information from the system. For offshore structures this is especially connected to reduction of the use of expensive diving inspectors and supporting outfits (diving systems, supply boats, etc.). Cost reductions may thus be achieved using remotely controlled or self-contained equipment like pigs for pipeline inspection and continuous monitoring systems or other automatized equipment, equivalent in performance to presently used methods or equipment for surveillance of structrue.

Further, cost savings are related to a more optimal use of inspection methods and the procedures to evaluate inspection results. This includes utilization of built-in structural redundancy present in steel jackets and concrete platforms.

Trends for the Future

The methods and equipment presently in use for offshore inspection purposes are fairly simple, but have to a large extent proven to give safe and reliable results.

To make inspection more cost effective and safe to divers and cover inspection needs not adequately covered at present, a development along the following lines is foreseen:

- Methods to replace manual diver inspection.
- Automatic recording equipment.
- Self contained (pigs) or remotely controlled equipment.
- More extensive use of ultrasonics, especially for corrosion mapping.
- Continuous monitoring systems with instrumentation.
- Wider application of photographic techniques.

Further, for cost effectiveness, more extensive inspection optimalizing systems have to be developed which require establishing better inspection efficiency data.

REPAIR OPERATIONS

Damage Evaluation

As a result of the inspection effort and overall surveillance, various types of damages will be observed on fixed offshore structures built in steel as well as in concrete.

Relative to the risks and values involved and with the extensive costs of offshore repair work, it is desireable to establish consistet guidelines for evaluation of the urgency and necessity of repair. Quality criteria for different methods should be established. Such guidelines and criteria will make it possible to develop rational contingency procedures which will minimize risks and unfavourable economic consequences.

This situation must be recognized in spite of the advanced methods by which present offshore structures are designed, analysed and constructed, and moreover the thorough independent control carried out through the design, construction and installation phases. Some offshore platforms will require some repair or maintenance during their lifetime, due to for instance unfavourable shapes of structural details, inaccurate analysis, occurrence of unforeseen loadcombinations, falling objects, ship impacts, explosions, fire, misuse, corrosion and other deterioration.

In most cases the recorded damages will be of less serious character. Nevertheless, in each particular case, the following aspects will have to be reviewed.

- reason for the detected damage
- the possibility of other undetected damages on the structure due to the same cause.
- possible need for further inspection
- the need for repair
- when should the repair be carried out
- possible consequences of postponement of the repair
- possible limitations in the operation of the platform before, during and maybe also after the repair has been carried out
- quality and suitability of adopted repair methods

One of the most crucial questions will be whether a damage by its present extent represents a critical reduction in the safety of the structure or if the damage will develop progressively towards such a state due to basic material characteristics, structural geometry, environmental conditions or the nature of the loads.

Further, it is important that the repair methods and procedures chosen are well fit and relevant for the individual damage, yielding integrity and durability in compliance with the environment and the loads on the structure. Such characteristics of the repair will, moreover, have to be obtained through very complicated sub-sea operations. Thus the need for advance testing of different repair and reinforcement methods is apparent.

Repair Methods for Steel Structures

The following repair and reinforcing methods have been used or proposed to date for offshore steel structures below water.

- Underwater wet welding.
- Dry welding in habitat.
- Clamping of prefabricated parts to the structure.
- Grouting of confining sleeve.
- Grouting between insert pipe and origianl member.
- Grouting and partly welded method.
- Grouted body inside tubular members.

Underwater welding performed by divers in a wet environment has been utilized, but with all the obvious difficulties involved the results are not to promising for structural welding purposes. There is a development in progress to provide better materials and procedures, but so far the experience with these new methods is rather limited.

Habitat welding is the preferred underwater welding technique for structural welds. The main requirements besides access and space in a watertight hyperbaric environment is the provision of power for preheating, drying of steel surfaces and for welding itself. Shield gas for welding, inert gas in the habitat, breathing gas and finally dry electrodes to the welder has also to be provided. Nondestructive destructive testing is fully possible to perform as a check of the welding operations. The method may be difficult to apply for complex joints and if the cracks are caused by inadequate material properties, the attempt to repair by welding is clearly not likely to succed.

Clamping methods may be used either to replace the functions of a damaged part of the structure by additional members adjacent to the damage. The problems here are related to the tight fit necessary to transfer the forces and the problems with loosening of the bolts. These problems may be overcome, but requires rigorous checking both during installation and later. The amount of underwater work is likely to be extensive due to the need of very accurate alignment and fit up.

Methods for repair and reinforcing based on grouting technique appear to be promising mainly because only a moderate amount of underwater work is necessary and that crack initiations and unfavourable residual stresses are not introduced in the structure. The application of grouted confining sleeves around defect joints, as shown in Figure 3 for a T-connection, also has its practical problems such as the varying and complicated geometry of the joints and the rather difficult task of securing tightness of the assembly during grouting.

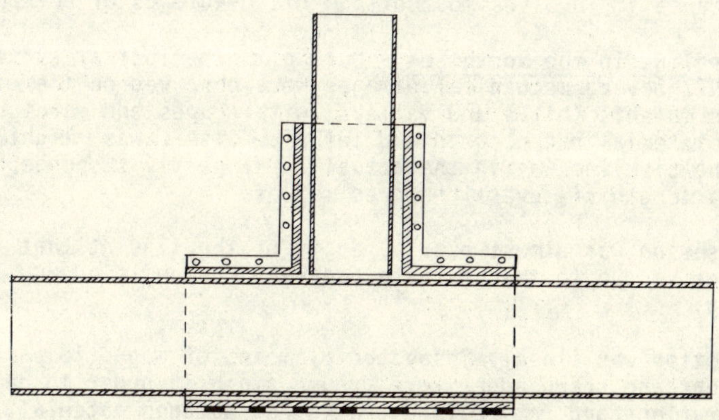

Fig. 3. T-joint with Grouted Sleeve

Grouting between an insert pipe and the original member, Fig. 4., is another alternative, but will have some limitations due to the need for placing an insert pipe. This might be a possible solution for reinforcing a leg, but for braces this alternative will be too complicated and most possibly replaced by a full internal grouting.

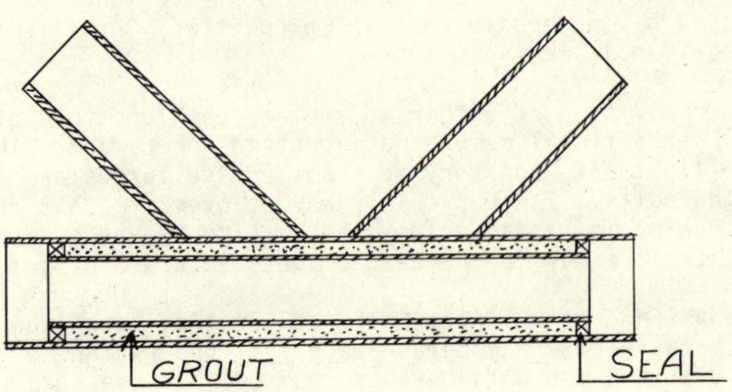

Fig. 4. K-joint with Grouted Insert Pipe.

The combination of grouted sleeve with welded sealing seems to be a less attractive method because it involves most of the disadvantages of both methods.

Repair on a pipeline in the North Sea - During the periodical survey of a pipeline in September 1977 severe mechanical damages were observed on the pipe, consisting of several deep indents (hills and valleys) with gouges and scratches. Excessive yelding of the material had occured and the pipe itself was slightly bent due to the damage. The pipeline was in the actual area partly suspended and was unburied over a length of several hundred metres.

As the summer season was almost over a repair of the line at that time would be extremely expensive due to the possible bad weather condition and lack of time for proper planning.

The damaged section was closely inspected by means of magnetic particles and all crack indications and sharp edges were ground smoth in order to decrease the stress concentrations and remove fine cracks and damaged material. The line was supported by sandbags in order to avoid vortex shedding and further damage due to fatigue. A finite element analysis supported the decision to keep the line in production during the winter season.

When the section was to be repaired, a water sludge with a pig at each end was pumped thorugh the line of the actual area prior to cutting in order to avoid oilspill. The damaged section of a length of about 50 m was cut out and a prefabricated replacement section was fitted. The pipes could have been connected by use of flange connections, hydrocouples or welding. Dry welding has proven to be the best method in the long run and this method was also chosen here.

To accomplish this a habitat was placed over each of the connecting points and sealed off against the pipe. Then the water was pumped out and the divers could carry out the welding in a dry room. The habitat also fitted with hydraulic equipment to keep the pipe in position and move the pipe ends in line for welding. The welding process was followed from the surface on video by the controlling authority.

A general requirement for qualifying welds on offshore pipelines is pressure testing. To perform this after the repair would have implied additional technical difficulties and extended the repair period, so it was decided that the welds should be tested by means of ultrasonics in addition to the common radiographic examination instead.

To satisfy the requirements from the controlling body with regard to acceptance of the ultrasonic testing a special procedure was developed. The testing of the weld should be performed by a qualified operator using an ultrasonic equipment with display. This equipment was connected to a similar slave equipment placed on surface on which the controlling operator could watch and interprete the signals obtained during the testing.

Repair of crack in jacket leg in 15 metres of water July 1977 - The damage ocurred during installation in connection with a doubler plate welded to the leg for fastening of one of the temporary float chambers. The weld betweeen this plate and the leg had cracked, and the cracks had propagated into the leg in a total length of about 4 metres. The leg was internally grouted. The doubler plate was removed and holes were drilled at the ends of all cracks to prevent propagation. It was then decided to remove the cracks and reweld the area in a habitat which was fitted around the leg.

The cracks which mainly went all the way through the wallthickness were ground out, a bevel made and the grout underneath removed. Before starting rewelding, the area was magnetic particle tested to confirm that existing cracks were completely removed. In addition drying/preheating of the leg was carried out for 15 hours in order to remove the moisture from the grout underneath. After the repair the welds were surveyed by an inspector through close visual inspection, ultrasonic and magnetic particle testing with positive result.

Strengthening of jacket by installation of additional stiffeners in 20 metres of water - A crack in a tubular joint was detected by magnetic particle testing. Previous cracking with following repair welding in this area had caused new cracking. Material tests taken were showing inadequate material properties in the z-direction and consequently cracking or lamellar tearing could be the result.

It was then agreed to strengthen the jacket by mounting of additional stiffeners. The stiffeners were then clamped to the actual braces and the leg. This required a perfect fit and therefore a template was made on site before the stiffeners with clamps were fabricated.

To do the work a special purpose diving vessel was used, supplying life support and decompression facilities for the divers. In addition, the equipment included lifting gear, hydraulic power tools, cleaning equipment and photo cameras to document the correct positioning of the strengthening members.

Repair Methods for Concrete Structures

Below water the following methods for repair of concrete structures may be used:

- injection of cracks with cementitious grout or epoxy
- concreting with tremie method or grout intruded aggregates, based either on cementitious materials or epoxy
- clamping and bolting of steel plates or profiles to the structure
- prestressing with high quality bars or strands

These methods are normally utilized in construction of concrete structures and it is the application offshore especially at greater waterdepths that will be the most noteable change.

Up to date 17 reinforced concrete platforms have been built since the first offshore placement in 1972. During a period of 40 installation years only five minor subsea damages have been reported. This, in addition to reports on the behaviour of shore based structures confirm concrete as being a well suited construction material for application in marine environments requiring very little repair or maintenance.

In short, the five damages comprise:

- Leakage to cell at 110 meters water depth during mouning of the steel deck. Repair by epoxy injection of cracks successful and no further leakage has been observed.
- Abrasion to shaft by a loose swaying pipe at 34 meters water depth. Repair by epoxy patching judged to be successful.
- Impact damage to a cell top by a dropped pipe. Repairs will probably be carried out but has not yet been performed.
- Leakage of a cell top due to improper sealing of a tension tie rod. The 26 cm dia. hole was repaired and no further leakage is observed.
- Damage to a breakwater wall. Partly repaired but further investigation will be necessary.

Other minor damages have been reported, but they are of negligible nature. None of the reported damages have been a threat to the integrity and safety of the structures. The two leakages were repaired to prevent water from leaking into the cells, whereas the other damages were repaired to prevent a probable future corrosion of the reinforcing steel.

The conclusions that can be drawn from an investigation of damages to offshore concrete structures are limited due to the short period for which they have been exposed.

However, the following points have been made:

- Damages of various extent do happen to offshore concrete structures.
- Accidents are much more frequent than deterioration as causes for damages in the early stages of a platforms being.
- The most frequent type of defects to be expected, are various forms of impact damages in the splash zone and on top of the cells (near the base of the shafts).
- Due to the small number of defects expected, unique procedure for carrying out repairs will have to be worked out after a damage is discovered.
- Due to the limited experience of suitable repair methods for use under water, procedure testing will be required in each repair case as a part of the certification scheme.
- There is a need for development of suitable repair materials including specifications for their use and limitations as well as documented performance of these in realistic environments.
- Different power tools exist capable of performing operations like sandblasting, waterjetting, grinding and pumping at various depths although some modifications may be required to satisfy certain repair material characteristics.
- Formwork and working platforms needed during the repair operation should be premade in order to minimize the need for underwater work. The equipment should

preferably not be dependent on materials and support systems on floating units due to unfavourable effects of wave action which will limit the available operaion period.

- Today purpose made equipment (shuttering, working platform, habitat, etc.) will be required in each case due to the uniqueness of the possible damages, geometry at the site and the extent of the damage.
- Special arrangements such as cofferdams, limpets and other protection systems will be needed in the splash zone due to heavy seas and strong currents.

Site Inspection and Testing

As described in the examples referred to the assessment of the quality of repair may result in rather high cost if normal inspection procedures are enforced. Thus it is essential to develop inspection procedures where the sum of the cost of site inspection and the cost of repair and subsequent surveillance is kept at a reasonable level and still giving the required quality of the repair.

Procedure testing and training of the repair crew under realistic conditions at one of the deep water testing facilities, such as here in Bergen, is in many cases to be recommended in order to minimize the possibility of rejection of the finalized repair work. Site inspection will in most cases be possible to perform with the same or similar methods as above water, but it is evident that remote viewing and recording has to be taken into account in quite many cases for surveillance and control. Also the production of redundant test pieces for retrieval and later laboratory testing may be used for checking the quality obtained.

Choice of Repair Method

The choice of the most appropriate repair method is involving economical and technical considerations and has to be combined with experience and a sound judgement to find an optimal solution.

Habitat welding and clamping are methods already in use and especially habitat welding will be used when the conditions are suitable. Development in underwater wet welding methods and equipment may make them also available for more qualified repairs int the future.

The application of grouted repair to steel structures will most certainly see a development in the years to come. This method has a possibility of relieving the most stressed parts of a joint and also of securing the continuity in a joint where a total failure is expected due to punching shear or fatigue. It also has the possibilities of stiffening dented or otherwise severed members and make them fit for service even after a damage. Other applications using grout or concrete may also be expected for steel structures.

Temporary or permanent prestressing may also be actual for both concrete and steel structures and thus providing means of relieving or reinforcing members. VERITAS' is involved in research and development activities directly related to damage detection and repair of underwater installations and will through its organization supported by laboratories and personnel both here in Bergen and at the head office in Oslo be able to take active part in this development to the benefit of safety at sea.

REFERENCES

Det norske Veritas (1977). Rules for the Design, Construction and Inspection of Offshore Structures 1977, Appendix i, inservice inspection.

Førli, O. (1979). Methods for inspection and monitoring. Veritas conference 27 - 28 November 1979, Høvik, Norway.

Humphrys, B.G. (1979). Underwater repair of concrete. Offshore Europe 79 Conference, OE-79. SPE 8168. 1-4.

Inglis, M.R. and North, T.H. (1979). Underwater welding: a realistic assessment. Welding and Metal Fabrication, April 1979, pp. 166-178.

Marshall, P.W. (1977). Desing strategies for monitoring, inspection and repair of fixed offshore platforms. Paper for presentation to the ASCE Committee on Reliability of Offshore Structures, San Francisco, October 22, 1977.

Sletten, R., Fjeld, S., and Røland, B. (1977). In-service Inspection of North Sea Structures. Offshore Technolgy Conference, OTC. 2980.

Sletten, R., Kristoffersen, K., and Eliassen, S. (1979). In-service Inspection, Cathodic Protection and Repair of Offshore Structures and Pipelines. International Symposium on Offshore Structures, RILEM-FIP-CEB October 8 - 12, 1979, Brazil Offshore '79, Vol. III-13.

Straube, P. and Carlin, B.S. (1979). Platforms Defects and Quality of Repairs. Progress report no. 1. Review of damages to concrete structures. VERITAS Report No. 79-0711.

APPLICATION OF ADVANCED STUDIES IN SAFETY ANALYSIS TO OFFSHORE PRODUCTION SYSTEMS: APPLICATION TO A RISER AND TO A MANIFOLD

A. Leroy

Total CFP (TEP/DP/MER), 204, Rond-Point du Pont de Sèvres, 92516 Boulogne-Billancourt Cedex, France

ABSTRACT

As industrial units grew increasingly complicated, the need of systems analyses became evident. In particular, advanced studies on the safety of systems were worked out. These safety analyses were initially used in industries, such as aeronautics or the nuclear industry where hazards have to be identified and, accordingly, minimized. Compagnie Française des Pétroles is now utilizing them to assess the safety of deep sea systems.

The main safety methods are briefly reviewed : Preliminary Risk Analysis, Failure Modes and Effects Analysis, Fault Tree Analysis. Commentaries are also included. The methods used by TOTAL-CFP are described.

An application to a production and to a manifold is given. Quantitative analysis was performed according to values gathered from ship's and platform's logs. Main results are discussed.

KEYWORD

Safety - Deep Sea - Manifold - Riser - Fault Tree Analysis

INTRODUCTION

As soon as safety levels demanded of a system reach high values, the engineer's know-how and the usual codes or standards are no longer sufficient. A complete safety analysis of the system has to be carried out. The aim is to identify every event or set of events which can lead to an accident. Then the probabilities of this accident must be computed in order to know if the hazard is acceptable or not. The safety program is just as much as the reliability program or quality control, a part of the project.

Safety analysis must be carried out for each system but, of course, is more profitable when applied to sophisticated or new ones. This is the reason why it was first developed in the aircraft and nuclear industries.

The latest deep seadrilling (1000 m water depth) performed by TOTAL-CFP has shown that such drilling is no longer a problem. The next step is to produce oil and gas at such a depth. A deep sea production program has been under study by TOTAL CFP - SNEA (P) - IFP research team since 1975.

Safety analysis was carried out for each main part of the deep sea production unit. By using such analysis, the project group wanted to check the feasibility of the concept, to lower, as much as possible, the program cost and, of course, to increase the security of the production system.

GENERAL PROCEDURES

A safety analysis follows the steps described below :

(1) Analysis of the system and its functions.

(2) Determination of accident risks.

(3) Development of possible accident scenarios.

(4) Assessment of the system's safety level.

(5) Comparison of this level to that set by specifications.

(6) Proposal of remedial measures to raise the safety level, if the comparison is negative.

MAIN SAFETY METHODS

Preliminary Risk Analysis (PRA)

This is a methodical work of classification and description designed to identify dangerous parts of the system and dangerous events.

The P.R.A. has to be carried out first, before the F.M.E.A. and the F.T.A. Its implementation is summarized in figure n° 1.

Failure Modes and Effects Analysis (FMEA)

The aim is the same for the F.M.E.A. and for the F.T.A. The difference consists in their respective scientific methods of approach : the F.M.E.A. is a deductive method and the F.T.A. an inductive one. Its implementation is shown in figure n° 2.

Fault Tree Analysis (FTA)

The first step is to draw up a list of the unwanted events. These are chosen in function of the seriousness of their consequences for the system (men, equipment or economic losses).

Starting from an unwanted event, the fault tree is drawn until elementary events are obtained (see figure n° 3). The symbols used for this graphic representation are shown in figure n° 4.

QUANTIFICATION

ELementary events can be expressed in the form of Boolean symbols as events can only occur or not. So, if their probabilities are known, the probability of an unwanted event can be computed by using the rules of Boolean algebra. These rules are summarized in figure n° 5.

The main problem is to obtain probabilities of occurrence of the elementary events. These can be obtained from investigation on similar equipment, from laboratory tests or can be estimated. An important word in Security or (reliability) terminology is "failure". A failure is the end of the capability of the system to do what it has been built for. The failure rate is the sum of failure per hour.

METHODS USED BY TOTAL-CFP

The riser and the manifold were analyzed by using the P.R.A. and the F.T.A. The F.T.A. is ideal for new systems such as deep sea production units, because only combinations of breakdowns leading to grave consequences for the system are considered. In addition, plotting in the fault tree enables the specialists in the different components of the system, to engage in effective dialogue.

SAFETY ANALYSIS OF THE RISER

Description

The riser is described in figure n° 6. It is designed to :

- carry up the oil and gas from the manifold to the floating production unit. The operation is performed through satellite lines ;
- carry down the processed crude through the central pipe. Crude is then dispatched through a pipeline to a floating loading unit.

Qualitative analysis

Some of the unwanted events studied are :

- maintenance impossible ;
- pollution ;
- central pipe fall ;

Figure n° 7 shows a part of the fault tree of the unwanted event "maintenance impossible". Dangerous elementary events can be easily seen by following paths including only OR gates.

Quantitative analysis

Failures rates λ of elementary events were obtained from investigation of the logbooks of TOTAL-CFP drilling vessels and platforms. Some values of λ, in failure per hour, are given herebelow :

- anchor rupture

$$1.4 \times 10^{-5} < \lambda < 4 \times 10^{-5}$$

- flange leakage :

$$10^{-5} < \lambda < 3.10^{-5}$$

- winch breakdown :

$$5.10^{-5} < \lambda < 2.10^{-4}$$

- human error in execution of procedure

$$3.5 \times 10^{-5} < \lambda < 6.10^{-5}$$

The computation of probabilities of occurrence of the unwanted events was carried out. For example, the probability (in an hour) of "riser connection impossible" obtained is : P = 0.0017.

By using these values, specifications were drawn up for the manufacture of the riser components.

SAFETY ANALYSIS OF THE MANIFOLD

Description

The manifold is described in figure n° 7. Its main functions are :

- to carry the crude coming from wellhead to satellite lines ;
- to carry back the processed crude ;
- to allow circulation of T.F.L. tools ;
- to allow wellhead maintenance without production stoppage ;
- blowing off and deconnection of the riser ;

Its components are :

- one permanent manifold ;
- one removable manifold ;
- removable production units ;
- one removable loading unit.

Qualitative analysis

Some of the unwanted events studied were :

- permanent manifold setting impossible ;
- manifold removal impossible ;
- pollution ;
- T.F.L. maintenance impossible.

As no similar equipment exists, it has not been possible to do a quantitative analysis.

CONCLUSION

Deep sea production system project

Safety analyses carried out on the deep sea production system have confirmed its feasibility. Numerous modifications were suggested and they were included in the final report. In addition to that, it will be possible to provide the prime contractor, with a complete set of specifications.

Deep sea production system development

Safety (and Reliability) analysis will be well used if the maintenance program is drawn up accordingly. Safety will be increased and development cost lowered.

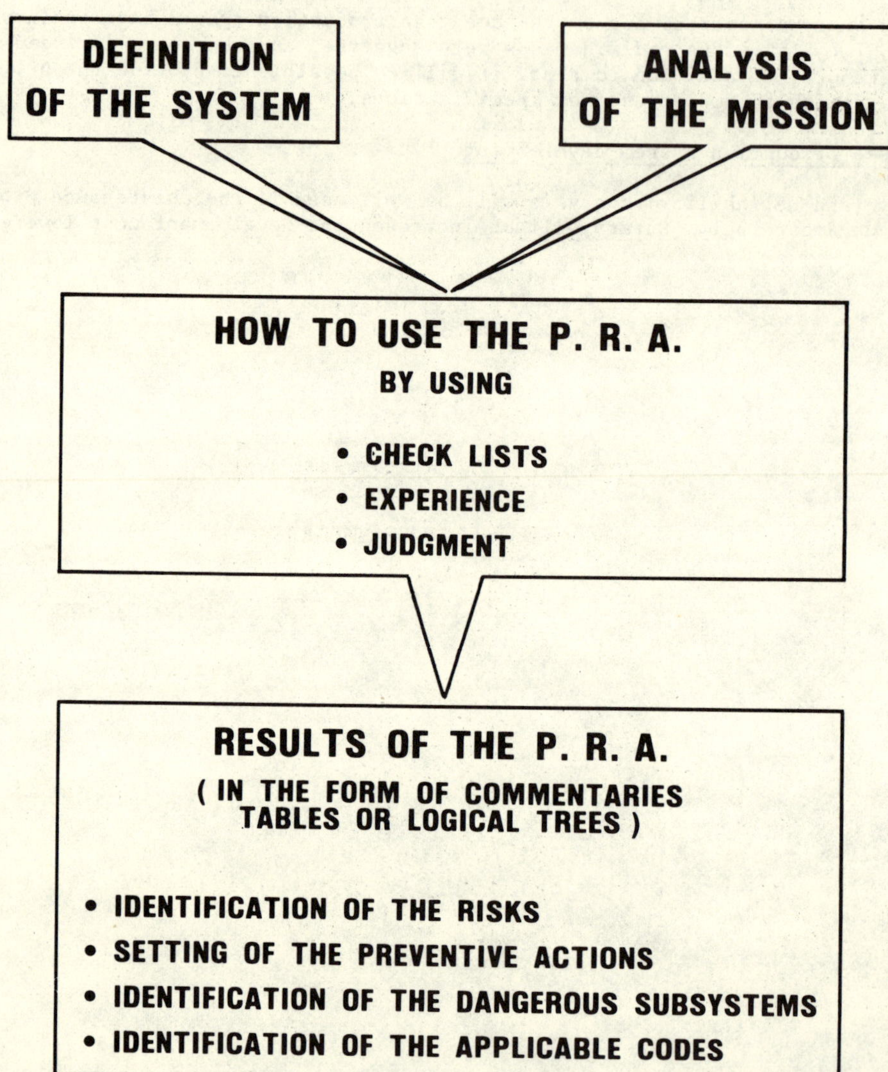
PRELIMINARY RISK ANALYSIS (P. R. A)
DEFINITION OF THE SYSTEM
ANALYSIS OF THE MISSION
HOW TO USE THE P. R. A.
BY USING
• CHECK LISTS
• EXPERIENCE
• JUDGMENT
RESULTS OF THE P. R. A.
(IN THE FORM OF COMMENTARIES TABLES OR LOGICAL TREES)
• IDENTIFICATION OF THE RISKS
• SETTING OF THE PREVENTIVE ACTIONS
• IDENTIFICATION OF THE DANGEROUS SUBSYSTEMS
• IDENTIFICATION OF THE APPLICABLE CODES

FIGURE N° 1

FAILURE MODES AND EFFECTS ANALYSIS

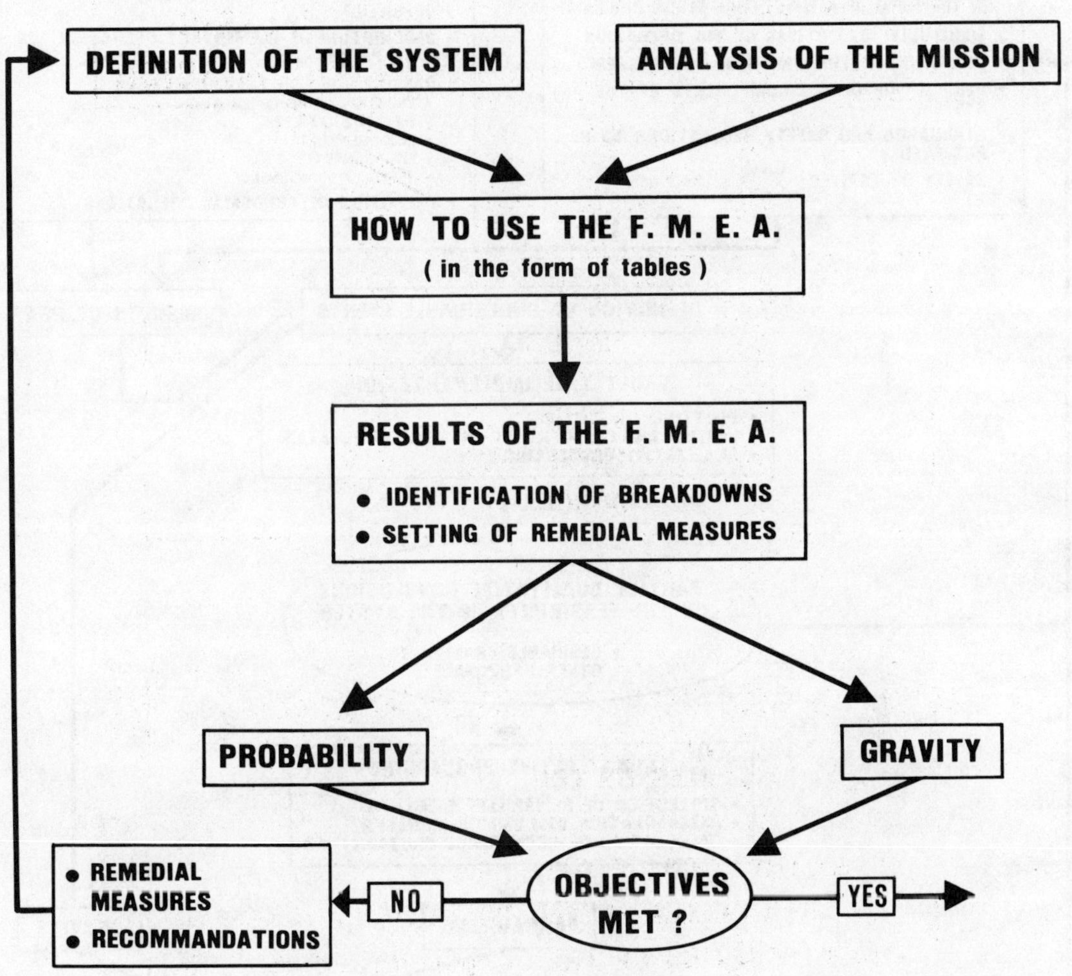

FIGURE Nº2

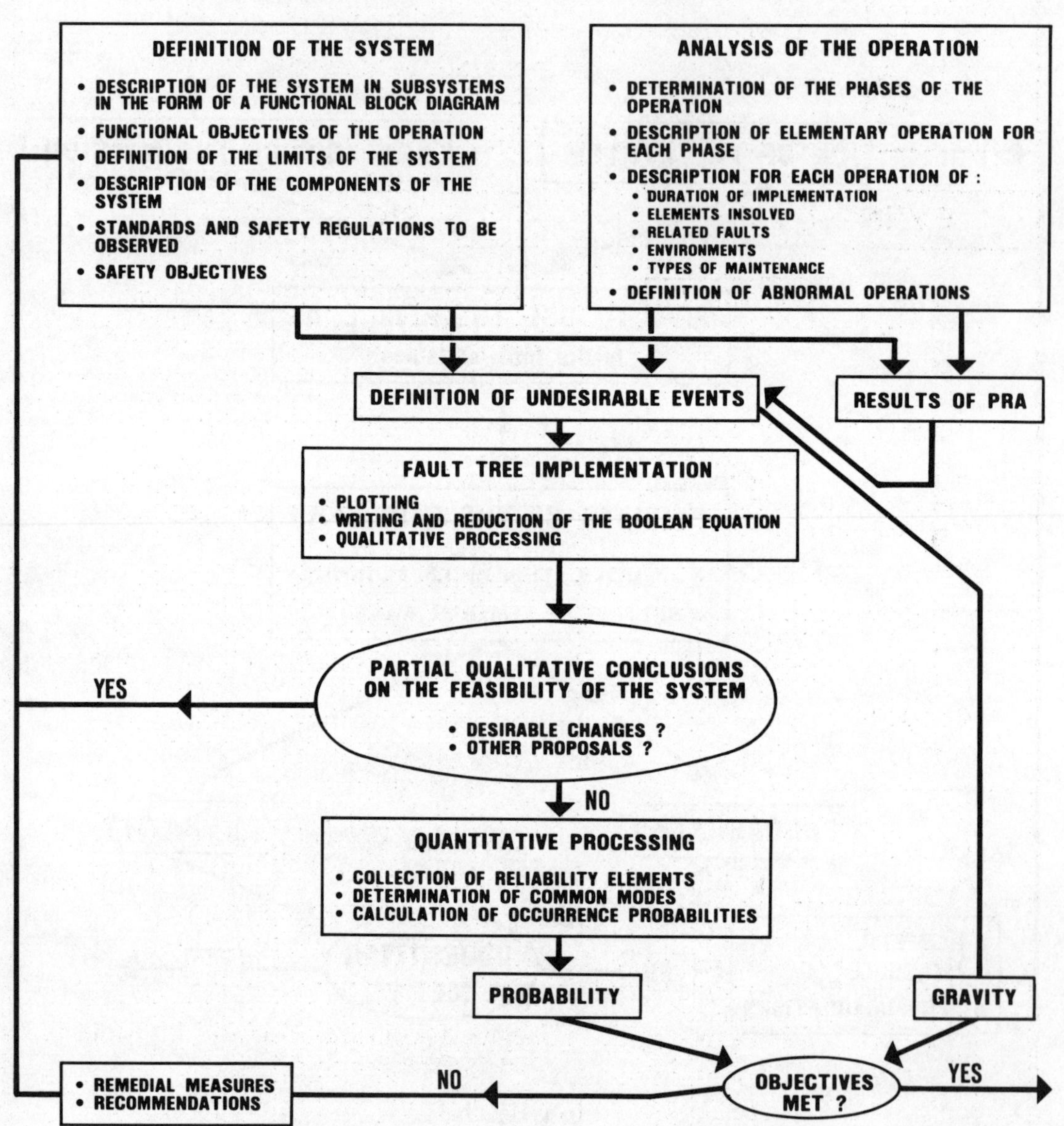

FAULT TREE ANALYSIS
DEFINITION OF THE SYSTEM
• DESCRIPTION OF THE SYSTEM IN SUBSYSTEMS IN THE FORM OF A FUNCTIONAL BLOCK DIAGRAM
• FUNCTIONAL OBJECTIVES OF THE OPERATION
• DEFINITION OF THE LIMITS OF THE SYSTEM
• DESCRIPTION OF THE COMPONENTS OF THE SYSTEM
• STANDARDS AND SAFETY REGULATIONS TO BE OBSERVED
• SAFETY OBJECTIVES
ANALYSIS OF THE OPERATION
• DETERMINATION OF THE PHASES OF THE OPERATION
• DESCRIPTION OF ELEMENTARY OPERATION FOR EACH PHASE
• DESCRIPTION FOR EACH OPERATION OF :
• DURATION OF IMPLEMENTATION
• ELEMENTS INSOLVED
• RELATED FAULTS
• ENVIRONMENTS
• TYPES OF MAINTENANCE
• DEFINITION OF ABNORMAL OPERATIONS
DEFINITION OF UNDESIRABLE EVENTS
RESULTS OF PRA
FAULT TREE IMPLEMENTATION
• PLOTTING
• WRITING AND REDUCTION OF THE BOOLEAN EQUATION
• QUALITATIVE PROCESSING
YES
PARTIAL QUALITATIVE CONCLUSIONS ON THE FEASIBILITY OF THE SYSTEM
• DESIRABLE CHANGES ?
• OTHER PROPOSALS ?
NO
QUANTITATIVE PROCESSING
• COLLECTION OF RELIABILITY ELEMENTS
• DETERMINATION OF COMMON MODES
• CALCULATION OF OCCURRENCE PROBABILITIES
PROBABILITY
GRAVITY
• REMEDIAL MEASURES
• RECOMMENDATIONS
NO
OBJECTIVES MET ?
YES

FIGURE N°3

LOGIC GATE SYMBOL	SIGNIFICATION
 OUTPUTS / INPUTS	AND
 OUTPUTS / INPUTS	OR

EVENT SYMBOL	SIGNIFICATION
 RECTANGLE	EVENT RESULTING FROM THE COMBINATION OF MORE ELEMENTARY EVENTS ACTING THROUGH LOGIC GATES .
 CIRCLE	PRIMARY EVENT
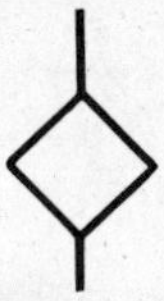 DIAMOND	EVENT DEVELOPED IN ANOTHER FAULT TREE WHICH CAN BE ASSUMED TO BE PRIMARY .
 TRIANGLE	SERVES TO TRANSFER THE PORTION OF THE TREE FOLLOWING THE SYMBOL :  TO THE LOCATION INDICATED BY THE SYMBOL : 
 HOUSE	EVENT RESULTING FROM NORMAL USE OF THE SYSTEM .

LOGIC SYMBOLS

FIGURE N°4

BOOLE ALGEBRA'S BASIC RULES

$$A \cdot A = 0$$

$$A \cdot 0 = 0$$

$$A \cdot 1 = A$$

$$A + A = A$$

$$A + 0 = A$$

$$A + 1 = 1$$

FIGURE N° 5

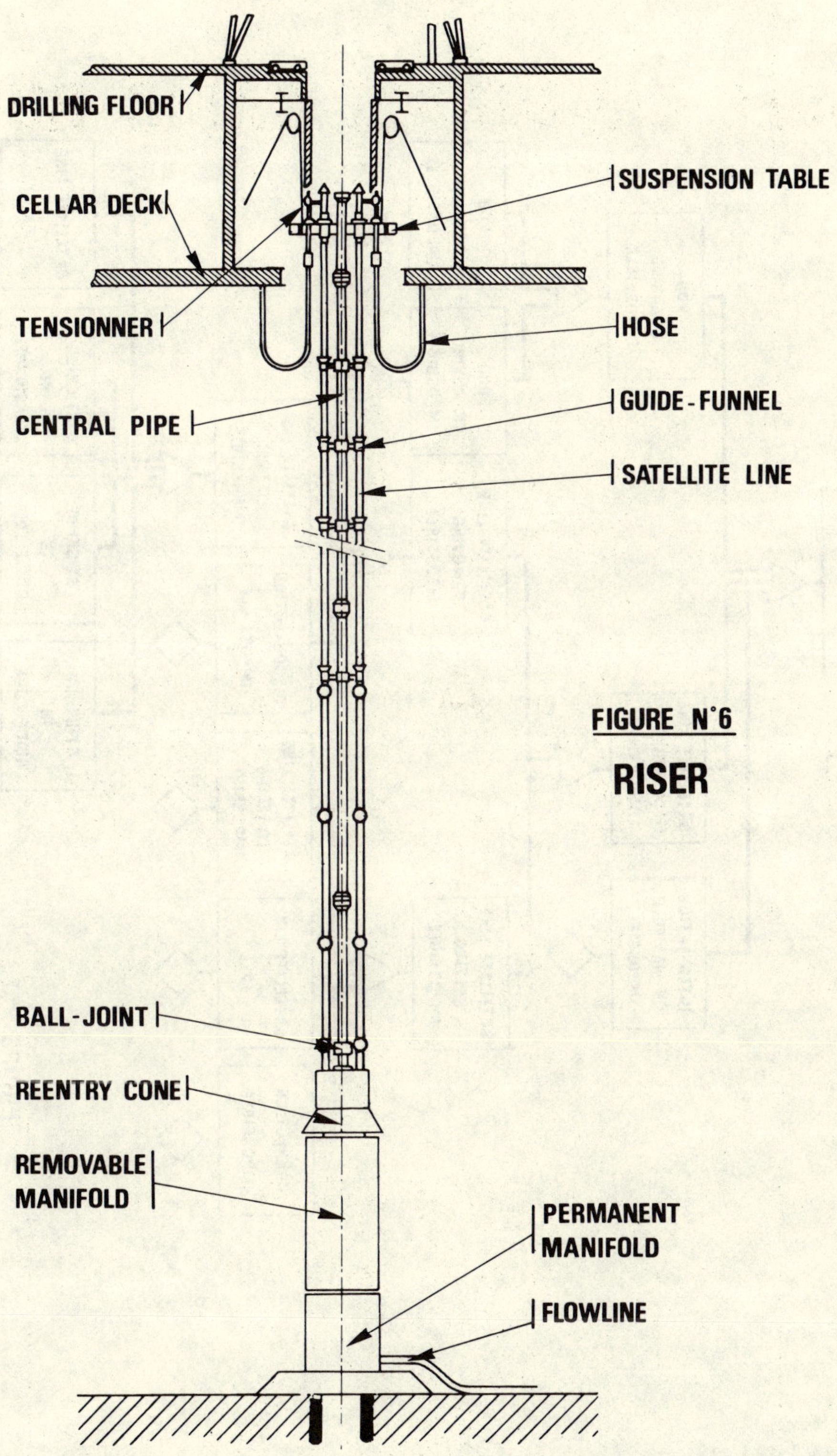

FIGURE N°6

RISER

MAINTENANCE IMPOSSIBLE
INTERVENTION ON MANIFOLD IMPOSSIBLE
SATELLITE LINE CHANGING IMPOSSIBLE
POD CHANGING IMPOSSIBLE
SATELLITE LINE SETTING IMPOSSIBLE
SATELLITE LINE REMOVING IMPOSSIBLE
POD REMOVING IMPOSSIBLE
POD SETTING IMPOSSIBLE
CONNECTION IMPOSSIBLE
SATELLITE LINE FALL
SATELLITE LINE LOWERING IMPOSSIBLE
DISCONNECTION IMPOSSIBLE
SATELLITE LINE LIFTING IMPOSSIBLE
STICKING IN GUIDE CONE
STICKING IN GUIDE TUBE
STICKING IN TENSIONNER
SATELLITE LINE FALL

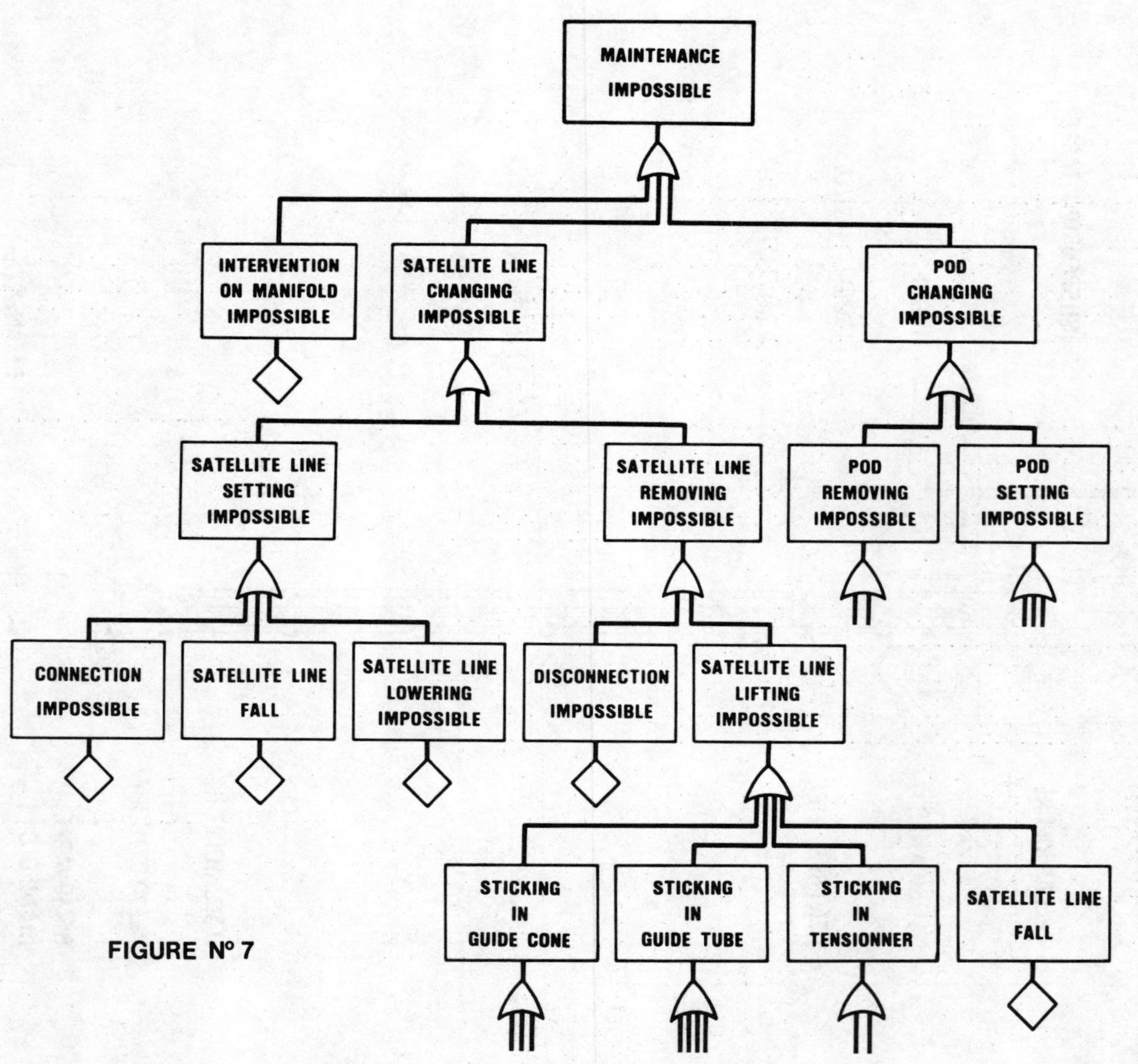

FIGURE N° 7

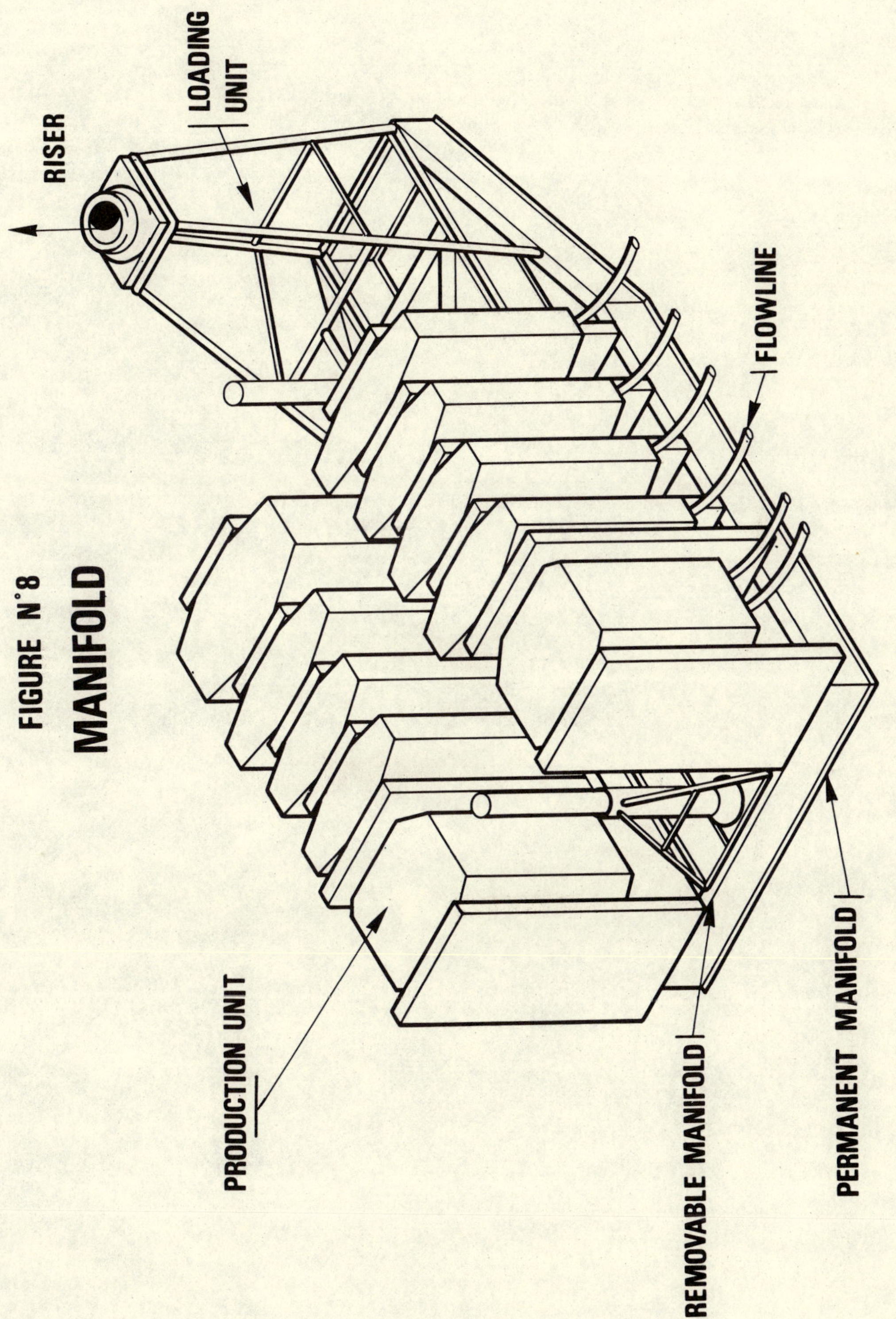

FIGURE N°8
MANIFOLD

SHOCK HAZARD IN THE HYPERBARIC ENVIRONMENT

A. Diesen

Norwegian Underwater Institute (NUI), Norway

ABSTRACT

This paper treats qualitatively the three main shock situations in the hyperbaric environment, where special attention is paid to electricity in water. Risk-distance between diver and electrical equipment is calculated for different equipment. The physiological/medical characteristics of electric shock are explained by means of diagrams and curves which show the effect of the shock current's duration and frequency.

KEYWORDS

Electric shock; electric fields in water; risk-distance in water; physiological shock effects; welding/cutting in water; cathodic protection with impressed current; electric cables in water.

INTRODUCTION

The contents of this paper are based upon results of the Norwegian research project, Electricity in water safety of divers, which was started in '77 and finished in '79. The purpose of this project was to investigate the effect of electric shock on persons in the hyperbaric environment and determine all the parameters involved in the shock circuit, including the diver. The aim was to establish criteria for safer use of electrical equipment, as well as for equipment having live parts in water (e.g. welding/cutting equipment, cathodic protection system with impressed current, etc.), as for enclosed equipment when insulation failure (earth fault, short circuit) occurs. In both cases, an electrical field is established in water. Because of the complicated situations, the electric field is, in addition to analytical calculations, determined by means of computer-calculations combined with dummy measurements. (For simplicity, the dummy measurements were carried out on a 1:6 scaled diver model, because the only intention was to investigate the dummy's relative influence on the electric field).

When choosing criteria for use of electricity in the hyperbaric environment where man is present, the human factor has to be a dimensioning contributor to the system. This must be in such a way, however, that we get an optimized system with regard to safety and technical solution. At present, many rules for use of electricity are both inflexible and unsuitable. An example is the rule of maximum voltage allowed inside diving bells. It reads as follows: All electrical equipment; maximum 30 Volt. Not mentioning if this is AC (Alternating Current) or DC (Direct Current). Designed in the wrong way, even 30 Volt may be dangerous, but in the right way, much higher voltages can be used. To-day we have extensive knowledge of the effect of electric shock on the human body. This knowledge, however, applies largely to atmospheric pressure on land, and therefore we have to transfer this knowledge to the hyperbaric environment. So, before going any further, a repetition of the "on-land" knowledge will be made, but without going too deeply into the subject.

First: The direct causes of death from electric shock can be:

A) Ventricular fibrillation
B) Deep burns
C) Respiratory paralysis

At present, we do not know of any fatal accidents in the North Sea, which were caused directly by electric shock (and we hope this will continue).

Second: The physiological sensitivity of electric shock is shown in Diagram 1, and it is especially important to regard the difference between 50/60 cycles AC and smooth DC. (The frequency dependency is also shown in Fig. 1).

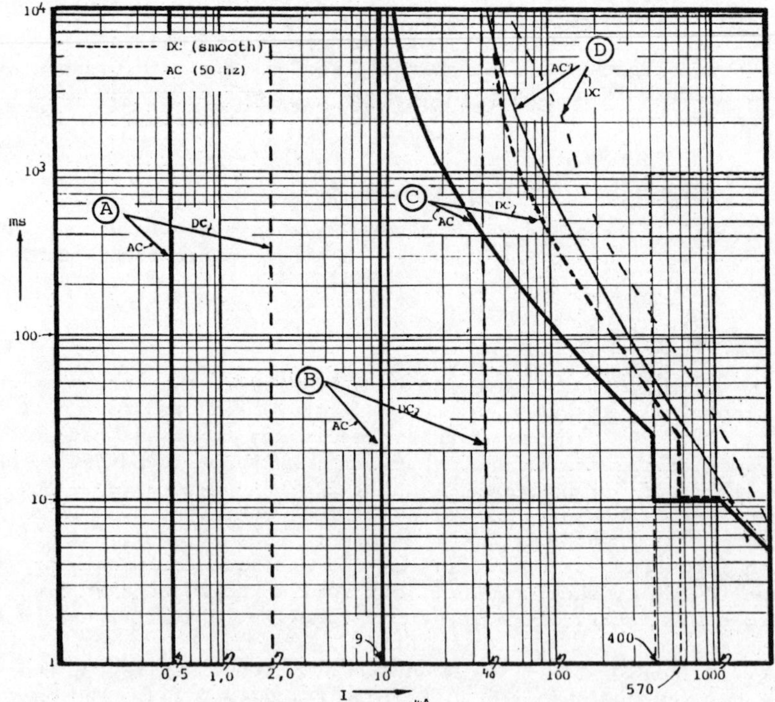

Diagram 1. Physiological sensitivity diagram of electric shock, showing: (A) — the perception level, (B) —let-go level, (C) — IEC-curve and Jacobsen's criterion, (D) — fibrillation level.

The three important threshold levels shown in Diagram 1, derived statistically, are:

1. The current magnitude at which a person begins to feel "electricity in his body" is called

Threshold of perception (curve Ⓐ)

2. The maximum at which a person is still capable of releasing a conductor by using muscles directly stimulated by that current is called:

Threshold of let-go (curve Ⓑ)

Let-go current is also shown in Fig. 1 as function of frequency, and it is important to note that the power frequencies (50 or 60 hz) are the most dangerous. When in - creasing the frequency, the let-go level will also increase, and a current

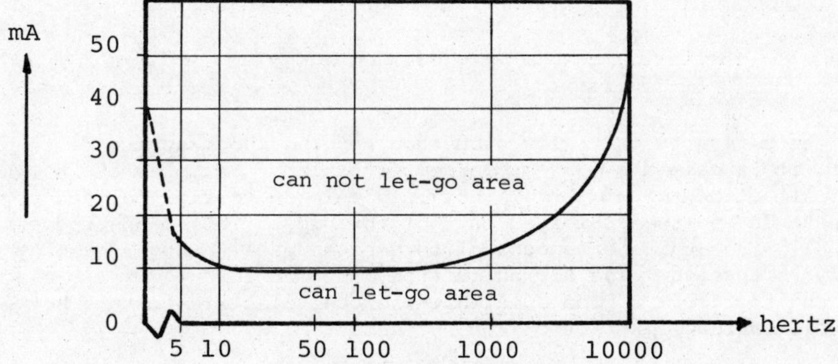

Fig. 1. Let-go current as function of frequency

with frequency equal to 7 khz has the same let-go current as smooth DC. (This means that if we increase the frequency in the system, we may tolerate higher amounts of shock currents to pass the body. Greater quantities of electrical currents in high frequency cases could cause heating and superficial burns of internal tissues). Supply based upon high-frequency distribution is often called shock-free supply.

3. The current magnitude at which the heart's pumping action stops, rhythmic contraction ceases and pulse disappears is called:

Threshold of ventricular fibrillation (curve Ⓓ)

and is one of the causes of fatal accidents.

To avoid a dangerous shock, the current magnitude between extremities must not exceed the time dependent IEC-security curve combined with Jacobsen's criterion as shown in curve C.

(Jacobsen's criterion is valid for shock duration in the interval between 10 ms - 1 sec and is the maximum safe current that will not induce ventricular fibrillation at whatever instant of cardiac cycle it occurs.)

In the hyperbaric environment, it is practical to classify the shock hazard situations into three main categories described below:

1. If both the person and the shock source are in a gaseous, humid environment, it is called a "dry" situation. This can be compared to the worst situation we may find on land; that is, in small humid rooms where it is easy to touch grounded metal. As on land, the current pathway will be between extremities in contact with the shock source and ground. We find a characteristic "dry" situation inside a welding habitat where the water level is below the habitat floor as shown in Fig. 2.

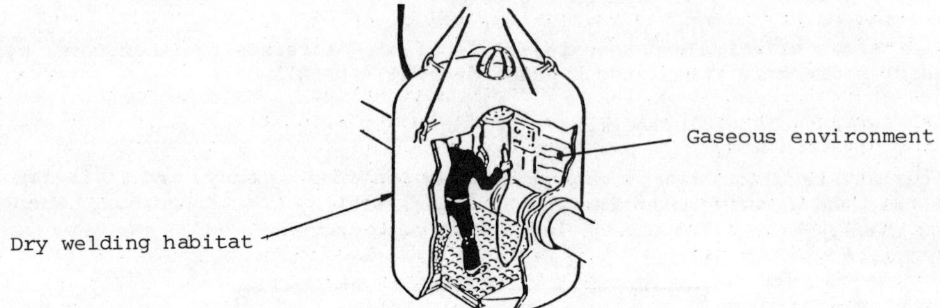

Fig. 2. Example of a dry situation. Dry, welding habitat

2. If the person is partially submerged and the shock source is in a gaseous environment, it is called a semi-submerged situation. Examples are shown in Fig. 3. In this situation, the current is not only allowed to flow between the hands but also between hands and the submerged part of the body. This means reduced body resistance, allowing a greater amount of current to pass through the body compared with the "dry" situation. The diving suit is included in the diver's electrical resistance, and therefore a good suit which reduces electric contact between body and water will increase protection against shock.

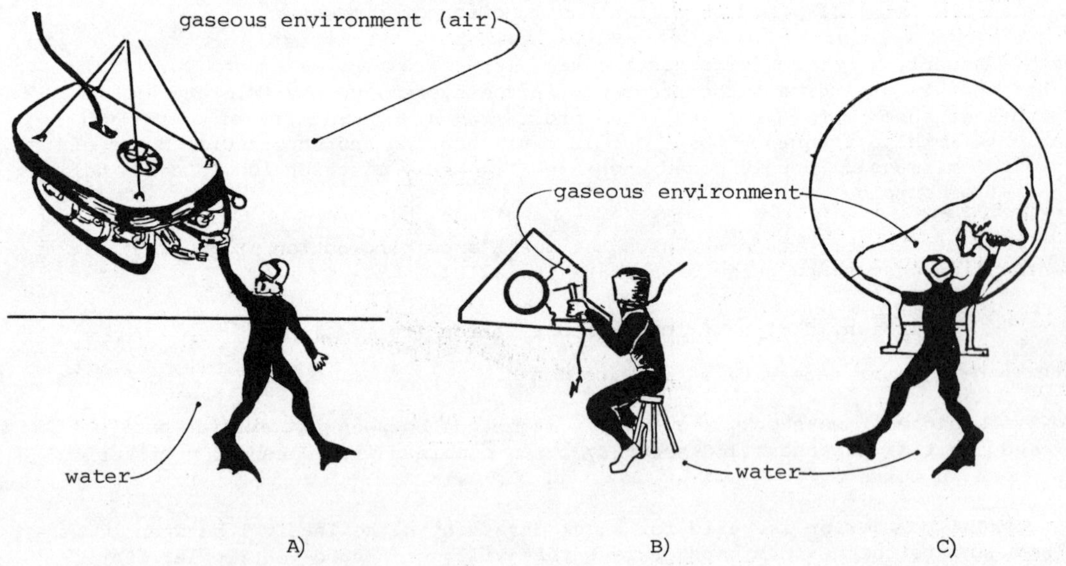

Fig. 3. Examples of semi-submerged situations:
A) handling of equipment above the sea level, B) welding in a hydrobox, C) touching equipment inside diving bells and DLO.

3. If both the diver and the equipment are submerged, it is called a submerged situation. A common theory is that the diver is swimming in a perfect conducting medium which will shunt the diver, making this situation even safer than on land. The situation is, however, more complicated; and the above conclusion is not generally accepted, even if the principle of the shunting effect is. This is because many parameters are involved, making this situation more complex than the other two described above. A closer study of this situation was therefore necessary. A potential shock hazard situation exists in water if an electric current field and potential field is established. The effect of this is that the diver is always touching the source via the conducting water. The electric potential field is shown in Fig. 4 as equipotential surfaces (with dotted lines); and depending on how many equipotential surfaces the diver is "breaking through", a voltage drop is established across the diver's body. This voltage drop will drive a current through the diver's body. The aim is now to find a method to calculate this current and then judge the medical and the physiological effects of it as functions of its frequency, duration and magnitude. On land, we are able to find the real current, because it usually follows a well defined pathway through the body. In water, however, the current may follow any route, making it very difficult to find the real current. One method of calculating the current is to introduce an electrical equivalent scheme of the diver, realized by an electrical minimum resistance R_t. This R_t is not only dependent on the diver himself, but also includes his clothing, i.e., his diving suit and mask.

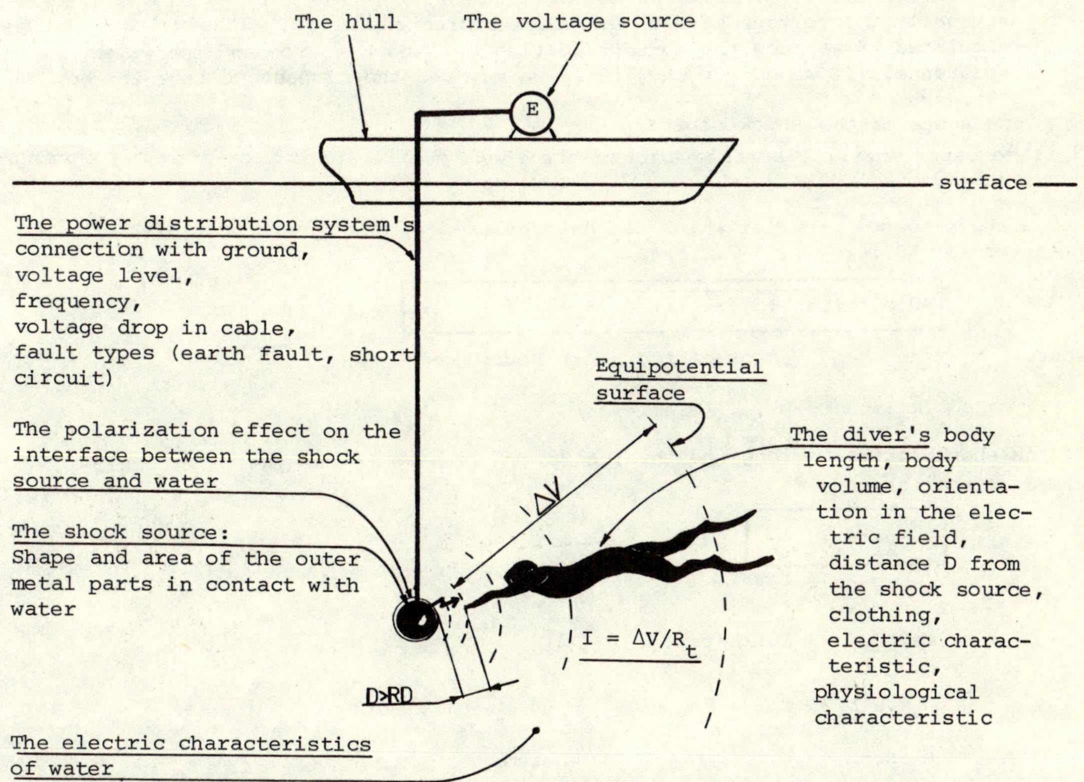

Fig. 4. Parameters involved in the shock circuit in water.
I = maximum current through the diver's body
ΔV = the voltage drop across the diver's body
R_t = minimum total resistance of the diver (clothing included)

Measurements have shown that a well clothed diver can be 4 or 5 times better insulated than a bad clothed diver. We therefore found it necessary to make a classification of the diver's clothing where R_t is included. The next parameter to be found is the voltage drop ΔV which is a solution of the electrical field. This voltage drop is dependent on many factors; where the most important are: A) the diver's length, body volume, orientation in the electric field and his distance from the shock source. B) the shock source's shape and area of the outer metal parts in contact with water. C) the distribution system's voltage level, frequency, cable length, fault types (earth fault, short circuit) and connection with ground. D) the polarization effects at the water/metal interface and the water's electric characteristics. The shock current is now equal to the ratio between the voltage drop ΔV and the minimum resistance R_t. (The basis for judging the physiological effects of this current is shown in Diagram 1 and Fig. 1.) Because we are able to find the current through the diver for a certain distance from the shock source, we are also able to find what we have called the risk-distance (RD). That is, if the diver keeps a distance from the shock source greater than the risk-distance, he is safe from getting a dangerous shock, because the current's magnitude and duration are below a tolerable maximum. RD is a function of:

1) The physiological characteristics of the human body
2) The electric characteristics of the human body
3) The diver's length and direction relative to the shock source
4) The diver's clothing
5) The voltage level. This means the voltage "across" the water volume and is usually the voltage between the shock source and earth. This voltage can be calculated if we know the current leaving the shock source and the water resistance. If we do not know this, we may use the highest voltage in the power system.
6) The shape of the shock source.
7) The outer metallic surface area of the shock source in contact with the surrounding water.

For a shock source in water which can be equalized with a sphere or a disk, the expression for RD is:

$$RD(m)=\tfrac{1}{2}(\sqrt{5.3+577\cdot k_s\cdot k\sqrt{F}V_o/R_t}-2.3)-0{,}28\sqrt{F}$$

where: k_s = the shape factor of the shock source

(k_s(sphere)=1, k_s(disk)=1.5)

If the shock source is long and thin, however, we have to use the more complicated expression;

$$e^m=n_1\cdot n_2/n_3\cdot n_4$$

where:

$$m=0.009\cdot R_t\cdot \ln(2L/t)/k\cdot V_o$$

$$n_1=\sqrt{\beta}+L/2 \qquad n_2=\sqrt{\beta+4.6\alpha+5.3}-L/2$$

$$n_3=n_1-L \qquad n_4=n_2+L$$

$$\beta=(L/2)^2+\alpha^2$$

L= the length of the shock source (m)
t= the thickness of the shock source (m)

If a short circuit between two (or more) terminals of a cable (or a connector) occurs where the water forms the short circuit path, a so-called electric dipole field is established in water. The expression for RD for such a field is:

$$\boxed{(2\alpha+2.3)/(\alpha(\alpha+2.3))^2=\beta}$$

where: $\beta=8000R_t/V_O\cdot d\cdot A\cdot k\cdot p$

d=diameter of terminals (mm)

A=center-center distance between two terminals (mm)

p=1 for a single phase system

p=1.5 for a three phase system

and the risk-distance:

$RD(m)=\alpha-D/2000$

D is the cable (or connector) diameter (mm)

For all the expressions above, apply:

V_O = the voltage between the shock source and ground or between phases. If V_O is to be calculated, the design of the power system has to be known when the water resistance is also included. Instead of the "real" V_o, one may use the highest voltage that can occur between the shock source and ground or between phases.

R_t = the equivalent resistance (ohm) of the diver when the clothing is also included.

Table 1 Physiological Shock Threshold Factor k of the Human Body

	AC(50/60 hz)	DC (<5% ripple content)
Perception level	k=18	k=4.5
let-go level	k=1	k=0.225
IEC-security level	$k=(1+1100/t)^{-1}$ 25 ms<t<10 s	$k=((1+1100/t)\log t)^{-1}$ 25 ms<t<10 s

Example 1 Cathodic Protection System with impressed Current using a Disc Electrode.

In this example, the shock source is one of the anodes in a cathodic protection system with impressed current. This impressed current is always rectified AC, supplied from the main power system via a transformer and a rectifier. (In this connection, one should be aware of the fact that an insulation breakdown may, in the worst case, lead the AC voltage into the water.)

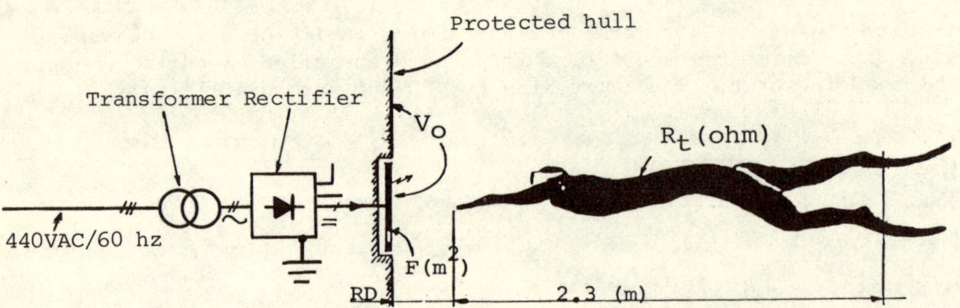

Fig. 5. Risk-distance between diver and a cathodic protection system with impressed current (disk-anode).

The diver is clothed in an open hot-water suit system and a well-insulated helmet. The resistance of the diver (including clothing) is R_t=230 ohm.

The diameter of the electrode is 0.72 (m).
The outer surface area of the electrode is:

$$F = \pi \cdot (D/2)^2 = \pi \cdot (0.72/2)^2 \simeq 0.4 \ (m^2)$$

In the expression for risk distance, we use the total surface area = 0.8 (m^2) = both sides.
The current rating of this electrode is I_{DC}=75 (A).
The voltage "across" the water volume:

$$V_o = I_{DC} \cdot R_W = I_{DC} \cdot 0.63 \cdot \rho_W/\sqrt{F} = 75 \cdot 0.63 \cdot 0.31/\sqrt{0.4} = \underline{23} \ (V)$$

R_W= the water resistance (ohm)

ρ_W= volume resistivity of water ≈ 0.31 (ohmm)

a.

If the risk distance is now based upon the perception threshold level (k=4.5 and k_s=1.5), we have:

$$RD(m) = \tfrac{1}{2}(\sqrt{5.3 + 577 \cdot 1.5 \cdot 4.5 \cdot \sqrt{0.8} \cdot 23/230} - 2.3) - 0.28\sqrt{0.8}$$

$$RD(m) = \underline{8}$$

b.

Based upon the let-go level (k=0.225)

RD = 1 m

Example 2 Cathodic Protection System with impressed Current using a long, thin Anode Bar.

In this example we have replaced the disk anode in example 1 with a long, thin anode bar as shown in Fig. 6 on the next page.

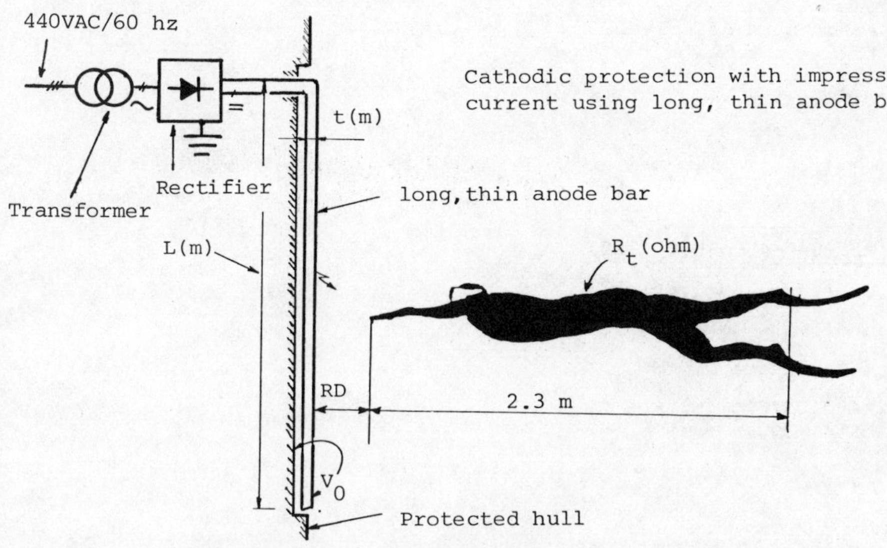

Fig. 6. Risk-distance to a long, thin anode bar.

The diver's equivalent resistance R_t=230(ohm) (see ex. 1).

The length of the bar = 2440 (mm) = L

The thickness of the bar = 30 (mm) = t

The current rating = 50 (A) = I_{DC}

The voltage "across" the water volume:

$$V_0=I_{DC}\cdot R_W=I_{dc}\cdot 0.32\cdot\rho_W\cdot \ln(2L/t)/L=50\cdot 0.32\cdot 0.31\cdot \ln(2\cdot 2.44/0.03)/2.44\approx \underline{11}\text{(Volt)}$$

R_W= the water volume resistance (ohm)

ρ_W= volume resistivity of North Sea water ≈ 0.31 ohmm.

a).

Risk-distance based upon the perception threshold level where k=4.5

$$m= 0.009\cdot R_t\cdot \ln(2L/t)/k\cdot V_0$$
$$= 0.009\cdot 230\cdot \ln(2\cdot 2.44/0.03)/4.5\cdot 11=0.2$$

$$\beta= (L/2)^2+\alpha^2=(2.44/2)^2+\alpha^2=1.5+\alpha^2$$

$$n_1=\sqrt{\beta}+L/2=\sqrt{\alpha^2+1.5}+2.44/2$$

$$n_2=\sqrt{\beta+4.6\alpha+5.3}-L/2=\sqrt{1.5+\alpha^2+4.6\alpha+5.3}-2.44/2$$
$$=\sqrt{\alpha^2+4.6\alpha+6.8}-1.22$$

$$n_3=n_1-L=\sqrt{\alpha^2+1.5}+1.22-2.44=\sqrt{\alpha^2+1.5}-1.22$$

$$n_4=n_2+L=\sqrt{\alpha^2+4.6\alpha+6.8}+1.22$$

and the expression to be solved:

$e^m = n_1 \cdot n_2 / n_3 \cdot n_4$

and

$e^{0.2} = 1.23$

$$\frac{(\sqrt{\alpha^2+1.5}+1.22)(\sqrt{\alpha^2+4.6\alpha+6.8}-1.22)}{(\sqrt{\alpha^2+1.5}-1.22)(\sqrt{\alpha^2+4.6\alpha+6.8}+1.22)} = 1.23$$

the solution of this equation is $\alpha=4$ (m)

and the risk-distance:

$RD(m)=\alpha-t/2=4-0.03/2\approx\underline{4\ (m)}$

b).

Risk-distance based upon the let-go threshold level (k=0.225)

$m=0.009\cdot 230\cdot \ln(2\cdot 2.44/0.03)/0.225\cdot 11= 4.25$

and $e^m = e^{4.25} = 70$

and the solution of the equation is:

$\alpha = 0{,}15\ (m) \mathrel{\hat{=}} 15\ (cm)$

and the risk distance

$RD = 15\ (cm) - 3\ (cm) = \underline{12}\ (cm)$

In these two examples the "real" voltage V_o across the water volume has been derived. However, one should be aware that the impressed current is fed via a rectifier transformer. Insulation breakdown between primary and secondary circuits may in the worst case, lead the main supply voltage into the cathodic protection system.

Example 3 Cable/Connector in water

A connector in operation is suddenly disconnected, and a diver finds himself close to this connector.

The electric field around the connector can be looked upon as a dipole field created by the short circuit current between the terminals. The situation is shown in the figure below.

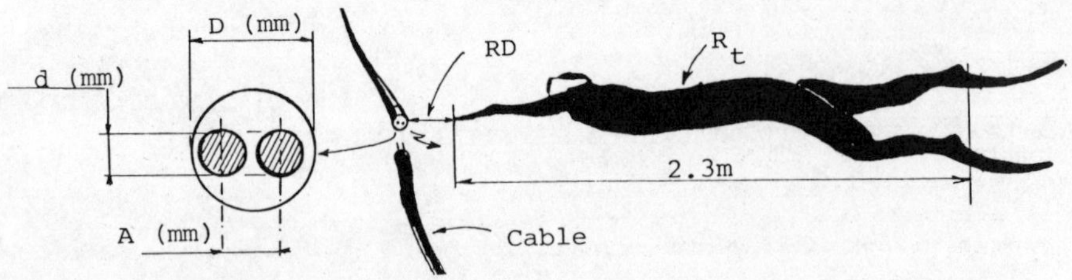

Fig. 7. Risk-distance to a connector or a cable

The diver is clothed in an open hot-water suit system and a well-insulated helmet. The resistance of the diver (including clothing) is R_t=230 (ohm).

The cross section of the conductor F=95 mm^2 and the diameter $d=2\sqrt{95/\pi}=11$ (mm) (single phase). The center-center distance between terminals A=13 mm. The cable/connector diameter is about D=30 mm. The voltage between phases is V_O=380 VAC.

a).

Risk distance based upon perception threshold level.

k=18

$\beta=8000\cdot R_t/V_0\cdot d\cdot A\cdot k=8000\cdot 230/380\cdot 11\cdot 13\cdot 18=1.9$

The equation to be solved:

$$(2\alpha+2.3)/(\alpha\cdot(\alpha+2.3))^2=1.9$$

and the solution is: $\alpha=0.47$ (m)

and the risk-distance RD(m)=α-D(mm)/2000

=0.47-30/2000=0.455m$\hat{=}$45.5cm

b).

Risk-distance based upon the let-go level (k=1)

$\beta=8000\cdot 230/380\cdot 11\cdot 13\cdot 1=34$

which gives the solution:

$\alpha=0.113$ (m)

and the risk-distance:

RD(m)=0.113-30/2000= .098(m)$\hat{=}$10(cm)

Example 4 Electric Welding/Cutting in Water.

Welding and cutting in water is usually based upon DC. From a safety point of view, this is advantagous. If the welding current, however, is rectified AC supplied from the main power system via a transformer and a rectifier, an insultation breakdown may, in the worst case lead the AC voltage to the electrode. This can be avoided if using a motor-generator set or an insulation monitoring system. In water, one should take care of a risk distance between body (hand) and open electrode in water. If the welding machine consists of a motor-generator set, the risk distance should be based upon the open-circuit-voltage (OCV) on the welding machine (no-load). The OCV is dependent on the voltage drop in the cables, which may be a problem in welding operations at great depths. The voltage drop in the cables must namely be compensated by a higher OCV, which means reduced safety. This problem can be eliminated by: A) increasing the cable dimensions, B) using sea as return conductor (sea-return) and C) installing automatic voltage reducing equipment. It is also important to keep the ripple content in the DC voltage as low as possible.

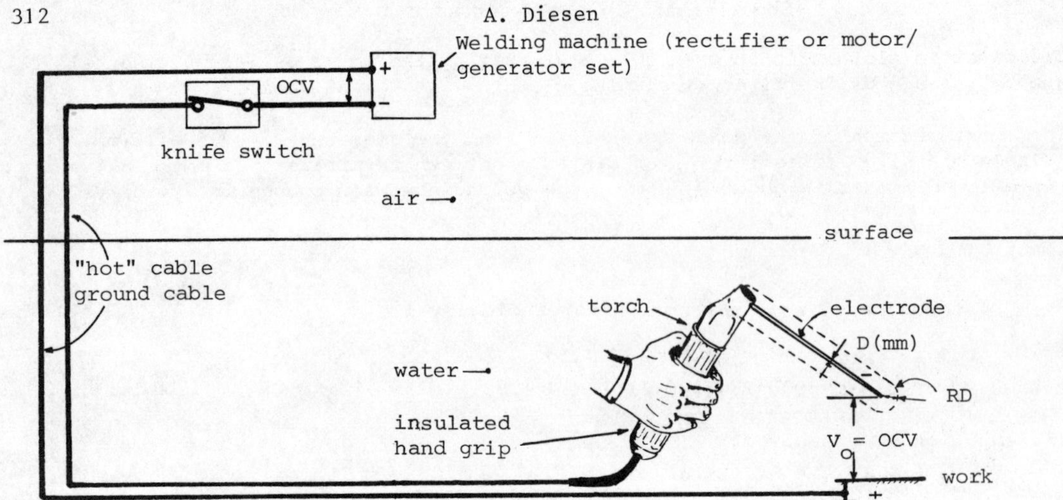

Fig. 8. Welding/cutting in water. Risk distance between the hand (or body) and the electrode. (Only the electric circuit is shown.)

The risk-distance:

$$RD(cm)=50(\sqrt{5.3+1.1\cdot k\cdot V_0\cdot D/R_t}-2.3)$$

The diver's equivalent resistance R_t=230(ohm) (see ex. 1)

V_0= open circuit voltage = 40VDC

D= Diameter of the electrode = ½"=13 (mm)

a) Risk-distance based upon perception level: K=4.5

$$RD(cm)=50(\sqrt{5.3+1.1\cdot 4.5\cdot 40\cdot 13/230}-2.3)=\underline{\underline{88cm}}$$

b) Risk-distance based upon let-go level: k=0.225

$$RD(cm)=50(\sqrt{5.3+1.1\cdot 0.225\cdot 40\cdot 13/230}-2.3)=\underline{\underline{6cm}}$$

Efforts which reduce the shock hazard are:

1) use of gloves
2) use of taped or enameled electrodes.

Example 5 Complex Electrical System.

An example of a complex electrical system is a tethered, remotely-controlled vehicle. A vehicle carries a lot of electrical equipment, the most important beeing the power cables, the hydraulic power pack, the propulsion system, the lights and the monitoring system. The supply voltage for such systems is usually high (500 VAC uptil 3500 VAC) and it is recommende that the diver keeps a risk-distance from the high-voltage cable/connector system. It is assumed that all metalic parts of the vehicle is well grounded and two defects are needed to constitute a shock risk, and where first fault is detected by the monitoring system. The worst situation occurs if two defects occur at the same time. This is realized if the power cable is suddenly disconnected from the connector, which will establish an electric (dipole) field in water. The situation is shown in Fig. 9

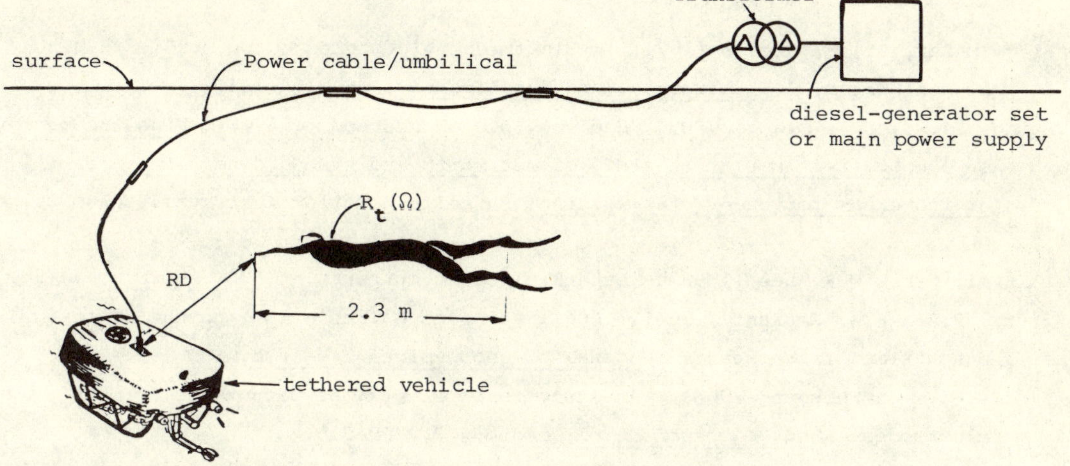

Fig. 9. Risk-distance from a complex electrical system. (Only the power system is shown in the figure.)

The resistance of the diver including clothing: R_t=230(ohm)

The voltage level V_0=3300VAC, 3-phase system

The conductor cross-section of the power cable is F=10mm^2.

Diameter of conductor $d(mm)=2\sqrt{F/\pi}=3.6(mm)$

center-center distance between phases A=5mm

Cable diameter: D=30mm

$\beta=8000 \cdot R_t/V_0 \cdot d \cdot A \cdot k \cdot p$

Risk-distance based upon perception level: k=18

$\beta=8000 \cdot 230/3300 \cdot 3.6 \cdot 5 \cdot 18 \cdot 1.5=1.14$

$$(2\alpha+2.3)/(\alpha(\alpha+2.3))^2=1.14$$

the solution to this equation is:

$$\alpha=0.605(m)$$

and the risk-distance RD(m)=0.605-30/2000(m)≙60(cm)

In conclusion, the hyperbaric environment creates worser electric shock hazard situations compared to equivalent situations on land (with the same voltage, frequency, protection system and shock source). The reason for this is mainly because of increased body contact with the surrounding, grounded environment, which will, in general, lead to reduced body resistance dependent on the person's clothing. The physiological basis for judging the effect of an electri shock is supposed to be the same as on land, but because of the hostile environment, the phychical effect may influence the situation more than on land. It is therefore important that the human factor is a strong dimensioning contributor when designing the electrical system, and it is also important to get rid of the fear for using electricity in the hyperbaric environment. This will make it possible, making maximum use of the electricity's advantages.

REFERENCES

Biegelmeier, G., K. Rotter (1971). Elektrishe Wiederstände und ströme in Menchlichen Körper. Elektrotech. und Maschinenbau, 88, 104-114.

Bridge, J.F., J. S. Glasgow, D.J. Hackman, N.C. Henderson (1969). Final report on comparative insulation study of wet suits and standard navy deep-sea diving dress for underwater welding. Battelle Memorial Institute, Columbus, Ohio.

Dalziel, C.F., W.R. Lee (1969). Lethal electric currents. IEEE Spectrum, 44-50.

Horne, R.A., R.A. Courant (1964). Application of Walden's rule to the electrical conduction of sea-water. Journal of geophysical research, 69, 1971-1977.

Mahler, J., J. Weinman (1964). An analysis of electrical properties of metal electrodes. Med. electron. biol. engng., 2, 299-310.

Mole, G. (1979). Underwater electrical safety - some guidance on protection against shock. CIRIA - underwater eng.group. Report UR 14. 79 pp.

Tuckermann, H. (1978). Stromtod - Durchströmung. ETZ-6, 30, 18-19.

Wright, A. (1978). How to protect against electric shock. Electr. Rev., 202, 31-33.

The contents of this paper is based upon the two reports below:

Diesen, A. (1980). Electricity in Water - Safety of Divers', Part 1: Fundamental Principles. NUI-Report 5-80. 64 pp.

Diesen, A., J. Giæver, P. A. Evjemo, H. Topp. (1980). Electricity in Water - Safety of Divers', Part 2: Practical Applications. DnV-Report 80-0238. 102 pp.

STANDARDISATION OF SONAR COMMUNICATIONS

J. K. Sear

Admiralty Marine Technical Establishment, Experimental Diving Unit, HMS Vernon, Portsmouth, Hants, UK

ABSTRACT

The paper states the need for a standard spot frequency for emergency sonar beacon on a lost manned underwater vehicle. Such a restriction will ensure that sonar search equipment can be made compatible on an international scale with obvious advantages to the rescue services.

Expanding this type of philosophy the need is discussed for a standard sonar warning frequency to be transmitted from obstacles such as oil rigs which might be hazardous to manned U/W vehicles.

In exploring the possible mutual interference with other sonar activities the need to restrict the operating band of those activities is proposed. Hence U/W speech could operate over a specified frequency band and dynamic positioning over another, whilst the spot frequency of echo sounders and active search sonar could be kept clear of those bands already allocated.

In order to arrive at some definite sonar frequency proposals the physics of sound transmission are reviewed. Recommendations for various frequencies of operation are made and ways of introducing the proposals are discussed.

KEYWORDS

Sonar Systems, Scenarios, Emergency Beacon, Diving Bell, Manned Submersible, Obstacle Markers, Programme of Introduction, Mutual Interference, Frequency Selection, U/W Speech, Recommended Frequencies.

INTRODUCTION

The first object of this paper is to demonstrate the need for commercial underwater operators to standardise the frequencies used for through-water sonar signal transmission.

Before exploitation of the mineral wealth on the sea bed, the commercial use of sonar was limited to conventional echo sounders and a few specialised fish finding sonars and side scanners. Now there are shallow echo sounders, deep probe echo sounders, dynamic positioning beacons, vehicle tracking beacons, various types of side scanner, doppler navigators and through-water speech communication between divers and to the surface support ship. All the time the list is growing. The transmission frequencies for these various systems are located in the sonar frequency range in an order which is dependent upon the characteristics of sound transmission in the area, and the requirement of each Sonar System.

Fig. 1 shows the spread which the various systems are located over the frequency range.

Fig. 1

THE ORDER OF FREQUENCY DISTRIBUTION

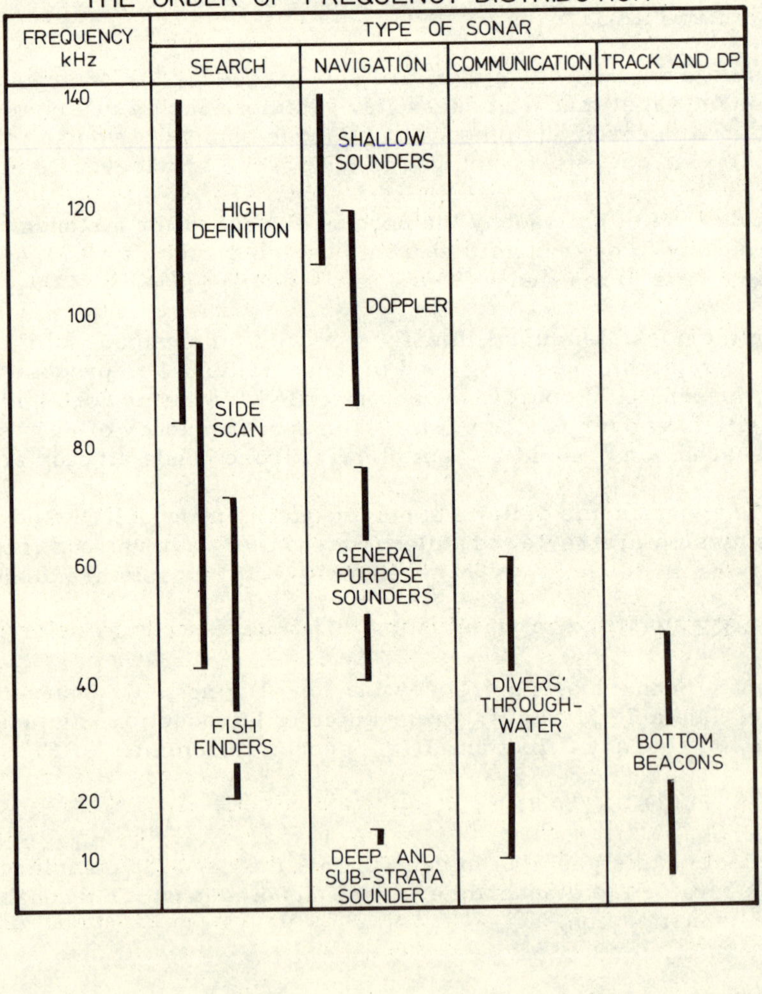

FREQUENCY kHz

TYPE OF SONAR

SEARCH | NAVIGATION | COMMUNICATION | TRACK AND DP

140
120
100
80
60
40
20
10

HIGH DEFINITION

SIDE SCAN

FISH FINDERS

SHALLOW SOUNDERS

DOPPLER

GENERAL PURPOSE SOUNDERS

DEEP AND SUB-STRATA SOUNDER

DIVERS' THROUGH-WATER

BOTTOM BEACONS

The second object of this paper will be to suggest that since there is a need for mutual co-operation between operators, it is necessary to be more precise about the use of the sea as a communication medium.

SCENARIOS

There are many scenarios which would illustrate the need for careful frequency selection. Prime examples are the emergency beacon for diving bells and the emergency beacon for a submersible and Sonar Markers for U/W obstacles.

Emergency Beacon for Diving Bells

The only reason for a beacon on a diving bell is to be able to locate it if dragged from the work site and if it becomes detached from the mother ship. The mother ship, might then have to locate the bell at considerable range with the navigational accuracy to pin point it within, say, 30 metres. Obviously another diving ship must take position over bell to effect a rescue. As the bell will then be remote from the worksite there will be no sea bed beacon transmission to assist the rescuer. Only the emergency transmission from the bell for location can assist the rescuer. The solution to this is for all vessels using sonar means for dynamically positioning to have an emergency reception channel in their system tuned to an emergency beacon frequency. Then if the emergency frequency is a standard for diving bells, any diving ship might position herself and her diving bell alongside the one to be salvaged. In these circumstances a rescue ship equipped with sonar DP would deploy beacons compatible with its system at the earliest opportunity. However, there should be if possible, no mutual interference between that DP system and the emergency beacon. Similarly any other sonar systems used in the proximity of the rescue site by the attending ship must not interfere with the emergency signal. If they do, they must be switched off.

In this simple scenario of a lost diving bell the following can be identified as the needs of an emergency bell location system.

1. To locate at long range.
2. To be compatible with other DP systems.
3. To be clear from mutual interference.
4. To be at a standard frequency.

It is possible to go a long way in fulfilling all of these needs by selecting the right frequency; the methods being discussed later in the paper.

Emergency Beacon for Manned Submersible

There is, at the time of writing this paper, a certain amount of timely standardisation for the emergency sonar transmission from the manned submersible. However, it is recommended that both the diving bell and the manned submersible should use the same emergency sonar frequency. There are instances when the dynamically positioning ship with an emergency bell beacon receiver would need to locate herself over a trapped submersible, in order to deploy divers to effect a rescue.

It would be convenient, therefore, to set the emergency frequency of the submersible to correspond to the existing channel for emergency on the DP ship. Obviously a manned submersible effecting a rescue on a trapped bell would have a receiver for that purpose. If the frequency was a universal standard, the same receiver could locate an immobilised unmanned submersible. Therefore there could be a rescue force of a bell or a submersible or a DP ship each able to act in accordance all at one frequency. It is easy to conceive of other types of search facilities once a standard of frequency has been agreed: e.g. helicopters with dunking sonar and, possibly, sonar buoys. Both of these facilities have the great advantage that as sonar platforms only they are quiet and therefore capable of achieving ranges which are limited by sea noise. This advantage, and mobility would enable large areas to be searched in a short time.

Sonar Markers for U/W Obstacles

The third scenario is based on the need to mark U/W obstructions i. e. oil rigs, drill ships etc., by means of sonar at standard, well publicised frequencies. The aim of a sonar marker on the U/W obstruction is two fold. Firstly it is to avoid the danger to the crew of a submersible and secondly it is to avoid damage to the static rig. This second aim might well be the most important to the human race in the long term.

There are two types of submersible at risk. One is a large military submarine which might hit a rig "en passant", another is a smaller working submersible which would normally be working in very close proximity to the obstruction. In the first case there is a need for a reasonably long range warning pinger with a frequency which is recognised internationally. In the second case pingers or transponders could be used for navigating around parts of the structure. As this would be localised the frequency would not be of international importance; but, of course, should certainly be known in case there is an accident. An important application of the first case is a long range marker beacon for the avoidance of submarine type accidents. The frequency for this must necessarily be low to achieve the range and the power output must be reasonably high; the available electrical source being adequate for this. The range required would be that necessary to give reasonable warning to the submarine in order to allow it to change course in time even at top speed. The frequency for the local navigation of a submersible in close proximity to an oil rig may well be high in order to achieve directional accuracy. The noise level in this area could be deafening as there is a fair chance that the operation of a transponder could be difficult. It is not easy to receive a signal when the receiver is in close proximity to the source of noise. It is far easier to transmit in the proximity of a noise source to a receiver in a quiet position.

STANDARDISATION

The Extent of Standardisation

To insist on total standardisation of all U/W transmissions would be too restrictive for a developing technology. It would be dangerous and quite wrong to compare the allocation of radio wavebands with the selection and standardisation of sonar frequencies.

The ranges of reception involved over the normal sonar bands are so small, compared to radio bands that many of the problems found in radio do not occur because of the "sea room" around most sonar operators. Standardisation is nevertheless required when interfacing sonar systems or when working at close range with other sonar systems.

A typical interface problem might be the ability of a diving ship taking part in a rescue to achieve through water communication with those trapped in another ship's diving bell. Standards of frequency in close range working are important for a rescue operation in the initial search stages. Here the only craft that can assist in the passive long range search are those equipped with an emergency receiving channel in the DP system.

Active search sonars e.g. fish finders, side scanners which could assist with the search may simply have to be switched off.

To establish a progressive programme for standardisation three categories of priorities have been identified relating to present and future needs. These are given in the following paragraphs.

It is suggested that the frequency standard that is <u>essential</u> is as follows:-

(i) Emergency lost beacon for diving bell and manned submersible effecting:
  a. bell beacon
  b. submersible beacon
  c. DP receiver channel for plotting purposes
  d. DP operation for temporary positioning
  e. submersible receiver for search
  f. restriction of use of all other transmitting sonar at the selected frequency

It is suggested that those frequency standards that are <u>desirable</u> are as follows:-

(ii) Emergency through water communications for Diving Bell and submersible effecting:
  a. bell transmitter (providing bell de-scrambler)/receiver
  b. submersible transmitter/receiver
  c. surface ship borne receiver/transmitter
  d. restriction of use of other sonar in frequency band selected

(iii) Marking of U/W obstacles at long range effecting:-

a. obstacle beacons

b. submersible receiver

c. eliminating the use of other sonars which transmit at the selected frequency

It is suggested that those frequency standards that are desirable only in the long term are as follows:-

(i) Marking of obstacles for close range working

(ii) All DP systems

(iii) All U/W Navigation systems

(iv) All search sonars both forward and side scan

(v) Close range through water speech

The above standardisations would allow all the named systems to be operated at the same time, in the same area or even on the same vehicle.

Limiting the Number of Emergency Frequencies

So far, in this paper, one spot frequency has been suggested as an essential standard and a further spot frequency as a desirable standard within a frequency band. Now it is in the interest of all users to keep the number of these standards to a minimum. So it is suggested that the emergency spot frequency for lost manned craft might be located in the emergency speech band. This has the advantage of simplifying the receiver design to a rationalised multi-purpose unit. It also puts less restriction on the use of other sonar frequencies. With regard to the mutual interference that would occur, it is logical to suppose that preliminary speech exchange could occur between the pulse transmission, and that the pulse transmission would be switched off once the position was fixed.

The problem of understanding helium speech in a high pressure bell atmosphere is best overcome by unscrambling the speech before transmission. It is then possible to use one carrier frequency for both one atmosphere and high pressure conversation. This step has the further advantage of narrowing the band of transmission thereby improving the signal to noise.

By the above reasoning only one standard 3KHZ band for all emergency transmission is required.

If the marking of all obstacles was restricted to one spot frequency and the long range identification of the obstacles was achieved by coding the position, then only one standard 3KHZ band and one spot frequency are required.

This type of approach, minimising the number of standard frequencies, should always be tried in order to reduce the varieties of transmitters and receivers that will be generated.

Mutual Interference

Interference from other sonars can destroy a video sonar picture or an U/W navigation computer. The severity of the interference largely depends on the type of processing in the effected system and the duration of the insonification of the locality.

Even the allocation of just one spot frequency, and a relatively narrow frequency band, is not all that easy because sonar transmission, being a mechanical process, produces mechanical side effects. Any transmission at 32 KHZ will give interference at 16 to 8 KHZ. So that an emergency beacon transmitting at say 32 KHZ might preclude the use of a DP beacon at 16 KHZ or any frequency in that region, depending on the bandwidth of the receiving system. As a general rule a frequency which is an even generation of the normal DP frequency is avoided.

With regard to other search sonar systems in a rescue situation, at the present time the simple rule is to switch them off if they transmit, and are not vital to the operation. If frequencies had been allocated and selected for various operations from the beginning of commercial sonar development then this problem would not exist now.

As a final comment on mutual interference problems I do think it is necessary that the frequency of transmission of all commercial operators should be published in a comprehensive form. An Appendix to an appropriate Ship Classification Society Register would ensure world wide coverage.

Summarising there are three rules relating to avoidance of mutual sonar interference:-

1. Do not use frequencies which are multiples of one another.

2. Switch off unnecessary transmissions.

3. Ensure that all transmission frequencies are published in comprehensive form.

Selection of Sonar Frequency

The frequency of an active sonar system is usually dictated by the range that has to be achieved. Sometimes, as in the case of higher frequency sonars, the detail of target definition is the overriding consideration in the selection of frequency. In other cases the practical size and shape limitation of the transducer must be the predominant influence on the choice of frequency e. g. the importance of a reasonably small transducer for a divers communication system.

There are, in a competitive commercial world, instances of systems being designed around an existing transducer in order to reduce cost.

Emergency Beacon

In considering the selection of the appropriate frequency of an emergency beacon it is necessary to take into account the range to be achieved and the size of transducer to be accommodated.

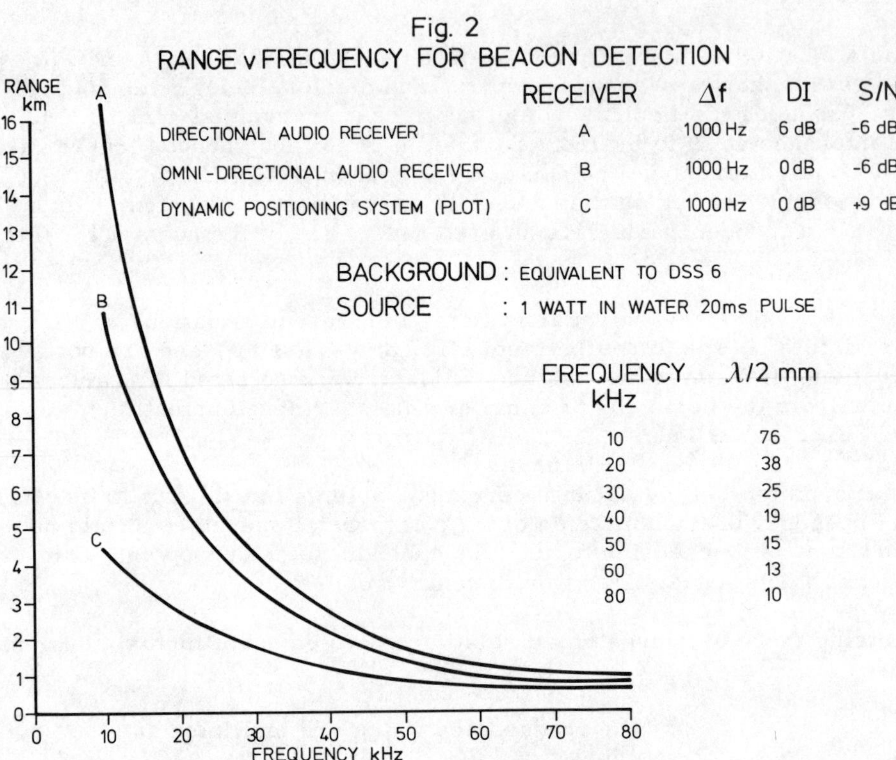

Figure 2, above, shows the detection range vs. frequency for a simple omni-directional source putting just 1 watt of power into the water, and being detected by three different kinds of receivers, the details of which are as follows:-

Curve A — This is based on the type of sonar that would be mounted on a submersible with a modest directivity index of only 6 db a bandwidth of 1000 HZ. It is assumed that the operator achieves audio detection 6 db below background, which is not unreason able.

Curve B It is a simple omnidirectional receiver, at the same bandwidth and with the same audio operator, used from the surface.

Curve C Is an approximation to a dynamic positioning system receiver where a signal excess of some 12 db is required to be able to plot the position of the beacon.

These three curves are idealised and whilst A is achievable from the platform of a submersible B might have a reduced performance due to the self noise of the surface ship. The ranges of C might also be reduced by self noise and would be further reduced by an increase in bandwidth and it should be noted that 1000 HZ for a DP system is somewhat small. However, the trends are clear and the lower frequency of operation is highly desirable in every case.

The half wavelength relationship to frequency is also shown and a transducer at 10 KHZ measuring some 76 mm in diameter would seem to be a reasonable size. At 10 KHZ the range looks good and the transducer is not too big. It is proposed that frequency for emergency beacons should be 10 KHZ. This will give a DP equipped ship with a 10 KHZ receiver channel the possibility of locating a manned vehicle at a range of 4 kilometres in ideal conditions but probably little more than 1 kilometre in practice.

Emergency Through Water Speech

It has been recommended that the helium speech of trapped divers should be de-scrambled at source and so reduce the bandwidth of the transmitted speech from 6 KHZ to 3 KHZ. It has also been suggested that the emergency signal should be in the band selected and that should be 10 KHZ. This may be regarded as an unconventional way to recommend the adoption of the existing Naval submarine rescue frequency carrier which would enable the Naval rescue service to assist. The carrier frequency recommended is 8.075 KHZ with plain speech, trans-mitted in a 3 KHZ band, between 8 KHZ and 11 KHZ. The recommended beacon emergency frequency is in the band at 10 KHZ. This has the advantage of alerting an initial detection on the speech receiver in a rescue craft. In addition it would be preferable for the pulse length of the beacon to be in excess of say 20 milliseconds to allow reasonable audio detection.

Hazard Marker

The sonar hazard marker should be capable of long range detection and must necessarily use a reasonably low frequency. It should also be at one spot frequency to avoid the difficulty of multi frequency receivers and should have a coded transmission. To ensure audible detection the pulse length should be in excess of 20 milliseconds. On structures that reach from the surface to the sea bed the markers should transmit at various levels to ensure reception in adverse bathymetric conditions. At a minimum, one transmitter should be located at 10 m depth to anticipate surface duct conditions and another at 100 m depth.

It must be appreciated that the detection system will have good directional discrimination so that coding techniques for transmission can be reasonably crude. I would use a combination of 20 and 200 MS pulses with PRF's of uneven numbers of seconds e.g. 7, 9, 11, 13, 15 etc.

For no other reason other than it is a low frequency (and hopefully not unlucky) 13 KHZ is recommended.

Recommended Frequencies

Summarising the recommendations so far:-

Emergency beacon for all manned U/W craft	.....	10 KHZ
Emergency through water speech	.....	8-11 KHZ
Hazard markers	.....	13 KHZ

In the future the following frequency for other sonar activities are recommended

Dynamic Positioning Beacons and U/W Navigation	.....	15-31 KHZ at 15, 17, 19, 21, 23 etc.
Close range markers	.....	35,37, 39 KHZ
Forward sonars	.....	20-40 KHZ at 20, 25, 30 & 40
Side Scanners	.....	50-110 KHZ at 50, 70, 90 & 110
Close range through water speech	.....	57-63 KHZ

Finally it would now be quite difficult to slot in such things as echo sounders and doppler navigators but it could probably be done in the long term as follows:-

Deep Echo Sounder	.....	14 KHZ
CP Echo Sounder	.....	45 KHZ
Doppler Navigator	.....	100 KHZ
Shallow Echo Sounder	.....	200 KHZ

These proposals are illustrated by Figure 3, on the following page.

Fig. 3
RECOMMENDED FREQUENCY DISTRIBUTION

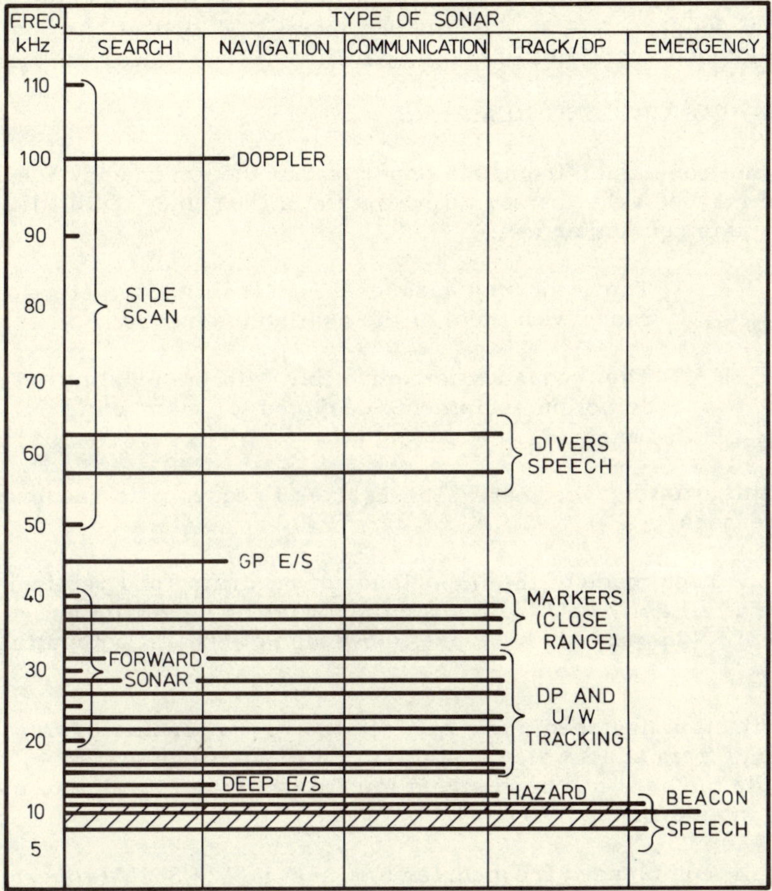

It is pertinent to add that all the recommendation frequencies are capable of being juggled into other positions and the recommendation serves mainly to show how the task could be tackled.

Implementation of Frequency Standards

The recommended frequencies for the emergency situation where human life is a risk must be agreed soon for the North Sea and also internationally. The national diving organisations of the countries concerned are giving serious consideration to this task and, possibly by the time this paper is published some directives will have been issued on the frequencies to be used.

The publication of all sonar transmission frequencies from rigs, survey vessels, diving ships etc., is the next important step. The best organisation to ask for this is the insurer, since this would minimise risk due to mutual interference. The frequency, source level, directivity of sonar should be published to an international public to assist planning of normal U/W operations and to form the basis of command decision in rescue operations.

CONCLUDING REMARKS

The important conslusion from this paper is that the emergency SONAR BEACON for trapped manned vehicles should operate at a frequency of 10 KHZ. The reasons for this conclusion are:-

a. The frequency ensures a reasonably long detection range with most of the available sensors.

b. The frequency is compatible with the existing detection equipments designed for submarine rescue.

The use of the existing Through Water Carrier Frequency is recommended for the same reasons.

Emphasis has been made of the importance of recording and publishing the frequencies of all active SONARS associated with RIGS, SHIPS and SUBMERSIBLES in order to have pre-knowledge of SONAR compatibility before an accident.

A clear distinction has been made between the requirements for long range SONAR alerts from underwater obstacles and close range markers. On the receiver side the need has been stated for DP system to be able to plot the position of an emergency beacon.

Finally, the recommended frequencies for most active SONAR operations has been made, primarily to motivate designers towards the concept of standardisation of SONAR frequencies.

ACKNOWLEDGEMENTS

I wish to acknowledge, with gratitude, the time given by the Admiralty Marine Technology Establishment to produce and publish this paper. I would also like to acknowledge the advice and assistance of friends and colleagues, during the preparation of this paper. Finally, I would like to thank the Board of this Conference for allowing me the opportunity to make public my thoughts on this subject.

ACOUSTICAL CAMERAS FOR UNDERWATER SURVEILLANCE AND INSPECTION

H. Heier and J. J. Stamnes

Central Institute for Industrial Research, Blindern, Oslo 3, Norway

ABSTRACT

For a number of years research has been going on at the Central Institute for Industrial Research to develop acoustical lenses. We are now in a position to produce large aperture lenses that combine high resolution with low weight and absorption, and in addition are cheap to manufacture. In this paper emphasis is placed on explaining how acoustic cameras based on such lenses can be successfully applied in various underwater operations. To mention a few examples, such cameras can be used for surveillance and guidance of work at the sea bottom in situations of poor optical transmission, mapping and inspection of paths for oil pipe lines, inspection and nondestructive testing of off-shore structures, and inspections to locate construction garbage at the sea floor.

KEYWORDS

Acoustical imaging; underwater imaging; holographic lenses; nondestructive testing; bottom mapping; robot vision.

INTRODUCTION

Together with the increase in underwater activities in recent years the need for efficient and reliable methods for surveillance and inspection of underwater operations has arisen. For reasons soon to become apparent, the commercially available methods that exist today for such purposes are of limited practical value.

Therefore, we present in this paper the concept of an acoustical TV camera that has a very wide range of underwater applications.

The paper is organized as follows: First we give a survey of various methods for underwater imaging and detection and compare their advantages and disadvantages. The main objective of the survey is to determine in what situations an acoustical TV camera should or should not be used.

Then the principle of the holographic lens, on which this camera is based, is described, and the advantages of holographic lenses over conventional lenses are presented.

The attention is then turned to the main parameters of the camera, such as, lens diameter, focal length, range, field of view, resolution, wave length, and their interdependence. Also, the properties of a typical holographic lens are presented, together with an idea of how a suitable detector could be made. The detector represents the element of uncertainty in our camera concept. Depending on the image size and resolution required, one may think of various ways in which the detector could be made.

Finally, we present examples of acoustical TV cameras for various applications, and summarize the main results of the paper.

SURVEY OF METHODS

To place our camera concept in some perspective we now give a brief survey of various methods that can be used for underwater surveillance and inspection. Although the survey is not exhaustive, we feel that the four methods presented below are representative for our comparative purposes.

Optical TV Camera

The most obvious surveillance method is to use an optical TV camera. The advantage of the optical TV camera is first of all that it is commercially available, secondly that it provides pictures with high resolution in clear water. The prime drawback of the optical TV camera is its limited range. Even in clear water the optical range is at most 10 m, and in turbid water, produced f.ex. by a working operation at the sea bottom, the range rapidly drops to a few centimeters.

Side Looking Sonar

The use of a side looking sonar is another surveillance possibility. The side looking sonar is commercially available, with a range of the order of a kilometer. Also, the long wave length sound waves of the side looking sonar penetrate somewhat into the bottom making it possible to obtain information not only about the bottom surface itself but also about the uppermost layers of the bottom. The disadvantage of the side looking sonar is mainly its limited resolution transverse to the beam, which is too low to make it attractive for surveillance of underwater working operations. Also, it must be towed continuously in order to build up a two dimensional picture.

Acoustical Holography

The prime advantage of acoustical holography lies in its ability to provide genuine three dimensional reconstructions. The range and resolution depend on the implementation. Long range (several kilometers) combined with relatively high resolution, can be obtained, but only at large expenses. As with side looking sonar penetration into the bottom for seismic purposes (Fitzpatrick, 1979; Ljunggren, Løvhaugen and Mehlum, 1980) is feasible, as well as penetration into materials for nondestructive testing purposes (Collins and co-workers, 1980; V. Schmitz and M. Wosnitza, 1980; Takahashi, Suzuki and Kanamori, 1980). The acoustical holography method is not commercially available for purposes as described here, and it requires expensive data collection as well as sophisticated data processing. Correctly implemented, the acoustical holography method is probably superior to all other methods from an information point of view, but a correct implementation is often economically prohibitive.

Acoustical TV Camera

An acoustical TV camera provides an image very much in the same way as an optical TV camera, except that light waves are replaced by sound waves. The range depends on both the wave length and the lens diameter. Generally speaking, the acoustical TV camera has a relatively long range and a high resolution, and as with acoustical holography penetration into the bottom for seismic purposes or into other materials for nondestructive testing purposes is possible. Except for the EMI Sokolov camera (Wardley, Brown and Crouchee, 1977), acoustical TV cameras for underwater imaging are not commercially available.

Finally, we mention that the method of phased acoustical arrays (Macowski, 1979; Kino, 1979) which has been used both in medical applications and in nondestructive testing, in some ways is similar to acoustical holography, and in other ways to an acoustical camera. For underwater applications, however, we feel that the phased array method is not competitive with acoustical TV cameras, as it will be significantly more expensive without providing significant advantages.

Comparing the descriptions given above, we find that it is advisable to use

- optical TV cameras for surveillance and guidance of working operations at the sea bottom in clear water at short range,
- side looking sonars at long range for large scale surveys where knowledge of small details is of less importance, f.ex. in large scale mapping of paths for oil pipe lines,
- acoustical holography when detailed knowledge of the three dimensional structure is very important, and when in addition the range is too long for an acoustical camera to be practical, f.ex. in oil prospecting (Ljunggren, Løvhaugen and Mehlum, 1980),
- acoustical TV cameras for surveillance and guidance of work at the sea bottom in turbid water at short range, and in clear and turbid water at intermediate ranges; for detailed investigations of paths for oil pipe lines; for inspections at short and intermediate ranges of oil pipe lines and off-shore structures; for nondestructive testing purposes, and for inspections to locate construction garbage at the sea floor.

From the considerations above it is clear that the various techniques to a large degree are complementary rather than exclusive. Thus, one can easily think of situations in which several of them could be used simultaneously or sequentially with great advantage. We mention a few examples:

When it comes to resolution it is hard to beat the optical TV camera. Thus, for surveillance and guidance of working operations in clear water one could think of using both optical and acoustical TV cameras, the former for close up inspections and the latter for providing overview scenarios from a more distant observation point.

As another example, in the mapping of paths for oil pipe lines, a side looking sonar can be used first for large scale surveys. Bottom segments that look particularly interesting can then be investigated in more detail by an acoustical TV camera.

Thirdly, one could combine a side looking sonar and an acoustical TV camera in the search for construction garbage or wrecks, in which the side looking sonar does the large scale search, and the acoustical TV camera does the close up inspections of interesting areas.

ADVANTAGES OF HOLOGRAPHIC LENSES

The acoustical camera we propose is based on so-called holographic lenses[1], a term we use here mainly to distinguish these lenses from conventional ones. A typical holographic lens for acoustic waves is shown in Fig. 1. It consists of a plane parallel plate of perspex with a concentric pattern of circular groves, which give the desired lens or focusing effect.

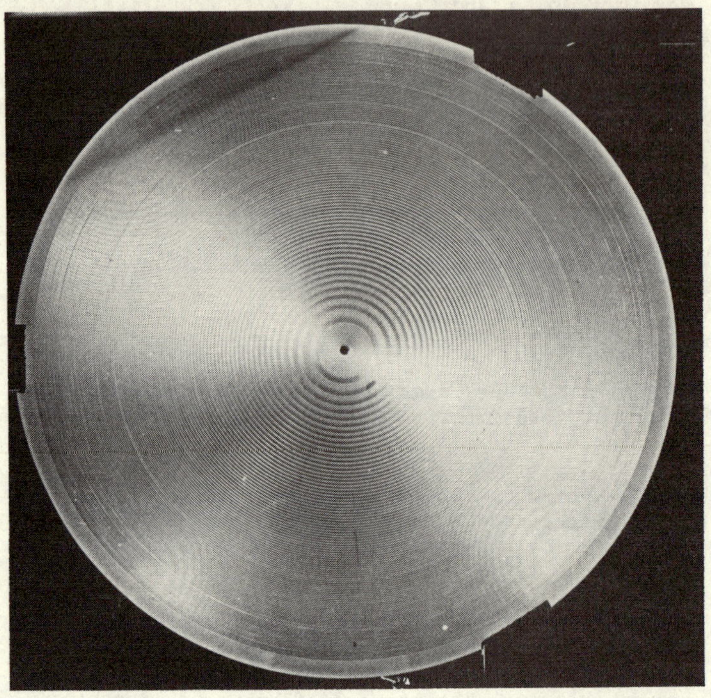

Fig. 1. Holographic lens for sound waves.

In contrast to holographic lenses, conventional lenses usually consist of suitable materials bounded by spherical surfaces. For large lens diameters an increase in diameter therefore must be accompanied by a corresponding increase in lens thickness. The advantage of holographic lenses over conventional ones is first of all that one can increase their diameters without simultaneously increasing their thickness, thereby avoiding the problems of absorption, weight and material costs. The thickness need not be larger than is necessary to provide mechanical support. Large lens diameters are needed to obtain lens systems that combine long range with high resolution.

Another advantage, which has to do with image sharpness, is that one without additional difficulty can produce holographic lenses which are equivalent to aspheric conventional lenses. Thereby one can decrease the number of lens elements needed to obtain a desired degree of aberration correction. We give an example of this later.

---

[1] The term "holographic lens" originates from the fact that the lens structure resembles the interference pattern of a hologram.

Thus, in summary, the advantage of holographic lenses over conventional ones is that one can obtain long range imaging systems of high resolution without running into insurmountable problems with weight, absorption or price.

THE CAMERA CONCEPT

Figure 2 shows a schematic view of an acoustical TV camera. In the figure the camera is placed at the bottom of a boat with several sound sources placed around it. The camera consists of a lens system and a hydrophone array mounted together in a camera body which is filled with water.

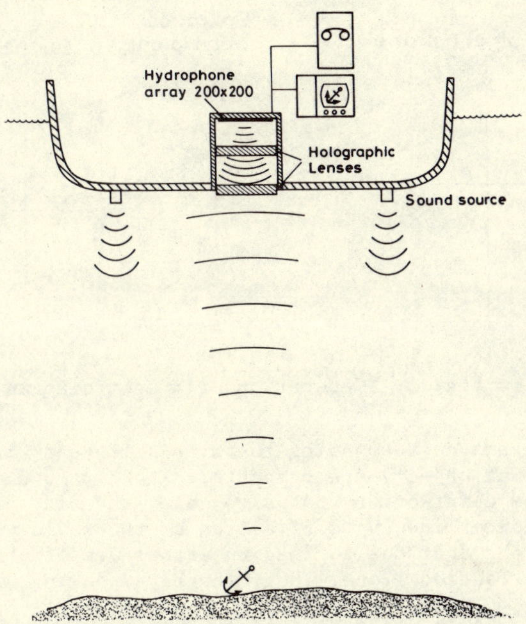

Fig. 2. Working principle of acoustical TV camera.

The acoustical waves emitted by the sound sources are reflected from objects in the water and the sea bottom itself back to the camera. The reflected waves are then focused by the lens system so that an acoustical image is formed on an array of hydrophones. The signals detected by the hydrophones are transferred to a TV monitor for display of the image. Also, the signals may be recorded on magnetic tape together with other information, such as f.ex. position. From this description it is clear that the working principle of an acoustical TV camera is analogous to that of an optical TV camera.

We now continue to describe in more detail the lens system, the main parameters of the camera, and the hydrophone array.

The Lens System

The acoustical lens system consists of two holographic lenses, as indicated in Fig. 2. Each lens has a concentric circular pattern of grooves, as indicated in

Fig. 1. A single lens does not provide an acceptable image sharpness over an extended field of view. Therefore a doublet is used. The patterns of grooves of the two lenses are optimized so that imaging errors play no practical role within a field of view of ± 20°. A more detailed description of this doublet will be published elsewhere (Heier, 1980).

To illustrate the quality of the holographic doublet we make use of spot diagrams to describe imaging errors. The generation of a spot diagram is shown in Fig. 3. The object point emits rays of sound that are bent by the lens to meet in the image plane. The display of the intersection points between the rays and the image plane is called a spot diagram.

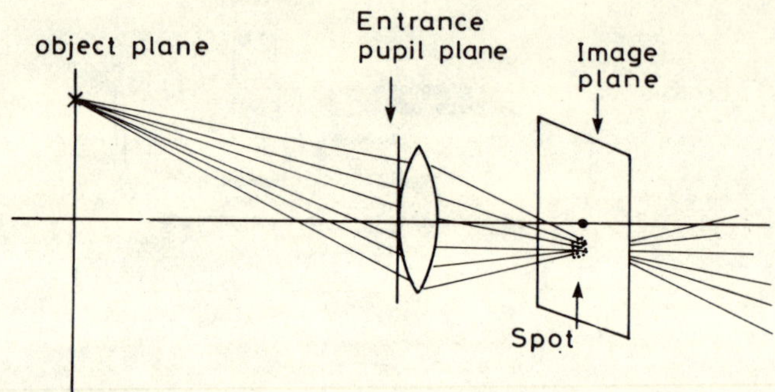

Fig. 3. Generation of a spot diagram.

In Fig. 4 spot diagrams corresponding to three different field angles (angular positions of the point object) are presented. The circle underneath the spot diagrams represents the diffraction spot size, i.e. the size of the theoretical lower limit of the spot, determined by diffraction of the sound waves around the lens edges. We see that the spot due to imaging errors are of about the same size as the spot due to diffraction alone. Thus, roughly speaking, imaging errors may be neglected within a field of view of ± 20°.

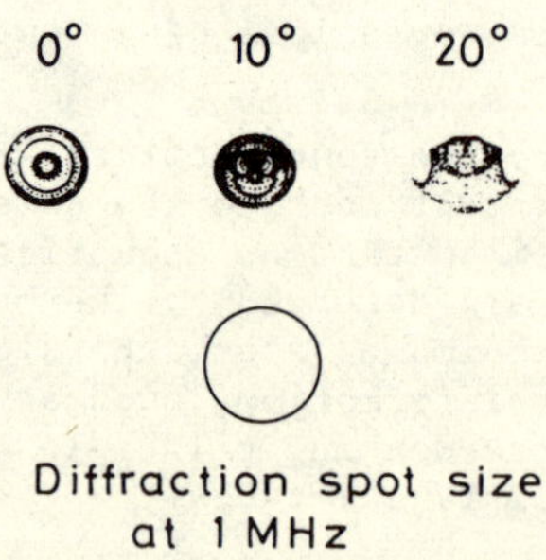

Fig. 4. Spot diagrams showing the image sharpness at various field angles.

In this example the lens diameter and the focal length are 1 m, and the sound frequency is 1 MHz. An acoustical camera with the same lens diameter as above, was developed at the Naval Coastal Systems Center in Florida about three years ago (Sutton, 1979). Their lens system consists of six elements of conventional type, and the camera has a field of view of $\pm 4.5^{\circ}$. Thus, with two holographic elements one obtains more than four times the field of view obtained earlier with six conventional elements.

Camera Parameters

The main parameters of the camera are the lens diameter, focal length, range, field of view, resolution, and wave length. In Fig. 5 it is illustrated how one derives the relationship between these parameters:

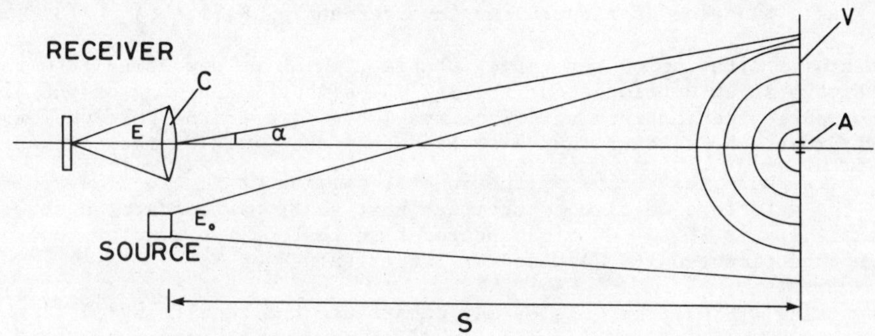

Fig. 5. Illustration relating to the derivation of the relationship between camera parameters [Eq. (6)].

A source emits a sound wave of energy E_o. This wave insonifies objects within the field of view of the camera so that the energy from the source is spread out over an area V given by

$$V = \pi(\alpha S)^2, \tag{1}$$

where α is the half angle field of view of the camera, and S is the range. Let an object element (corresponding to the image through the lens system of an image element) have an area A and reflection coefficient R. This element is assumed to reflect the sound over a solid angle 2π, so that the energy E collected by the camera opening, of area C, is given by

$$E/E_o = 10^{-2aS} R\cdot(A/V)\cdot((C/S^2)/2\pi), \tag{2}$$

where a is the attenuation coefficient. In sea water at 15 ^{o}C, the attenuation depends on the wave length in the following way (Kinsler and Frey, 1962)

$$a = 0.06(1/\lambda)^2/(1/\lambda^2+10000) + 2.4\text{x}10^{-7}/\lambda^2 \quad [\text{db/meter}]. \tag{3}$$

The area A of the object element may be written

$$A = \pi(S\Theta/2)^2, \tag{4}$$

where Θ is the angle subtended by the object element at the camera opening. In the case of diffraction limited imaging, which we henceforth assume, Θ is equal to the angular resolution of the camera, i.e.

$$\Theta = 2.44\lambda/D, \tag{5}$$

where D is the lens diameter. Substitution from (1), (4), and (5) into (2) yields

$$S^2 10^{2a(\lambda)S}/\lambda^2 = 611KR/\alpha^2, \tag{6}$$

where α is the half angle field of view in degrees, and $K = E_o/E$ is the inverse damping of the signal between the source and the receiver. If the reflection coefficient R and the field of view α are known, then (6) may be solved numerically to give λ as a function of range S and inverse damping K, i.e. $\lambda = \lambda(S,K)$.

A fundamental limitation on the range is noise, which we now bring into the calculation through an assumption about the maximum allowable damping of the signal between the source and the receiver. From available data on the EMI Sokolov camera (Wardley, Brown and Croucher, 1977) we find that the limit of the allowable damping corresponds to a value of the inverse damping of $K = 10^{11}$. Available numbers for the detectors we plan to use (see next section) indicate, however, that their sensitvity is 10 to 100 times better than that of the Sokolov tube. For that reason we have computed $\lambda(S,K)$ also for larger values of K.

After computing $\lambda(S,K)$ from (6), we may substitute the result in (5) to obtain the lens diameter D as function of angular resolution Θ, with the range S and the allowable inverse damping K as parameters. The resulting curves for R = 0.5 and $\alpha = 20^o$ are displayed in Fig. 6. We see from the figure that some uncertainty in the K value does not lead to dramatic shifts of the curves, a result that is to be expected since the damping is exponential.

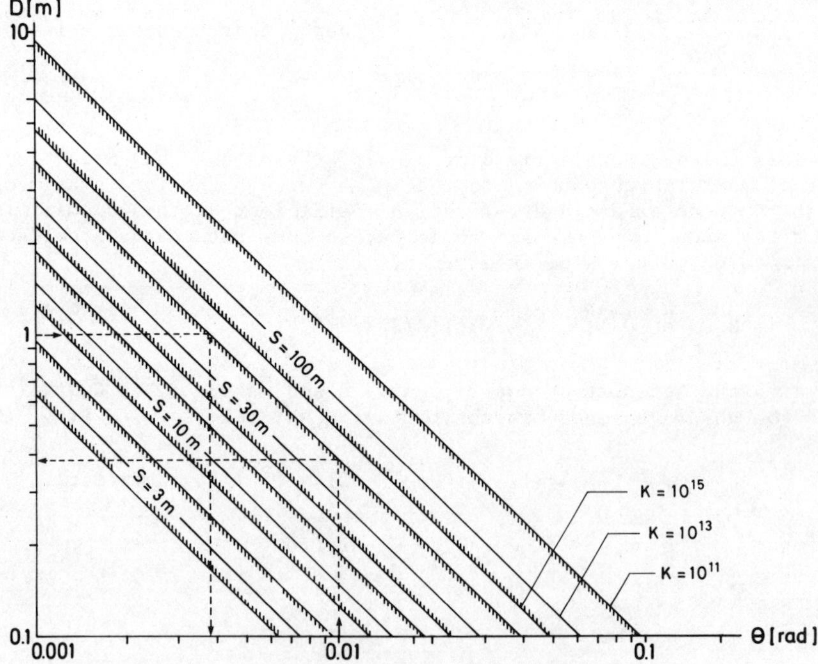

Fig. 6. Lens diameter D as function of angular resolution Θ, with range S and allowable inverse damping K, as parameters.

From Fig. 6 we see that a lens of 1 m diameter and with a range of 30 m has an angular resolution of about 0.004 radians. We also see that the EMI Sokolov camera which according to available data (Wardley, Brown and Croucher, 1977) has a diameter of 22 cm, a range of 10 m, and an angular resolution of about 0.009 radians, corresponds to a K value of 10^{11}, as asserted above.

In comparison, the 1 m diameter lens system of the Naval Coastal Systems Center (Sutton, 1979) is said to have a maximum range of 150 m and the same angular resolution as that of our 1 m diameter camera. We do not know enough about the NCSC camera to tell what causes the 5 times larger range than in our case, but it may be due to a much larger allowable K value than 10^{11}. Thus, it looks like we have been conservative in our assumption about K.

Also, we have been conservative in our formula for the angular resolution [Eq. (5)], which probably may be improved by a factor of 1.5 to 2. A factor of 2 improvement in resolution would increase the range of a camera with 0.004 radians resolution from 30 m to about 100 m.

The Hydrophone Array

As mentioned in the introduction, the hydrophone array represents the element of uncertainty in our camera concept.

One may think of various ways in which an acoustic detector array could be made, depending on the required image size and resolution. Nevertheless, we present now an idea of how to make a detector array that possesses many attractive features, such as high resolution, high sensitivity, fast read out, and low production costs. Although the idea is based on established technology, a minimum of a year of research and development would be needed to produce an array prototype.

The detector array would consist of pressure sensitive MOS transistors with a piezo electric layer of zinc oxide underneath the gate electrode, as illustrated in Fig. 7. The array would consist of modules, each module having a typical size of 2 x 2 cm^2. To cover larger image sizes one would need to place several such modules side by side, so that the manufacturing process would have to be adjusted to the image size.

This modular construction is illustrated in Fig. 8, where several detector elements are integrated on the same module, and the modules coupled together on a thick film substrate by means of the well known flip-chip technique.

The acoustical signal from each detector element would be used to charge a capacitor, and the charges read out by means of shift registers. Thus, we have a detector that directly measures the intensity of the signal.

The total number of detector elements required is equal to the image size divided by the transverse resolution in the image. Thus, the complete detector array would consist of this number of detector elements spread out evenly over an area corresponding to the image size.

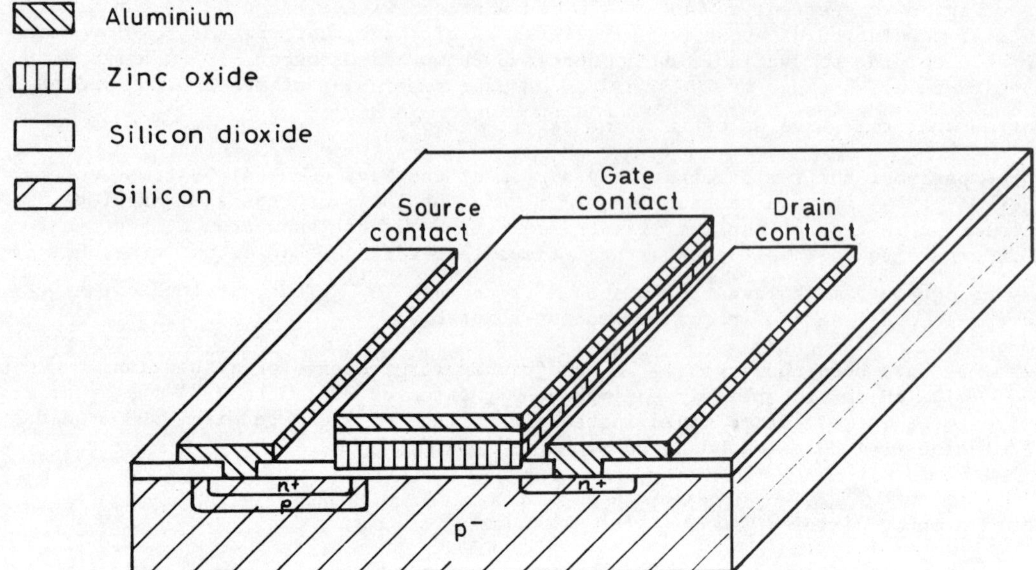

Fig. 7. Pressure sensitive MOS transistor with zinc oxide underneath the gate electrode.

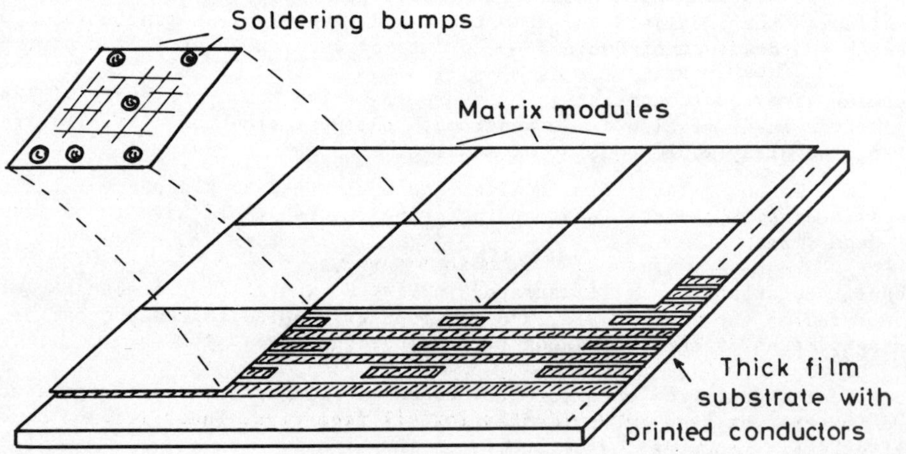

Fig. 8. Modular coupling by means of flip-chip technique.

CAMERAS FOR VARIOUS APPLICATIONS

As an illustration, we now present specifications for acoustical TV cameras to be used under various circumstances. To that end, we first classify the cameras after range, and make use of Fig. 6 to determine the relationship between range, lens diameter, and angular resolution. After presenting the specifications, we then give examples of applications for cameras of various ranges.

As mentioned earlier, we are able to design diffraction limited holographic lenses with a half angle field of view (α) as large as 20°, and with an f-number (focal length divided by lens diameter) as low as 1. Therefore, we consider now cameras for which $\alpha = 20^{\circ}$, and f-number = 1.

First we choose the desired range S, and angular resolution Θ. The remaining specifications are then given as follows:

Lens diameter : D (follows from Fig. 6 with $K = 10^{13}$),
Wave length : $\lambda = \Theta D/2.44$,
Frequency : $\nu = C/\lambda$; C = 1500 m/s,
Transverse object coverage : $H = 2S\tan\alpha$,
Transverse object resolution : $R = 2S\tan(\Theta/2)$,
Depth of field : $FD = \pm 2n^2\lambda/(1+n\Theta/1.22)$; $n = S/D$,
Image size : $H' = 2S'\tan\alpha$; $S' = SD/(S-D)$,
Image resolution : $R' = 2S'\tan(\Theta/2)$
Number of detector elements : NxN; $N = \tan\alpha/\tan(\Theta/2)$.

Numerical examples of the quantities specified above are given in Table 1 for various values of the range S, and the angular resolution Θ. As mentioned earlier, we have been conservative in our assumptions, so that it may turn out that the lens diameters need not be as big as indicated in Table 1.

As an example, we may suggest the following underwater applications for cameras of various ranges:

1. Ranges shorter than 1 m: Nondestructive testing of oil pipe lines or platforms.

2. Range 3 m: Guidance of robots or manipulators; surveillance of small scale working operations.

3. Range 10 m: Inspection of oil pipe lines; surveillance of medium scale working operations.

4. Range 30 m: Investigations of paths for oil pipe lines; surveillance of large scale working operations.

5. Range 100 m: Search for wrecks; inspections to locate construction garbage at the sea floor; mapping of bottom topography.

The necessary size of the lens diameter to obtain a desired resolution at a given range follows from Table 1. F.ex., if we want a transverse resolution of 15 cm at 30 m range, then we need a lens of 55 cm diameter.

TABLE 1 Examples of imaging parameters for various values of the range and angular resolution. The table gives the lens diameter (D), wave length (λ), frequency (ν), object size (H), object resolution (R), depth of field (FD), image size (H'), image resolution (R'), and number of detector elements (NxN) corresponding to a given range (S) and angular resolution (Θ).

Θ	S (m)	D (m)	λ (mm)	ν (MHz)	H (m)	R (cm)	FD (m)	H' (m)	R' (mm)	NxN
0.005 radians	3	0.15	0.3	4.9	2.2	1.5	+ 0.26 − 0.22	0.11	0.8	146 x 146
	10	0.30	0.6	2.4	7.3	5.0	+ 1.54 − 1.17	0.23	1.6	
	30	0.55	1.1	1.3	21.8	15	+ 8.4 − 5.3	0.41	2.8	
	100	1.1	2.3	0.7	72.8	50	+ 60 − 28	0.81	5.6	
0.002 radians	3	0.40	0.33	4.6	2.2	0.6	± 0.04	0.33	0.9	364 x 364
	10	0.70	0.57	2.6	7.3	2.0	± 0.23	0.55	1.5	
	30	1.30	1.07	1.41	21.8	6.0	+ 1.18 − 1.10	0.99	2.7	
	100	3	2.46	0.6	72.8	20	+ 5.78 − 5.18	2.25	6.2	
0.001 radians	3	0.70	0.29	5.2	2.2	0.3	± 0.01	0.66	0.9	728 x 728
	10	1.30	0.53	2.8	7.3	1.0	± 0.06	1.09	1.5	
	30	2.6	1.07	1.4	21.8	3.0	± 0.29	2.07	2.9	
	100	6	2.46	0.6	72.8	10	± 1.36	4.64	6.4	

SUMMARY

We have presented a concept of an acoustical TV camera, which can be used for many different underwater purposes. The camera lens is a low cost, low absorption, light weight holographic doublet with a diffraction limited resolution over a field of view of $\pm 20^{o}$. The detector promises to be very sensitive, and can be manufactured by integrated circuit technology.

ACKNOWLEDGEMENT

The idea of how to make a suitable hydrophone array is due to J. Bakken at our institute.

REFERENCES

Collins, H. D., R. P. Gribble, T. E. Hall, W. M. Lechelt, J. T. Luebke, J. Spalek, E. M. Sheen, and A. Stankoff (1980). Acoustical Holography Matrix Array Imaging System for the Underwater Inspection of Offshore Oil Platform Weldments, In A. F. Metherell (Ed.). Acoustical Imaging, Vol. 8, Plenum Press, New York.

Fitzpatric, G. L. (1979). Seismic Imaging by Holography. Proc. IEEE, 67, 536-553.

Heier, H. (1980). To be published.

Kino, G. S. (1979). Acoustic Imaging for Nondestructive Evaluation, Proc. IEEE, 67, 510-525.

Ljunggren, S., O. Løvhaugen, and E. Mehlum (1980). Seismic Holography in a Norwegian Fiord, In A. F. Metherell (Ed.). Acoustical Imaging, Vol. 8, Plenum Press, New York.

Macowski, A. (1979). Ultrasonic Imaging Using Arrays. Proc. IEEE, 67, 484-495.

Schmitz, V., and M. Wosnitza (1980). Experiences in Using Ultrasonic Holography in Laboratory and in the Field with Optical and Numerical Reconstruction, In A. F. Metherell (Ed.), Acoustical Imaging, Vol. 8, Plenum Press, New York.

Sutton, J. L. (1979). Underwater Acoustic Imaging. Proc. IEEE, 67, 554-566.

Takahashi, F., K. Suzuki, and T. Kanamori (1980). Digital Signal Processing in Acoustical Focused Image Holography, In A. F. Metherell (Ed.), Acoustical Imaging, Vol. 8, Plenum Press, New York.

Wardley, J., P. H. Brown, and R. C. Crouchee (1977). The design and performance of an improved ultrasonic image converter tube. Ultrasonics Int. Conf. Procs., 121-124.

Kinsler, L.E., and A.R. Frey (1962). Fundamentals of Acoustics, J. Wiley & Sons, Inc., New York.

OPTICAL FIBRE WITNESS DEVICES FOR MONITORING THE INTEGRITY OF OFFSHORE STRUCTURES

B. S. Hockenhull*, J. Billingham*, G. Christodoulou* and K. F. Hale**

*Cranfield Inst. of Technology, Cranfield, Bedford MK43 0AL, UK
**National Maritime Institute, Feltham, Middlesex TW14 0LQ, UK

ABSTRACT

Specially treated optical fibres which fracture at particular predetermined strain levels have been incorporated into devices for monitoring structural integrity offshore. By attaching the fibres firmly to the structure the opening or progression of any cracks along the surface can be monitored. The devices can be attached to the structure subsea by the use of specially developed adhesives and an optical monitor can interrogate the device from a position on the deck and hence determine the strain level and, therefore, structural integrity, at the point in question.

Laboratory studies have been used to monitor the development of strain and the movement of cracks within both steel and concrete samples and this work has been currently supplemented by field trials offshore on the Christchurch Bay Tower test facility. The technique is capable of detecting both opening cracks and the movement of cracks along a surface such as those occurring at nodes in offshore structures.

The apparatus and techniques used presently are inexpensive and simple to operate and understand and for these reasons the technique appears most promising for use in an offshore situation. It also offers the possibility of being able to accurately locate as well as size the cracks within the structure.

KEYWORDS

Offshore structures; structural integrity monitoring; fatigue; fracture; optical fibres.

INTRODUCTION

Offshore structures are required to retain their structural integrity for long periods in a hostile environment. Although structural integrity is related to the degree of redundancy in design, there is no doubt that localized plastic deformation and the growth of cracks by fatigue are important modes of degradation for such structures. The detection and location of any such degradation at the earliest possible stage is important if the structural integrity and related personnel safety are to be maintained.

The most common means of assessing structural integrity is through a conventional inspection programme using visual inspection, ultrasonics and magnetic particle inspection (Mainwaring-Davies, 1975).

This involves the use of divers, submersibles or remotely controlled vehicles (Pass, 1978) and requires the clearing away of any marine growth on the surfaces of the structure together with relatively good weather conditions. The cost of such inspection, particularly related to crack detection, has been said to be about five times more expensive than onshore inspection (Goodfellow, 1976) and indeed the information obtained may be less reliable in association with the high risk levels involved in all diving operations.

In view of the concern regarding the effectiveness of current inspection programmes, there is considerable interest in examining alternative means of assessing structural integrity. One approach is to continuously monitor the integrity of the structure using remote monitoring systems based on vibration analysis (Brown and Huckvale, 1978), or acoustic emissions techniques (Parry, 1977). In both cases, transducers are placed at or close to critical regions of the structure and characteristic signals are monitored for changes indicating that some event, such as crack growth or member failure, has taken place. Such systems are costly to instal, complex in operation, usually involving extensive computing for example, and doubts have been expressed regarding their capability to detect other than major structural damage.

This paper describes an alternative technique being developed at C.I.T. under the sponsorship of the N.M.I. The technique involves the use of specially prepared optical fibres which are capable of detecting small permanent changes in strains (Hale and colleagues, 1978, 1980). These fibres, bonded to a structure, can detect plastic strain in the surface at specific points (Hale, Hockenhull and Christodoulou, 1979). The technique is capable of operating as a continuous monitor, interrogated from the surface or in a semi-continuous rôle where particular critical areas are examined periodically using diver inspection techniques. The method has the potential to provide a cheap, robust and reliable technique for structural integrity monitoring.

BASIC PRINCIPLES

If an optical fibre is used to transmit light then a continuous signal may be received, unless the fibre is broken, when a transmission loss will occur. An optical fibre will fracture when the strain in the fibre exceeds a critical value. This may be utilized in a device by bonding the fibre to a surface undergoing elastic or plastic strain or if indeed cracking were to occur at the surface. Such a system depends upon the strain transfer from the surface to the fibre occurring satisfactorily through the bonding agent. High quality, low loss fibres have been developed for the telecommunications industry and therefore the light source and receiver can be of almost any kind, depending upon the system to be used. In addition to the detection process, it is feasible to apply time domain reflectometry to a system in which the reflected signal from the fracture in the optical fibre is timed to determine the exact position of the fracture along a particular fibre length.

EXPERIMENTAL WORK

Laboratory Tests

The first stage undertaken was to degrade essentially perfect silica fibres, which exhibit fracture strains of about 7%, to lower values, in a reproducible manner. The fibre principally used was a commercial telecommunications fibre of 125 μm diameter, coated with a silicone resin. The fibre is of the step index type in which the outer silica coating controls the total internal reflections which allow transmission. The surface of samples of this fibre was degraded by immersion in an acid solution for times ranging from 1 to 30 hours. Fibres thus treated were then made into tensile specimens by bonding the ends of the fibres onto aluminium strips with an epoxy resin. These specimens were then tested in tension and fracture strains were calculated from the fracture load and Youngs modulus of the fibre or directly from the strain over the gauge length.

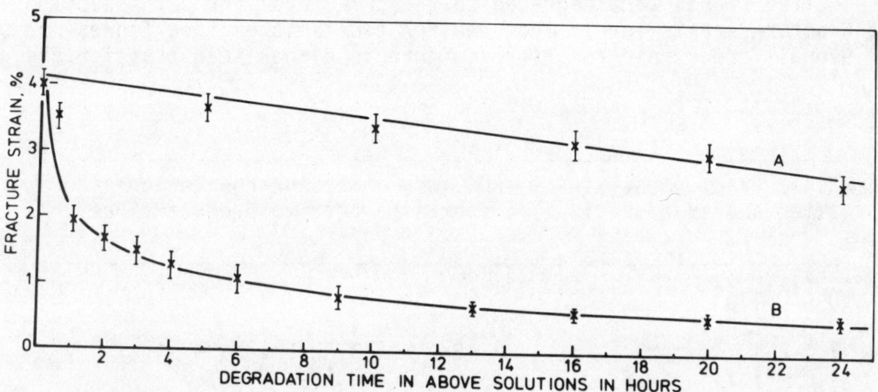

Fig. 1. Fibre fracture strain as a function of degradation time

Figure 1 shows the results of a number of tests. The error bars show the typical range of fracture strains encountered, namely about 25% for all strain values. It was not possible in these tests to measure fracture strains above about 5% due to fracture or pull out in the grips. The surfaces of typical fibres in the normal and degraded conditions were examined in a scanning electron microscope. Figures 2 and 3 show the surfaces obtained. The degraded fibre appears to be notched by the

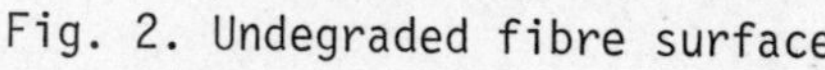

Fig. 2. Undegraded fibre surface

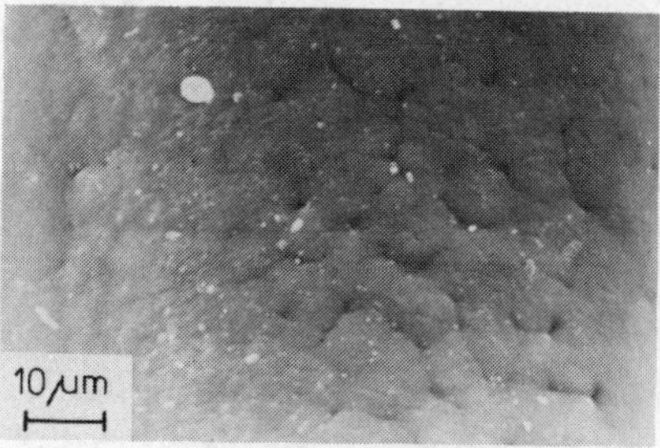

Fig. 3. Degraded fibre surface

chemical attack. No significant change in diameter occurred. It may be reasonably concluded that fibres of a known fracture strain range can be produced in a reproducible manner.

After these preliminary experiments, suitable adhesives were sought to bond the fibres to a steel surface. A variety of surface preparations were used and a number of commercial resin bonding agents investigated. A surface condition produced by abrasion to between 240 and 80 grit was found to be most suitable and was used in all subsequent experiments. A commercial adhesive for underwater application (designated 1941) was finally chosen for most of the experiments.

One set of fibres in 150 mm lengths were bonded to steel strip samples placed in sea water in the laboratory. Some were bonded under water. After seven months, some 10% of all fibres were detached, namely due to crevice corrosion effects, but the remainder showed good bond strength.

A further set of fibre lengths were bonded to steel strip tensile samples in groups. In each case, the fibres were degraded to the same level for each specimen. Typical resulting fracture strain levels measured for four sets of five fibres are shown in Table 1. Overall the strain for fibre failure is higher than that for the degraded

TABLE 1

Fibre fracture strain after degradation	No. of fibres	Strainrange in tensile test to produce fibre fracture
0.8 - 1.0	5	1.2 - 1.6
0.8 - 1.0	5	1.3 - 1.5
1.0 - 1.2	5	1.40 - 1.42
1.0 - 1.2	5	1.50 - 1.52

fibre but by no more than a factor of two and generally less. Figure 4 shows a typical stress-strain curve for a steel specimen to which fibres were bonded. It

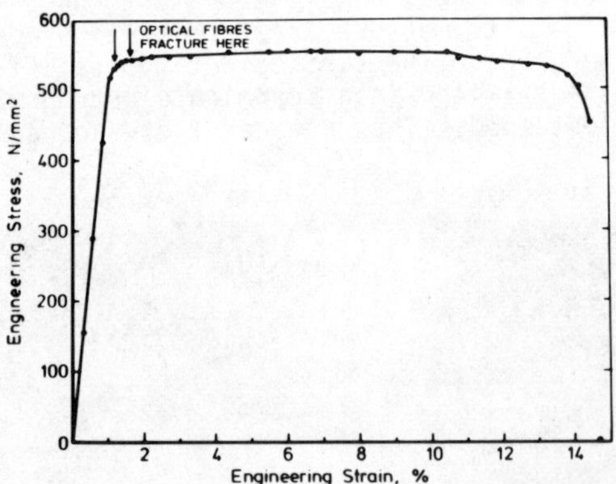

Fig. 4. Stress-strain curve showing fracture strain range of bonded degraded fibres

was noted that when the fibres failed, no light signals were transmitted. It was also noted, that where the fractures occurred, beneath the bonding resin, light was emitted from the illuminated, fractured end. Figure 5 shows this 'bleeding light' effect clearly. In general, more than one fracture took place in each fibre

Fig. 5. Bonded degraded fibre failure in tensile tests

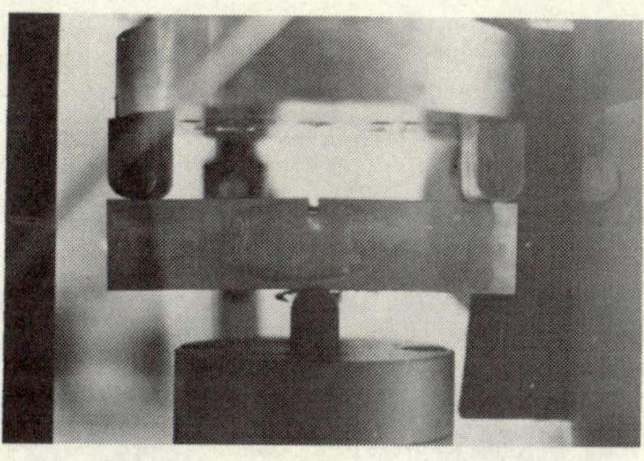

Fig. 6. Bonded degraded fibre failure in fatigue test

and these fractures were distributed along the bonded length of the fibre.

In order to assess the capability of the technique to monitor slow fatigue and growth, degraded fibres were bonded to the side surface of a notched three-point bending fatigue test specimen. When a fatigue crack was initiated by applying a cyclic stress it then grew in a direction normal to the bonded fibres, that is, the crack opening exerted tensile forces on the fibres. The crack growth was monitored using a simple travelling microscope with a resolution of 10 μm while concurrently the fibre integrity was monitored by passing light along the fibre. Fibre fracture occurred in sequence when the crack progressed some way through the sample and this gave rise to transmitted light loss and to a 'bleeding light' effect as shown in Figure 6. One factor which should be noted is that when the crack opening is small, the separation of the fractured ends of the fibre may also be small and under these circumstances the transmission loss is not total but will still be significant and readily detectable. Further fatigue tests have been undertaken using several specimen geometries and with fibres degraded over a range of 0.3% to 1% fracture strain. The results were similarly satisfactory in each case and the effects are reproducible.

To apply the systems practically, there are two requirements, firstly a set of fibres in a suitable geometrical arrangement or 'package' and secondly an optical transmitter and receiver system. Sets of parallel fibres have been prepared using a simple frame with the free ends mounted in ferrules fitted into electrical type connectors. Great care is necessary in the preparation of these connectors, which have to be geometrically correct. The fibre ends also require polishing but the result is a connection system of comparatively low loss (< 4dB/joint). Several optical power monitor units have also been built. These consist of transmitter and receiver units both of which are built into an electrical type connector and appropriate amplifier and signal conditioning systems. The signal level received by the unit provides an output to a printer unit. This is arranged to print the output level at fixed time intervals or alternatively to respond to a sudden change in signal level and print out accordingly. Each printing event is recorded with date and time to within 1 minute.

Optical fibre cables are used to join the input and output to the degraded fibre package. Figure 7 shows the general arrangement of the system set up in the laboratory with two 20m lengths of cables joining the electronic systems to the fibre package. It should be noted that outside the electronic systems, all of the signals are optical and are not subject, therefore, to electrical interference.

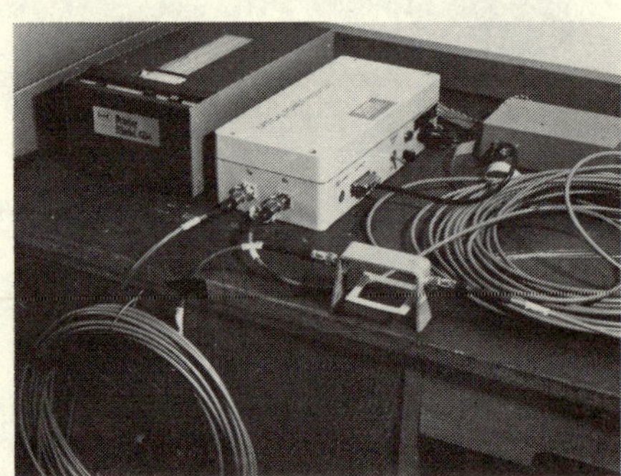

Fig. 7. General laboratory layout of optical fibre monitoring equipment

Fig. 8. The Christchurch Bay Tower

FIELD TRIALS

For preliminary proving tests, unstressed steel plates, with a representative arrangement of fibres for crack growth detection, attached to the plates with adhesive and associated interconnectors have been exposed about half a metre below the low tide level of the sea on the NMI Christchurch Bay Tower, shown in Figure 8, which is located in about 8.5 metres of water about one mile offshore between Christchurch, Hampshire and the Needles, Isle of Wight. The tower is a research facility used for relating wave action to forces on tubular members to facilitate the design of safe yet economical production platforms. Over one hundred environmental parameters are continuously monitored and the data stored on tape recorders for subsequent analysis. Some of the more immediately useable data such as those concerned with wave height are telemetred ashore for management purposes. The tower is therefore ideal for relating operational performance of oceanographic types of instrumentation to environmental conditions.

Figures 9 and 10 show the first plate before and after exposure for 28 days. No failure or debonding of the fibres occurred although 4 metre waves and force 8 - 9 winds were recorded during this time. A second plate was installed for a further month, again without loss of optical continuity, but in this case one of the connectors became detached from the plate. Modifications were made to the design to include clamps on the connectors to strengthen these points and a small sacrificial

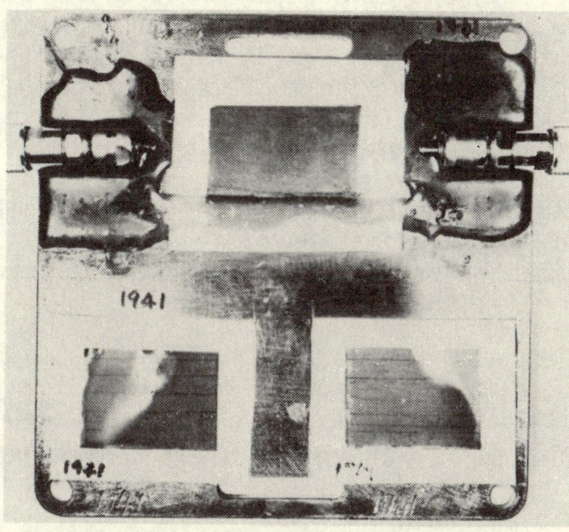

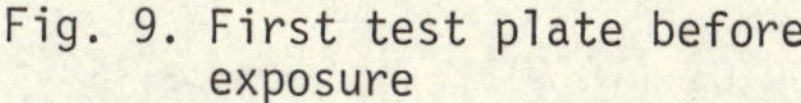
Fig. 9. First test plate before exposure

Fig. 10. First test plate after 28 days' exposure

anode for corrosion control. The third plate was diver installed at 6m depth. Optical continuity is still present and this plate is essentially as-new. It has been inspected and reinstalled after 2 months' exposure, again by a diver using gloves. This demonstrates the feasibility of making satisfactory optical joints under water.

DISCUSSION

The need for structural integrity monitoring is self evident and the technique described has the potentiality to fulfil this rôle in the near future. It has been clearly demonstrated in the laboratory that fibres can be degraded in a controlled manner to fracture strains of about 0.5% or less. Such fibres can be bonded to a metallic substrate and, when this is subjected to tensile loading, strain transfer to the fibre takes place, in a reproducible manner, causing the fibres to fracture at a strain fairly well related to that expected of the degraded fibres. Strain transfer causing fibre fracture also occurs when the fibres are bonded across a region through which a fatigue crack grows. The fibre fracture in each case causes a loss of light transmitted and this may be used as the basis of a structural monitoring system. So far, the behaviour in offshore conditions of the underwater components of the system, that is the optical fibre package bonded to the structure and the connectors and cables, has only been tested for a comparatively short time and not on a large structure, but the indications so far are encouraging.

The system offers a combination of essential simplicity, comparatively low cost, both of the package and the electronic system and the advantage of the continuous monitoring of those parts of the structure which are most critical, such as welded nodal joint regions. The monitoring system at deck level could be used to scan the signals from a large number of degraded fibre packages and offers a simple read-out in which a significant signal reduction indicates either cracking or plastic strain in an unequivocal manner requiring no special skills in interpretation. All of the signals below deck level are optical and therefore free from electrical interference. It is possible to speculate on the potential of the system used in location mode where a long length of fibre showing a fracture could be interrogated using time domain reflectometry. Such optical equipment with a satisfactory resolution capability is currently under development within the telecommunications industry. Alternatively, monitoring a matrix of fibre groups could give some location capability depending on the density and distribution of the fibre packages. The next stage proposed in the experimentation is to use the system to detect fatigue crack growth on a large nodal joint fatigue test at the National Engineering Laboratories and subsequently as large a scale trial as can be mounted.

The basis of the technique is simple but clearly there is much to be learnt yet about the detailed engineering technology of the application of the system, which can only be achieved by larger scale experimentation with an appropriate supporting laboratory programme.

REFERENCES

Brown, D.R. and Huckvale, S.A. (1978). European Offshore Petroleum Conference, 80, London.

Goodfellow, R. (1976). Offshore Engineer, p.23, July.

Hale, K.F., Hockenhull, B.S. and McCarron, V. (1978). SUT Meeting, Structural Integrity Monitoring, April.

Hale, K.F., Hockenhull, B.S. and Christodoulou,G. (1979). Conference, British Society of Strain Measurement, Bristol, September. (To be published in Strain Journal).

Hale, K.F., Hockenhull, B.S. and Christodoulou, G. (1980). NMI Report R.72. OT-R-8006 January.

Mainwaring-Davies, D. (1975). International Conference on Underwater Construction. Cardiff, U.K.

Parry, D.L. (1977). Proceedings, Offshore Technology Conference, 467 Vol. II.

Pass, H. (1978). Petroleum Review, p. 78, July.

TESTING

HYPERBARIC WELDING AT 320 MSW, DEVELOPMENT OF ADEQUATE WELDING PROCEDURES

H. O. Knagenhjelm

Materials Technology Section, Norsk Hydro Research Centre, Norway

ABSTRACT

The paper will review the activities necessary to obtain a suitable welding procedure for joining pipelines of API 5LX-65 quality at 320 MSW.

The development started in 1975 performing the first simulated dive at Taylor Diving and Salvage in New Orleans USA. During three more such simulated dives, filler metals and welding methods were tested on different pipeline steels within the API 5LX-65 specification. These tests were carried out under simulated conditions in a pressure chamber system. Paralell to these tests a number of detail problems were solved by research work carried out mainly at SINTEF, Dept. 34, Trondheim, and at Inst. de Soudure, Paris. This included testing of thirteen brands of electrodes in a minisimulator and establishment of the effect of pressure on weldmetal chemistry. An arc stability study of TIG arcs was carried out in Paris. SINTEF, Dept. 34, was special consultant on welding metallurgy.
Two different procedures were developed paralell to each other, one based on using TIG for root runs and MMA for fillers, the other using TIG for both root and fillers. The results of the qualification complied with DnV's "Rules for the Design Construction and Inspection of Submarine Pipelines and Pipeline Risers" for transport of sweet hydrocarbons. The TIG procedure can be modified to meet the stricter hardness criteria for sour hydrocarbons (260 HV_5). This does not seem possible for the MMA welding. The mechanical properties of the full TIG weld were of almost parent material quality.

The program ended carrying out a full scale Fjord Test using the MMA procedure. The results confirmed those of the simulated tests and the procedures were finally accepted by DnV. Regarding choice of welding procedures for future tie-ins at 320 MSW one would prefer the full TIG procedure as it gives a greater margin of safety, the MMA procedure may cause problems if not followed very carefully.

Norsk Hydro today have adequate procedures and knowledge to specify and supervise tie-ins at 320 MSW.

KEYWORDS

Hyperbaric welding; Tie-in; Deepwater; Tungsten inert gas welding; Pipelines; Procedure development.

INTRODUCTION

Norsk Hydro a.s as Operator on behalf of the Petronord Group and Statoil submitted in Arpil 1976 a report to the Norwegian Department of Industry on the technical feasibility of landing gas from the Frigg Area to Norway.

This report indicated a number of critical operations in connection with a pipeline project in such deep waters. Among these operations were pipeline repair as this had not been carried out at the depths found in the Norwegian Trench. The report therefore concluded that deep water pipeline repair is crucial for the landing of oil and gas in Norway and that a full scale repair test would be desirable. The undertaking of such a full scale test was thereupon taken up within the Petronord Group.

As a similar situation had in the meantime arisen in connection with the Statoil/Mobil Group's Statfjord Transportation System Project (STSP), Norsk Hydro approached Statoil, as the representative of the Statoil/Mobil Group, in order to evaluate the possibility of performing a test jointly.

Of the various alternative techniques which could be developed, the method considered most promising was welding under pressure by divers (hyperbaric welding). This method had so far been successfully employed at depths down to 150 MSW. Analyses carried out by both groups, with the assistance of experts from companies experienced in this field, showed that there was great probability that this method would also be feasible in the depths found in the Norwegian Trench.

On this background the hyperbaric welding method was chosen for the Deepwater Hyperbaric Welding Test (DHWT). The aim of the DHWP was to demonstrate that it is possible to repair pipelines under the depth and bottom conditions present in the Norwegian Trench, by modifying the methods which have been used extensively in the North Sea.

In the Deepwater Hyperbaric Welding Program (DHWP) Norsk Hydro a.s acted as operator on behalf of the Statoil/Mobil Group and the Petronord Group (incl. Statoil). Main Concractor for the DHWP has been Brown & Root Offshore N.V. (B&R NV) with Taylor Divning & Salvage Co., Inc. (TD&S) as responsible for all the underwater operations.

Very little data and knowledge existed for hyperbaric welding this deep (320 MSW). One therefore decided to start with conventional methods and see how they worked. From the problems that arose, a lot of research work had to be carried out to solve problems of welding technical art. Norsk Hydro, Materials Technology Section, was responsible for this work, which will be described in the following. SINTEF, Dept. 34, Trondheim acted as a special consultant on welding metallurgy.

FIRST SIMULATED DIVE

This dive was performed in may 1975. Two coupons of 813 mm O.D. x 19 mm WT pipe were welded under a pressure of 30 bar. The objective of this test was mainly to test out "normal" electrodes of basic type, and to test out welding of the root using TIG. The power supply used was a normal motor/generator DC type. The welding procedures are outlined in Table 1.

Results of TIG Welding

TIG welding of the root caused serious problems. The arc was difficult to ignite and was unstable. The shielding gas of pure argon was too heavy at this pressure and did not give adequate coverage in the overhead and vertical position. A mix of 85% Ar and 15% He solved this problem. Most of the root was welded at 1 bar to be able to do filler passes using electrodes.

TABLE 1 Welding Procedures First Simulated Dive

Pipe dimensions	813 mm O.D. x 19 mm W.T.
Pipe material	Sumitomo, C = 0.05%
Bevel preparation	Angle: 75°, Land: 1.6 mm, gap: 4 mm
Electrodes	Root pass: 3.2 mm TIG Oxweld 65 Fillers : 2.4 mm Atom Arc 7018 2.4 mm Atom Arc 8018-C3 3.25 mm Tenacito 70B-1c 2.5 mm ESAB OK 53.05
Power supply	Motor Generator (MG)
Preheat	100°C/ No preheat

Results of MMA Welding

One of the points was to compare American electrodes to European ones. It seemed that the diver/welders had some diffuculties in handling the European types. This was found to be a function of limited experience and electrode diameter. It was extremely difficult to handle the puddle of the 3.25 mm Tenacito electrode, especially in the overhead position, here the weld metal would easily form "grapes". Due to the pressure the puddle is rather thin and shiny and freezes faster than under atmospheric conditions, this also increases the danger of undercuts. Even with 2.4 mm electrodes it takes skill and care to put in the beads smothly and even.
The full weave technique was used for all electrodes except OK 53.05 which was welded using stringer beads. Preheat was only used for electrode Atom Arc 7018. The rapid freezing of the puddle causes poor flow in the weld process, this forms the typical rippled surface of the weld, see Fig. 1.

The testing of the mechanical properties was performed at SINTEF Trondheim and Institute de Soudure Paris. The results are given in Table 2 and 3, were weld metal chemistry and mechanical properties are listed.

It can be seen that a carbonpickup of 0.1-0.16% takes place and that manganese and silicon are burnt off to a variable extent. The oxygen content is raised to 800-1100 ppm, which is considerably higher than under atmospheric conditions.

The impact values where rather low in the root region for all electrodes except Tenacito. In the rest of the weld the values are acceptable, but did not offer a great margin of safety. It was felt that an impact value of average 70 J at -10°C would be a safe level. The DnV requirements were 47 J (min. 34 J) at -10°C.

The hardness varied a lot, but preheat did not seem to have any influence. The level stayed below 300 HV_5, but did not meet the requirements for transportation of sour hydrocarbons, max. 260 HV_5.

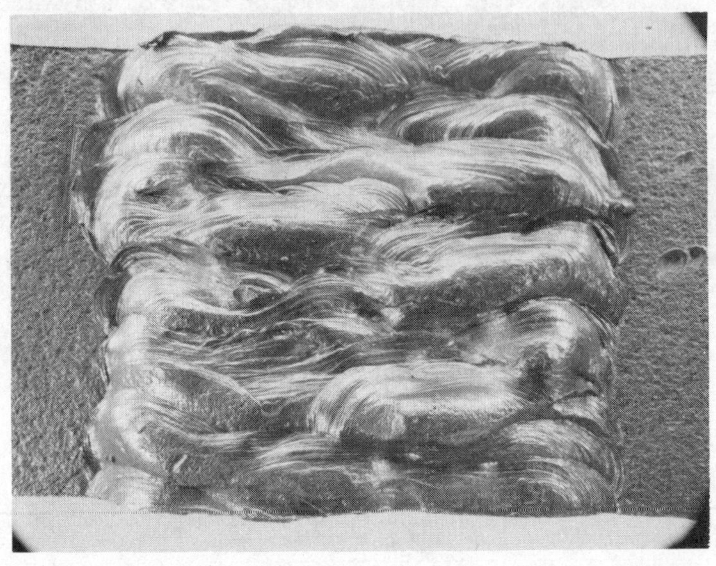

Fig. 1. Typical surface of MMA welds

TABLE 2 Chemistry of Welds from First Simulated Dive

Electrodes	C %	O ppm	Mn %	Si %	Ni %	(C%)(O%)·10^3
Tenacito 70 B 1 bar	0.033	n.d.	1.64	0.32	2.26	-
" 30 bar	0.12-0.146	620-790	0.66-0.9	0.16-0.23	1.5-2.0	7.4-11.5
Atom Arc 8018 1 bar	0.05	n.d.	0,97	0.24	1.0	-
" " " 30 bar	0.18	610-690	0.81	0.19-0.23	0.79-0.95	11-12.4
Atom Arc 7018 1 bar	0.04	n.d.	0.98	0.34	0.03	11.9-22.6
" " " 30 bar	0.176-0.20	677-1130	0.61-1.01	0.27-0.40	0.03	
OK 53.05 1 bar	0.1	n.d.	0.8	0.4	-	
" " 30 bar	0.225	560	1.46	0.44	0.03	12.6

TABLE 3 Mechanical Properties of Welds from First Simulated Dive

Electrode	Hardness HV_5		Charpy-V(J) $-10^{o}C$		Tensile MPa	Sidebend
	HAZ	Weld	Top weld	Bottom weld	-	-
Tenacito 70 B 30 bar	175-286	201-286	49-67	54-79	585	12.5% Minor-defects
Atom Arc 8018 30 bar	161-232	192-257	39-47	35-46	587	No defects
Atom Arc 7018 30 bar	192-232	185-265	41-55	22-42	565-584	25% gross defects
OK 53.05 30 bar	192-232	210-303	55-65		-	-

Conclusions after First Simulated Dive

Welding of the root pass seemed to be the main problem and one had problems with arc stability. It was decided to try to improve arc stability.

Acceptable joint quality was obtained using MMA welding, Tenacito 70 B-lc gave the best weld metal properties. However, impact values of the root were rather low. It was clearly demonstrated that welding procedures used for welding at 150 MSW could not readily be applied at 300 MSW, adjustments and research had to be carried out.

SECOND SIMULATED DIVE

Second simulated dive was performed in March 1976. Three coupons of 406 mm O.D. x 19 mm W.T. pipe were welded under a pressure of 32 bar.

The main purpose of this dive was to evaluate.

- Electrode Tenacito 70B-lc with smaller diameter (2.5 mm).
- The effect of increasing the number of TIG passes in the weld root (especially the effect on Charpy-V values).
- Possible improvement of arc stability using a high frequency electronic power supply (EPS).
- Possible improvement of arc stability using MMA for welding of the root pass.
- Different types of pipe material.

The welding procedures are outlined in Table 4.

Results of TIG Welding

The new electronic power supply (EPS) improved the arc stability, and welding of the root no longer caused any serious problems. The main results of the TIG welding are given in Table 5.

The impact values of 6 and 8 layers represent a 100% TIG weld, 2 and 4 layers have a part Tenacito 70B-lc weld metal. The trend is quite clear; welding of the root with more than 1 layer of TIG improves impact values. If one compare impact values in root positions using TIG or MMA, values are increased from 31-38 J at -10°C for MMA to 58 J for TIG. 58 J is about the level of MMA weld metal for fillers. One other important feature is that no significant change in carbon content or oxygen content takes place in the TIG weld, this means reduced risk of cracking of the root.

The following typical weld metal chemistry of TIG and MMA illustrates this clearly:

TIG - 0.063% C, 130 ppm O
MMA - 0.145% C, 695 ppm - 1100 ppm O

It was clear that a full TIG weld would be very attractive from a weld quality point of view.

Small problems were still encountered with respect to welding torches etc., but as a whole the TIG welding functioned very well.

TABLE 4 Welding Procedures Second Simulated Dive

Pipe dimensions	406 mm O.D. x 19 mm W.T.
Pipe materials	Mannesmann, C = 0.07% and C = 0.14%
Bevel preparation	Angle: 75°, land: 1.6 mm, gap: 4 mm Angle: 60°, land: 3.2 mm, gap: 3.2 mm
Electrodes	Root pass: 3.2 mm Linde Mi88 2.5 mm Phillips 36 D 2.5 mm BOC Fortrex Fillers : 2.5 mm Tenacito 70B-lc 2.5 mm Murex Fortrex 8018-Cl 3.2 mm TIG Linde Mi-88 (evaluation of 2, 4, 6 and 8 passes in root)
Power supplies	Motor Generator (MG) Electronic Power Supply (EPS)
Preheat	100°C/ No preheat

TABLE 5 Mechanical Properties of TIG Welds

Number of passes	Charpy-V -10°C Weld root, (J)	Hardness HV_5	
		HAZ, mean	Weld, min-max
2	54-60	232	244-263
4	79-82	210	263-286
6	144-154	232	224-257
8	124-155	221	232-271

Results of MMA Welding

Welding of the root gave operational problems although the results using Phillips 36 D were passable. Using BOC Fortrex caused cracking of the root. As some hydrogen cracking was observed in connection with MMA welding of the root, combined with carbon contents of 0.18%, it was concluded that welding the root using TIG was a lot safer and produced nicer looking welds as well.

Preheat seemed to have no effect on either hardness nor impact values. Preheat on site will however, assure a dry groove and might also lower the hydrogen level. The smaller size electrode (2.5 mm) made it possible to weld rather easily in all positions, although the puddle is rather difficult to handle and undercuts occur easily. It takes a lot of training for the welders to be able to weld properly.

A summary of the mechanical properties is given in Table 6.

Weld metal chemistry was about the same as in first simulated dive. That is %C in the range 0.12-0.15 for the weld. In the root pass using electrodes carbon contents of up to 0.19% was observed. This is concluded to be the reason for the hot cracking of the root using BOC Fortrex 2.5 mm electrode.

TABLE 6 Mechanical Properties of Welds Second Simulated Dive

Electrodes		Preheat	Pipe	Tensile	Chappy-V - 10°C (J)		Hardness HV_5		
					Weld metal				
Filler	Root/ layers	°C	% C	MPa	top	root	weld top min-max	weld root min-max	HAZ Mean
Tenacito 70B-1c	Philips 36D/1	150	0.07	576 HAZ	61-62	31-43	182-225	232-251	263
"	"	NO	0.07	504 WM	60-64	25-37	192-234	251-289	282
"	TIG/1	150	0.07	560 HAZ	48-59	48-67	199-227	225-289	265
"	"	NO	0.07	561 PP	58-65	57-60	185-210	192-277	263
"	Philips 36D/1	100	0.14	652 HAZ	57	28-35	196-221	221-232	227
"	TIG/1	100	0.14	630 HAZ	50-54	54-60	183-192	221-286	217
-	TIG/1	NO	0.14	707 HAZ	-	52-63	201-210	244-257	243
Tenacito 70 B-1c	TIG/4	NO	0.14	634 PP	55-60	79-82	192-221	263-286	224
"	TIG/2	NO	0.14	595 HAZ	55-60	54-60	210-232	244-263	232
"	TIG/8	NO	0.14	643 HAZ	48-52	125-155	192-205	232-271	223
"	TIG/6	NO	0.14	634 HAZ	52-58	144-154	192-201	224-257	232

PP: Parent plate WM: Weld metal
HAZ: Heat affected zone

Conclusions after Second Simulated Dive

- Acceptable weld metal quality was achieved at a carbon content of 0.14%, a lower level is desireable.
- Root welding using MMA did not improve impact values, TIG root welding, one layer, improved impact values in the root to MMA level for fillers. More layers of TIG improved impact values further, and all TIG weld metal had about twice the impact values of MMA weldmetal.
- Preheat at $100^{o}C$ seems to have little effect on impact values or hardness.
- An acceptable joint quality using one layer of TIG for the root and Tenacito 70B-lc for fillers was demonstrated. However, a higher margin of safety regarding impact values would be desireable.
- Hardness level does meet requirements for sweet hydrocarbons, max. 300 HV_5, but not for sour, max 260 HV_5.
- A further evaluation of electrodes will be needed to raise impact values, and possibly reduce hardness.
- Arc stability of TIG was very good using the electronic power supply.

EVALUTION OF ELECTRODES AND OTHER STUDIES

As mentioned, one were not satisfied with the impact values obtained using MMA electrodes. In order to optimize the choice of electrodes and to learn the basic mechanisms of weld metallurgy at 32 bar pressure, the following activities were carried out:

- Design and construction of 8 l minisimulator at SINTEF, Trondheim for testing of electrodes and weld metal chemistry.
- Testing of 13 brands of electrodes at SINTEF, Trondheim.
- Evaluation of effects of pressure on weld metal chemistry. SINTEF, Trondheim.
- Evaluation of transformation behavior (CCT-diagrams) for the pipe steels to be used. SINTEF, Trondheim.
- Arc stability studies of the TIG process at Institute de Soudure, Paris.
- Evaluation of risk of hydrogen cracking, registration of cooling rates when welding in the habitat.

Minisimulator Testing

The minisimulator is shown in Fig. 2.

In order to check whether the welding in the simulator was comparable to simulated dives, welding was first carried out according to procedures used in second simulated dive. It was concluded that the results were quite similar, a slightly lower carbon content, higher impact values and lower hardness were observed, but the differences were quite small. Testing therefore commenced evaluating different brands of electrodes and studying the effect of pressure on weld metal chemistry and hydrogen pickup.

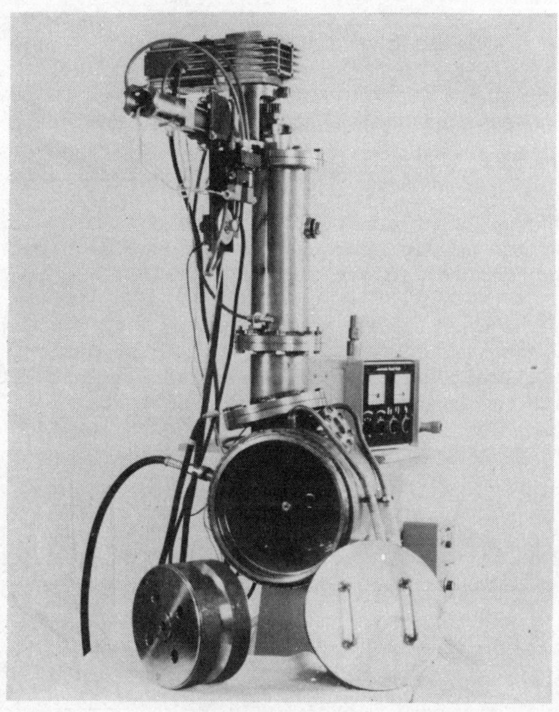

Fig. 2. 8 liter minisimulator

Effects of pressure on weld metal chemistry and hydrogen pickup. Both a theoretical and experimental study was carried out at SINTEF, Dept. 34, Trondheim. For reference see Christensen (1976, 1977a, 1977b).

The chemical reactions taking place in the weld metal under helium + 2% oxygen pressure is not yet fully understood. Some empirical formulas have been derived based on the effect of the observed constriction of the arc on the reactions. One would expect the [C%]·[O%] to increase in direct proportion to pressure,

$$m_p = m_1 \cdot P \quad (1)$$

where m_p is [C%] [O%] at P bar
and m_1 is [C%] [O%] at one bar

In the same way hydrogen pickup would be expected to follow Sievert's law which easily could be expressed

$$[H]_p = [H]_1 \sqrt{P} \quad (2)$$

where $[H]_p$ is the hydrogen content in ml/100 g deposit at P bar
and $[H]_1$ " " " " " " " at one bar

The experimental results however, showed that $[H]^2$ and [C%] [O%] did not increase in direct proportion to pressure. For the hydrogen this could be logically explained by the constriction of the arc as pressure increases which results in only a partly coverage of the arcplasma over the puddle. A part of the puddle is therefore exposed to a non hydrogen containing atmosphere which means that some hydrogen will be ventilated off to the surrounding atmosphere. It was found that the hydrogen content under pressure can be expressed as:

$$[H]_p = [H]_1 \quad \sqrt{P/f} \quad f = \frac{m_1}{m_p} \cdot P \tag{3}$$

f has been determined experimentically. This means that by knowing the hydrogen potential at 1 bar one can predict the hydrogen pickup at P bar. At 30 bar this means that the hydrogen content in the welds are two times higher than at 1 bar.

Regarding the reaction between carbon and oxygen in the weld metal, it is found that the increasing pressure supresses the formation of carbonmonoxide. This means that more oxygen is available for burnoff of manganese and silicon which is also observed. It also leads to an increase in oxygen content in the solidified weld metal approaching a limit of approximately 1000 ppm.

The experimental results are shown in Fig. 3, 4, 5, 6 and 7.

The carbon content is increased by about 0.1% to 0.15% at 30 bar, manganese is reduced by ≈ 0.4%, silicon by ≈ 0.2%. This means that these losses must be compensated in the composition of electrodes for hyperbaric use.

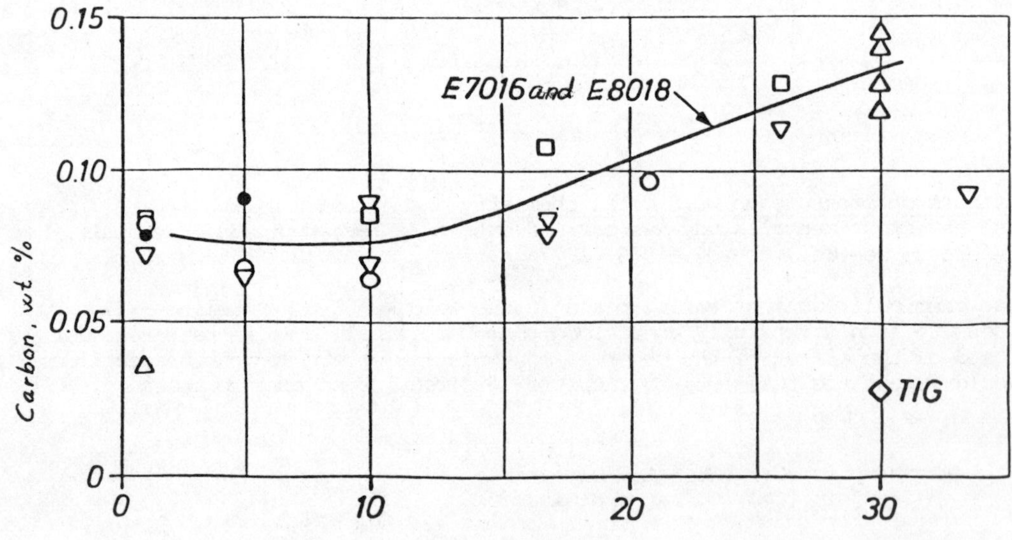

Total pressure P, bar

Fig. 3. Carbon content as a function of pressure

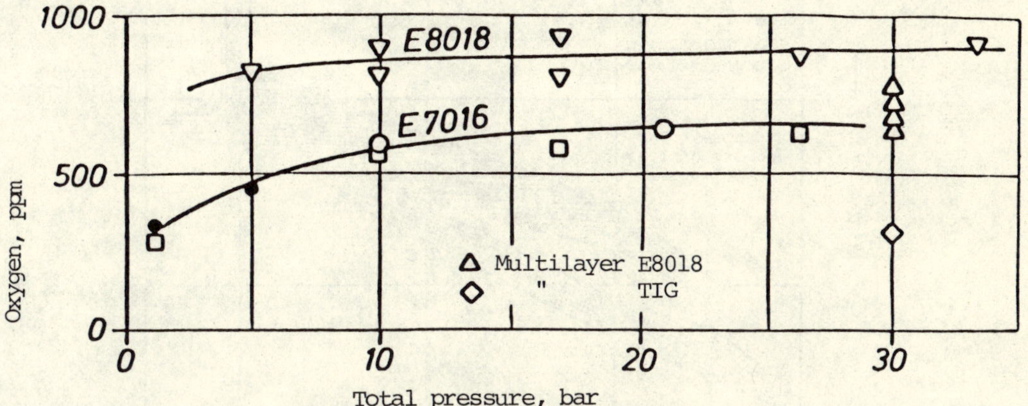

Fig. 4. Oxygen content as a function of pressure

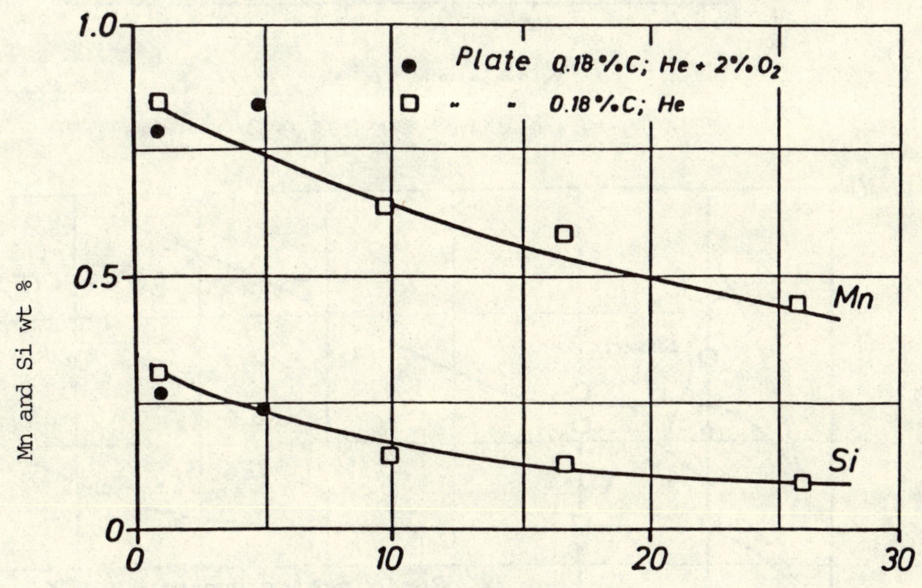

Fig. 5. Manganese and silicon content as a function of pressure. Electrode E7016

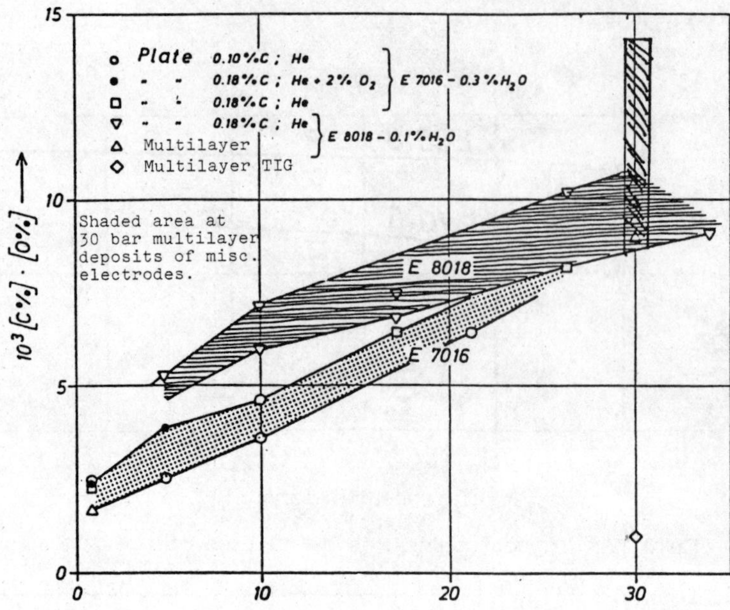

Fig. 6. Effect of pressure on [C%]·[O%]

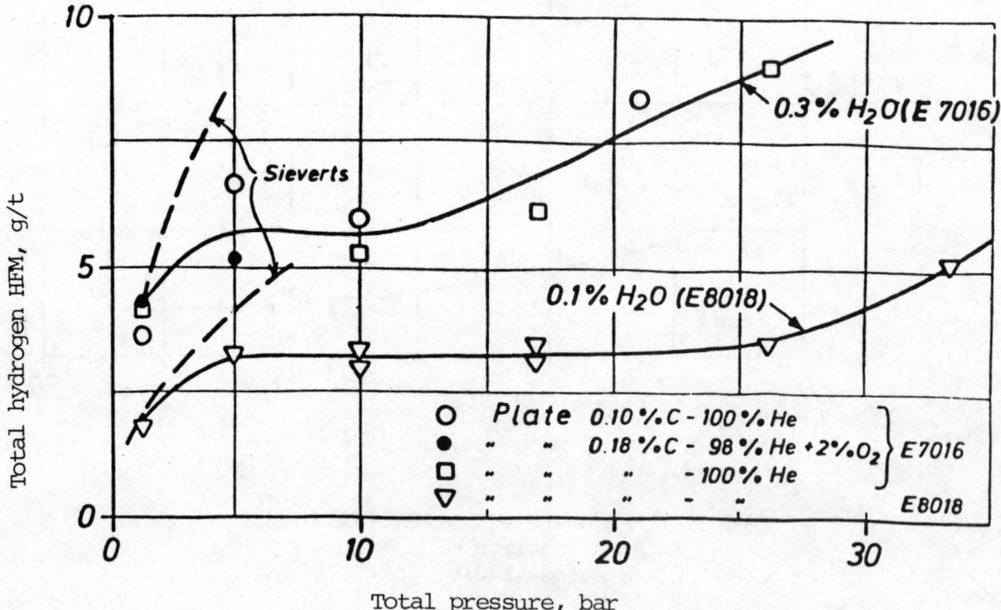

Fig. 7. Hydrogen content as a function of pressure

<u>Testing of electrodes.</u> In order to pick out suitable electrodes for testing a roundtrip was made to different electrode manufacturers to have their points of view.

In the end 13 electrodes were picked for testing, they were of the following main types:

- Basic low-alloy type.
- Basic 2.5% Ni type.
- Basic 1.0% Ni type
- Cellulosic low-alloy
- Basic austenitic type
- Basic austenitic/ferritic type
- Nickel base type

These electrodes were tested in the minisimulator with respect to welding performance, carbon pickup and notch toughness at $-10^{\circ}C$. Four electrodes were picked out for further testing in manned tests:

- 2 of basic 2.5% Ni type
- 1 of basic austenitic type
- 1 of austenitic/ferritic type

<u>Effect of Environment Moisture</u>

It was concluded that the risk of hydrogen pickup by infiltration of habitat gas and high humidity into the arc is not serious. By proper treatment of the electrodes one assumed that max 2 hours exposure to the habitat atmosphere would not result in any dangerous water absorption in the cover. The electrodes were baked at $350^{\circ}C$ for 1 hour and stored at $150^{\circ}C$. Electrodes for use were to be pressurized in dry helium and kept in containers holding $70^{\circ}C$ before use.

However, later it became known that the waterabsorption in the cover would rise rather dramaticly by raising RH from 70% to 90%. These data were published for atmospheric conditions by Oerlikon. The main trend is shown in Fig. 8.

The rules says max 2 hr exposure to the habitatatmosphere of 90% RH. Normally the electrodes will only be exposed to the atmosphere for about 10 min before used, but it is seen that an almost immidiate pickup of 0.5% takes place at 90% RH. This would result in the order of 13 ml/100 g deposit hydrogen, which might cause problems. One does not know whether the absorption mechanisms are the same under atmospheric conditions and at 30 bar, a project will be initiated to establish water pickup as a function of time and pressure. It is known that some diving companies have very strict rules for exposure times of electrodes to the high humidity habitat atmosphere.

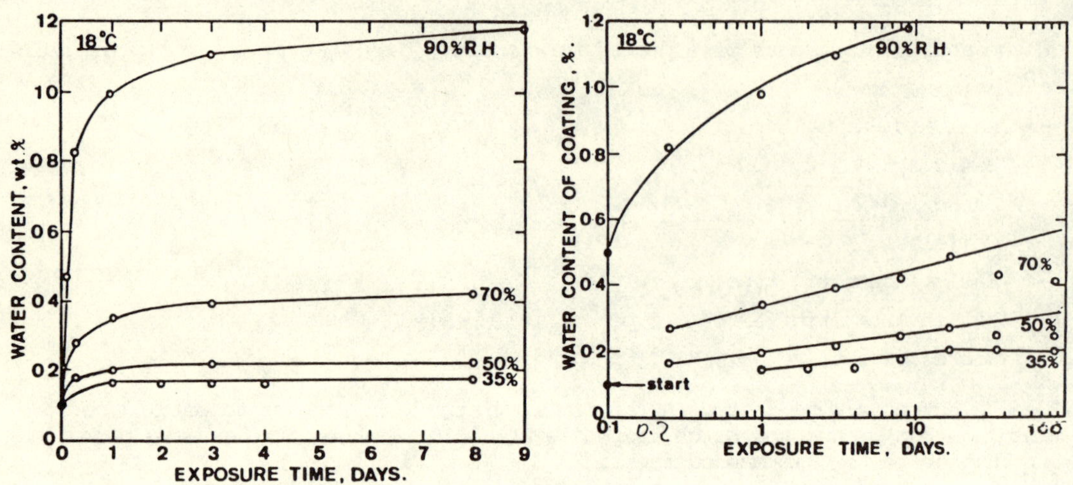

Fig. 8. Absorption of water in cover of Tenacito 70B-lc type (Oerlikon)

Cooling Rates under Hyperbaric Conditions

To be able to relate expected hardness and welding parameters, one measured cooling rates nearby the weld from the inside just underneath the groove surface. All measurements were treated by a computer. Heat inputs and temperatures were registrated, the cooling time between 800 and 500°C was also calculated. This registration was done during third and fourth simulated dive and a summary of the results are given in Table 7.

TABLE 7 Distribution of cooling time 800 to 500°C

Interval seconds	Number of events TIG	Number of events MMA
1-5	3	0
5-10	0	6
10-15	5	6
15-20	5	4
20-25	2	7
25-30	3	1
30-35	4	0
35-40	2	0
40 —	2	1

It can clearly be seen that TIG welds give a somewhat slower cooling. The slowest cooling of 25-40 s has been obtained during filler passes with TIG at rather high current. The very fast cooling rates origins from root welding with TIG. Coupled with CCT diagrams for the steel, expected max hardness in the HAZ may be estimated.

Other Studies

In addition to the above mentioned, Inst. de Soudure studied stability of TIG arcs in Ar and He and different additions to these. The conclusion was that argon gave the most stable arc and that there was no realistic way to improve this. As mentioned 15% He has been added to the Ar for flow reasons, but this does not influence arc stability.

SINTEF, Dept. 34, Trondheim acted as consultant on a regular basis in welding metallurgy and also established CCT diagrams for the pipe steels involved. The following items were also studied:

- Effect of postheat
- Effect of preheat
- Effect of restraint

THIRD SIMULATED DIVE

Third simulated dive was performed in July 1977. The purpose was to compare covered electrodes selected from the minisimulator testing program at SINTEF, and to evaluate the power sources to be used in the DHWT, (EPS and RPS). Second simulated dive showed that excellent weld quality was obtained by TIG welding with high frequency EPS. It was felt that increasing the welding current and consequently the weld metal deposit rate, TIG welding would be compatible to MMA welding with electrode diameter 2.5 mm. A comparison between electrode Tenacito 70B-1c, Esab OK 67.52, Philips 75 and Linde-Mi88 for filler passes was therefore planned for (Philips RSP-B was left out because it produced a slag that was difficult to remove). Second simulated dive showed that by increasing the number of TIG passes the Charpy-V values in the root were improved, and 3 passes were selected for third simulated dive. Second simulated dive also showed some peculiar HAZ hardness results, giving higher values for the 0.07% C steel than for the steel containing 0.14% C. It was found necessary to look more closely into the effect of steel composition, and for third simulated dive four different steel qualities were chosen. At this stage the Statoil-Mobil group had joined the project and the choice of pipe dimensions and steel qualities were influenced by the planning of the STSP.

Two coupons of 914 mm O.D. x 25.4 mm W.T. pipe were welded under a pressure of 32 bar. The welding procedures are outlined in Table 8.

TABLE 8 Welding Procedures Third Simulated Dive

Pipe dimensions	914 mm OD x 25.4 mm W.T.
Pipe material	Mannesmann , C = 0.06% Nippon-Kokan , C = 0.08% Bergrohr-Thyssen, C = 0.10% Italsider , C = 0.14%
Bevel preparation	Angle: 60°, land: 1.6 mm, gap: 4 mm
Electrodes	Root pass, hot pass and first filler: TIG 3.2 mm Linde-Mi88 Fillers: 2.5 mm Tenacito 70B-1c 2.5 mm Esab OK 67.52 2.5 mm Philips 75 3.2 mm TIG Linde Mi-88
Power Supplies	Electronic Power Supply (EPS) Rectified Power Supply (RPS)
Preheat	100°C No preheat

Results of TIG Welding

As in second simulated dive TIG welding caused no serious problems and the EPS was found superior to RPS for the root pass, for the filler passes no great difference was observed between RPS and EPS, but RPS power supply for TIG welding demands higher skill than the EPS. No special training in welding TIG at high amperage had been carried out prior to the test. One had anticipated a welding current of about 200 A in quadrant one (flat to vertical position). The ceramic nozzle melted at this amperage and the maximum welding current used was 150 A. The weld was completed after 26 passes taking 580 min welding time. Welding on quadrant using MMA for fillers took 463 min. In addition one had problems with tungsten electrodes getting stuck and blunting. It was therefore not welded a full quadrant in addition to the first as planned, but half of this groove was filled up with Tenacito 70B-1c.

TIG welding was more timeconsuming than MMA, but this was partly due to unsufficient heat resistant nozzle material and limited training in handling high current TIG and optimizing welding technique as planned. Arc stability and welding performance were superior to MMA welding. The results of the mechanical testing are showed in Table 9.

The mechanical properties of the TIG weld metal and HAZ are very near parent plate values and oxygen content is very low, 1/10 of MMA.

The hardness level has possibilities of meeting the requirements for transportation of sour hydrocarbons, 260 HV_5 or lower.

The COD value of 0.74 mm indicates a very ductile weld free of microdefects. The chemical analysis is given in Table 10.

TABLE 9 Mechanical Properties of Welds from Third Simulated Dive

Filler	Preheat ^{o}C	Pipe % C	Tensile MPa	Charpy-V (J) -10°C, min-max values			
				WM	F.L.	F.L.+2	F.L.+5
TIG	No	0.06	527 P.P	249 285	281 294	- 294	- 294
TIG 15 layers + Tenacito 70B-lc	"	0.06	534 F.L.	71 84	55 80	165 294	- 294
Tenacito 70B-lc	"	0.08	581 P.P	-	-	-	-
Philips 75	"	0.08	601 F.L	66 73	71 83	159 165	150 213
Tenacito 70B-lc	100	0.1	-	75 77	71 87	126 189	193 224
Philips 75	No	0.1	548 F.L.	60 67	69 96	95 199	175 204
OK 67.52	100	0.14	570 P.P	96 116	109 120	137 163	167 195
OK 67.52	No	0.14	555 P.P	85 93	98 110	114 194	147 182

Table continued next page.

Filler	Preheat °C	Pipe % C	Hardness HV_5, min-max values					COD mm
			WM Top	WM root	HAZ top	HAZ root	P.P	
TIG	NO	0.06	244 266	244 254	232 260	234 262	201 232	0.74
TIG 15 layers + Tenacito 70B-1c	"	0.06	229 239	232 234	246 299	223 241	206 225	0.274
Philips 75	"	0.08	206 274	241 251	236 329	236 260	201 234	0.188
Tenacito 70B-1c	100	0.1	178 190	246 280	214 239	239 293	214 223	0.22
Philips 75	No	0.1	206 257	251 274	254 321	249 283	223 234	0.157
OK 67.52	100	0.14	236 241	214 262	219 329	187 341	188 225	0.444
OK 67.52	No	0.14	187 208	206 251	223 299	195 260	192 232	0.303

WM : Weld metal HAZ : Heat affected zone
F.L. : Fusion line F.L. + 2 : 2 mm from fusion line
P.P. : Parent plate F.L. + 5 : 5 mm from fusion line

TABLE 10 Chemical Composition of Welds from Third Simulated Dive

Filler	C %	Mn %	Si %	Cr %	Ni %	0 ppm	$[0\%]\ [C\%]\cdot 10^3$
TIG	0.049	1.2	0.37	-	1.67	106	0.52
TIG + Tenacito 70B-1c	0.05 0.14	1.18	0.37	-	1.62	108	0.54
Philips 75	0.15	0.43	0.23	-	2.03	1330	20.0
Tenacito 70B-1c	0.13	0.69	0.21	-	1.91	1320	17.2
Philips 75	0.14	0.41	0.21	-	2.08	1340	18.8
OK 67.52	0.23	6.4	0.63	15.9	8.4	560	12.9
OK 67.52	0.23	6.5	0.76	16.6	8.7	735	16.9

Results of MMA Welding

Type of power source, RPS or EPS, was found to be of small importance to arc stability and weldability of MMA welding. All three electrodes performed well and gave little difficulties. The results of the mechanical testing and chemical analysis are given in Table 9 and 10. OK 67.52 showed strange micro cracking and the selective corrosion aspect (austenite-ferrite)and the high carboncontent of 0.22% caused a decision not to develop this electrode any further. The Tenacito 70B-lc and Philips 75 were quite equal in performance and test results. However, Tenacito 70B-lc was specific composed taking into concideration carbon pickup and burning off of Mn and Si. Tenacito 70B-lc gave slightly better impact values and COD values. Due to more experience using Tenacito 70B-lc, this electrode was chosen for the procedure qualification test. The mechanical properties were found to have a proper margin of safety against the DnV requirements.

Conclusions after Third Simulated Dive

TIG welding is superior with respect to quality. The welding was 25% slower than MMA, but this can be overcome by change of equipment and welding technique. One decided to evaluate the process further as backup procedure in fourth simulated dive. The test proved that 3 layers of TIG for the root and fillers with Tenacito 70B-lc would meet the DnV requirements for transportation of sweet hydrocarbons.

FOURTH SIMULATED DIVE

The fourth simulated dive was performed in November 1977. The purpose was to qualify the welding procedures to be used in the DHWT, and to test the incapsulated power sources. Based on the experience from the previous work, the procedure listed in Table 11 was selected as the main procedure.

TABLE 11 Main Welding Procedure for the Qualification Test

Pass no	Welding Process	Electrode	Power supply	Bevel angle	Preheat
Root pass hot pass 1. filler	TIG	Linde-Mi88	EPS	75°	$100^{\circ}C$
Remaining fillers and cap	MMA	Tenacito 70B-lc			

Two coupons made from 914 mm O.D. x 25.4 mm W.T. pipe were welded with this procedure, one made from Mannesmann steel with 0.06% C, the other from Bergrohr-Thyssen with 0.10% C.

0.10% C was considered to be a realistic carbon content for a future pipe line. However, third simulated dive indicated that the DnV hardness requirements might be the most difficult to meet, especially if sour hydrocarbons are present (maximum 260 HV_5) and the 0.06% C pipe was selected to evaluate if this requirement could be met by using an extra low carbon content. It was also decided to weld a third coupon to evaluate different procedures which might reduce the hardness values. These procedures are listed in Table 12.

For the DHWT the RPS was intended to be used as backup for the EPS, and this power supply was evaluated on the third coupon.

Third simulated dive showed that high quality welds could be obtained by all TIG welding, and the procedure outlined in Table 12 quadrant 1 and 3, were evaluated as a back up procedure. Prior to the test, the diver/welders went through a training program. The purpose of this program was to ensure familiarity with the welding procedures, especially the high current TIG welding technique, which had been developed at Norsk Hydro Materials Technology Section. At the end of the training program each welder made a performance test on the actual procedures.

TABLE 12 Alternative Welding Procedures for the Qualification Test

<table>
<tr><th>Quadrant</th><th>Pass no.</th><th>Welding Process</th><th>Electrode</th><th>Power supply</th><th>Preheat</th></tr>
<tr><td rowspan="2">1 and 3</td><td>Root pass
Hot pass
1 filler</td><td>TIG</td><td>Linde-Mi88</td><td>RPS</td><td rowspan="2">100°C</td></tr>
<tr><td>Remaining fillers and cap</td><td>TIG</td><td>Linde-Mi88</td><td>EPS</td></tr>
<tr><td rowspan="3">2 and 4</td><td>Root pass
Hot pass
3 fillers</td><td>TIG</td><td>Linde-Mi88</td><td rowspan="3">RPS</td><td rowspan="3">100°C</td></tr>
<tr><td>Remaining fillers
Cap</td><td>MMA</td><td>Tenacito 70B-lc</td></tr>
<tr><td>Washing of cap</td><td>TIG</td><td>None</td></tr>
</table>

TIG Welding

Prior to the test the diver/welders had performed a testing program to perfectionate high-current welding technique. A new heat resistant nozzle had been made and the torch had been modified. In order to make the torch more heat resistant and avoid sticking of the electrode due to overheating of the collar, NH wanted water cooled torches. This was not possible due to short time. As one of the main items of this test was to evaluate the TIG process's potential, some special tests were performed:

- Registration of welding time etc., time studies.
- Deposition rate testing in air and under hyperbaric conditions for both MMA and TIG welding.

Time studies and deposition rates. The duty cycle at 1 bar was 0.5-0.6 both for MMA and TIG welding. The deposition rates showed 1.07 kg/archour for MMA welding. For TIG welding it varied with welding current, ranging from 0.37 kg/archour at 160 A to 1.15 kg/archour at 280 A. This clearly demonstrated the ability of TIG welding to compete with MMA welding on speed, if one uses the right technique. Under hyperbaric conditions at 32 bar for full TIG welds the duty cycle was about 0.32 against 0.45 for MMA. This was due to some problems with the torches due to overheating and extreme heat to the welders hands. The average deposition rate for

MMA welds were 0.81 kg/archour against average 0.88 for TIG welding. The maximum obtained by the most skilled welder was 1.27 kg/archour. This is 50% more than MMA welding. The spesific deposition rate test gave the results listed in Table 13.

TABLE 13 Deposition Rates of TIG Welding at 32 Bar

Welding current (A)	Deposition rate kg/archour
115	0,31
160	0,68
200	1,41
220	2,14

Melting of cups was not a problem anaymore.
Total welding time for a half coupon was 17 hr and 6 min using TIG for fillers. Using MMA welding with a somewhat wider groove (75°) it took 17 hr and 12 min for a half coupon. The conclusion is that TIG welding does not take significant longer time than MMA welding.

Welding performance. The arc was stable. EPS was superior to RPS for root welding. The maximum current the welders could handle in all positions was 160-195 A. Maximum welding current was 215 A in quadrant 1. (flat to vertical).

Heat problems were quite severe to welders hands and welding torches. Feeding of the filler wire caused contamination of the tungsten which therefore had to be changed rather often. This problem could be avoided by automatic wire feed (Semiautomatic TIG torch). It was concluded that for further development after DHWT, a semiautomatic TIG-torch, watercooled should be tested to solve these problems.

Mechanical properties. The mechanical properties are listed in Table 14. The weld metal chemistry is listed in Table 15. The results confirm those from third simulated dive, the impact values in the weld metal being a little lower. The new welding technique and handfeeding of filler wire caused a fully non recrystalized structure in the weld, but due to the cleanliness of the weld metal, impact values were still very high. The hardness mainly comply with the DnV rules for pipelines carrying sour hydrocarbons, 260 HV_5 in HAZ, the weld metal is a bit too hard, but this can be overcome by lowering the carbon equivalent to about 0.4% against 0,51% for the actual Linde MI-88 wire used.

TIG-washing of the cap did not seem to lower the maximum hardness significantly. 0.06% C in the pipe material seems only to give marginal improvement in hardness.

TABLE 14 Mechanical Properties from Procedure Qualification

Procedure	Pipe % C	Tensile MPa	Charpy-V (J)-10°C Average		
			WM	F.L.	F.L.+2
Main procedure	0.1	592-601 F.L.	top 64 root 44	88 Top	174 Top
Main procedure	0.06	539-557 P.P.	65 Top	87 Top	135 Top
TIG	0.1	605 P.P.	-	-	-
TIG	0.06	536 P.P.	155 top	260 top	259 top

Procedure	Pipe	Hardness HV_5		min - max		COD
	% C	WM top	WM root	HAZ top	HAZ root	mm
Main procedure	0.1	150 187	257 299	227 371	206 265	0.172-0.232
Main procedure	0.06	175 195	244 280	221 329	214 283	0.162-0.267
TIG	0.1	246 257	219 232	232 274	206 226	0.508
TIG	0.06	251 271	219 234	225 262	223 234	0.562

F.L.: Fusion line, HAZ: Heat affected sone. WM: weld metal
F.L.+2: 2 mm from fusion line. P.P.: Parent pipe material

TABLE 15 Weld Metal Chemistry of Welds from Procedure Qualification

Procedure	% C	% Mn	% Si	% Ni	Oppm	[% O] • [% C]
Main procedure	0.11 0.14	0.54 0.67	0.17 0.21	1.41 1.63	1020 1260	$11.2 \cdot 10^{-3}$ $17.6 \cdot 10^{-3}$
TIG	0.043 0.062	1.24 1.31	0.17 0.34	1.49 1.75	90 100	$0.38 \cdot 10^{-3}$ $0.62 \cdot 10^{-3}$

Results of MMA Welding

The two coupons for procedure qualification were welded without any mentionable problems. The results of the mechanical testing are given in Table 14, and the chemical analysis in Table 15.

The RPS was found satisfactory as a back up power source.

The results meets DnV's requirements for pipelines carrying <u>sweet</u> hydrocarbons at a design temperature of $0^{o}C$.

Conclusions after Procedure Qualification

- The main procedure and the back up procedure were qualified for welding at 320 MSW by DnV.
- Using the TIG high current technique, the time for filling up the groove was approximately the same as for MMA.
- For the DHWT it was recommended that the welding current should be increased for welding of hot pass and first filler as this seemed to have a beneficial effect on the hardness values.
- TIG washing of the cap layer in order to lower the HAZ hardness in top of the weld showed no beneficial effect.
- The quality of the all TIG welds was superior. The hardness measurements indicated that the requirement of 260 HV_5 might be met. However, due to the divers' lack of experience with high current TIG welding and the rather low field reliability of the welding equipment, this technique was not recommended for the main procedure.
- It was decided to work further in developing the TIG procedure into a fully mechanized system.

THE FJORD TEST

Due to a diver fatality the full scale test in Skånevik was stopped before the welding had started, but after having performed the most difficult part of the tie-in operation. The work carried out prior to the accident was:

- Removal of concrete
- Rough cutting of pipe
- Aligning of pipes
- Placing the SPAR
- Lowering of the habitat
- Piging of the pipes
- Retrieval of SPAR

Taylor Diving and Salvage inc. decided to carry out the actual welding by their own in Scotland. They used the main procedure and one coupon 902 mm OD·25.4 mm WT was produced and presented to Norsk Hydro for testing and final evaluation by Det norske VERITAS.

Test results

A summary of the results is shown in Table 16. The results as a whole comply with the results of the procedure qualification.

However, one found some hydrogen cracks in a certain portion of the weld, this was also reflected in the side bends which failed at a low angle of bending. After tempering at 100°C most of the sidebends passed the test. The fracture faces of the tensile tests showed some strange decohesion although the strength was high enough, these dimples were also observed on the sidebend specimens. Electron microscopy identified manganese-silicate slag balls in the dimples, these were also observed on fracture faces from earlier tests.

It was concluded that these slags combined with hydrogen had caused the low bending resistance. It was later confirmed that the electrodes used in this specific layer, where hydrogen cracks were found, had been exposed to the habitat atmosphere for 12 hours. This would result in approximately 19 ml/100 g deposited metal hydrogen in the weld metal which is very high.

DnV found the procedures suitable and they were accepted by DnV in a letter of 9.6.1978.

TABLE 16 Mechanical Properties of Weld from the Fjord Test

Tensile MPa	Charpy-V (J) at -10°C		Average max - min.	
	WM root	WM top	HAZ root	HAZ top
582 P.P, 587 P.P 576 P.P, 556 F.L.	48 45-51	63 64-80	55 140-23	47 80-34
COD Fusion line mm	Hardness	HV_5	Average max - min.	
	WM root	WM top	HAZ root	HAZ top
0.082, 0.138 0.190, 0.151	260 275-244	181 232-156	238 284-216	249 311-204

P.P.: Parent pipe material, WM: weld metal, HAZ: Heat affected zone, F.L: Fusion line

TABLE 17 Results of Sidebend Testing

Treatment	Plunger diam.	Tot.number of test	Number accepted	% Accepted
None	36 mm	14	3	21
250°/24 h	36 mm	2	1	50
650°/45 min.	36 mm	2	2	100
100°/24 h	36 mm	10	6	60
100°/24 h	75 mm	8	7	88

Conclusion after Fjord Test

The main procedure is suitable to join pipelines API5Lx-65 down to 320 MSW. However, welding metallurgical aspects and hydrogen pickup causes risks of low mechanical properties if the procedure is not followed very closely.

MMA welding has reached its maximum depth where applicable.
The back up procedure was also qualified.

ADDITIONAL WORK

The promising results of the backup procedure from fourth simulated dive forced the decision to do some more work to sort out some of the problems experienced during the simulated dives. One also felt the need to do a state of the art study summing up hyperbaric welding prosesses ability in general.

State of the Art Study

Inst. de Soudure Paris and Cranfield Inst. of Technology were asked to do a review of all the actual welding prosesses for hyperbaric welding, and to evaluate their potential in quality and speed. In addition to this NH summed up all procedures known to us qualified in the North Sea and the mechanical properties achieved.

The three processes mainly used are MMA, TIG or MIG flux cored wire. One tried to evaluate these prosesses against another, and against our own high current TIG procedure. The MIG fluxcored wire has a deposition rate of approximately 2 kg/archour under pressure in all positions. The mechanical properties of MIG fluxcored wire are about the same as MMA. Whether the method is capable of meeting the requirements of maximum hardness of 260 HV_5 is not quite clear. As the prosess is a high heat input prosess this should be possible. The independency between filler wire speed and heat input is the greatest advantage of the TIG prosess beside being an inert gass prosess. This enables an optimizing of heat input to deposition rate, which again makes it possible to obtain very low hardness in HAZ.

After this review the following conclusions were drawn:

- The weld quality produced by the high current TIG welding procedure qualified at 320 MSW is superior to that obtained with ohter procedures, even in moderate or shallow waters.
- The efficiency of the high current TIG procedure is comparable to MMA welding with possibilities of further reducing the welding time significantly by increasing the welding current, mechanizing the prosess and apply "narrow groove" welding.

Further Work on the TIG Procedure

Some details caused problems and delays during fourth simulated dive welding at high current. These were connected to:

- Overheating of torch (gas-cooled).
- Blunting of electrodes.
- Problems in feeding the filler wire steady.
- Somewhat high hardness in weldmetal.

One suspected that the blunting of the electrodes were caused by droplets of filler wire. A study on lifetime of electrodes at 32 bar at currents up to 300 A, 50 V was undertaken at SINTEF, Dept. 34, Trondheim. A lot of basic information came out of this study, and the conclusion was that lifetime and arc stability did not cause problems under the conditions tested. Arc ignition could easily be obtained without torch start using HF or a ball of steel-wool.

To test out possibilities of other filler wire types, three types were compared to Linde-Mi 88. 2,5% Ni type, 1% Ni type and unalloyed stabilized.

The welding was performed in a special way to simulate cooling rates under hyperbaric conditions. This testing was carried out using a prototype semiautomatic TIG-torch (water-cooled), feeding the filler wire automatically at the edge of the puddle. This prototype functioned very well and made it a lot easier to do the welding especially in the overhead position. The work pointed to solutions to all the problems experienced during fourth simulated dive, the following conclusions were made:

- 2,5% Ni and 1% Ni types give impact values in the range 130-280 J at -10°C, 2,5% Ni giving better results than 1% Ni.
- The hardness in HAZ and weld metal is well below 260 HV_5 by choice of adequate welding parameters.
- Independence between heat input and deposition rate gives possibilities of optimation of hardness/ductility in HAZ and weld metal.
- Depositon rates of 1-2 kg/archour can be obtained in all positions.
- A duty cycle of 50-85% can be obtained by using a semiautomatic TIG-torch.

CONCLUSION

Norsk Hydro as operator has been responsible of developing two welding procedures for hyperbaric welding at 320 MSW. The MMA procedure has been tested in the sea and both procedures are qualified for use by DnV.

The TIG process has been furhter developed into an efficient manual process with a high potential for mechanization. The mechanical properties of this welding procedure is practical of parent material values.

Serious work has already started to develop the TIG prosess for mechanized TIG welding down to 400 MSW.

ACKNOWLEDGEMENT

The author would like to express his thanks to Head of Section K. Gundersen, Project Engineer M. Tystad and Prof. N. Christensen of SINTEF, Trondheim for valuable help during preparation of the paper.

REFERENCES

Christensen, N., Simonsen, T., Gjermundsen, K. and co-workers (1975-1979). Confidential research reports. SINTEF Dept. 34, Trondheim.

Christensen, N., Gjermundsen, K. (1976). Effects of pressure on weld metal chemistry. 11 W Doc. 212-384-76.

Christensen, N., Gjermundsen, K. (1977a). Effects of pressure on weld metal chemistry. 11 W Doc. 212-395-77

Christensen, N. (1977b). Metallurgisk forskning ved NTH. Jernkontorets analer nr. 5.

Gaudin, J.P., (1975-1979). Confidential research reports. Institute de Soudure, Paris.

Gundersen, K., Knagenhjelm, H.O. (1975-1979). Internal reports. Norsk Hydro Materials Technology Section.

Weibye, B., Tystad, M., Andersen, K. and co-workers (1975-1979). Internal reports Norsk Hydro Offshore Engineering Subsea Group.

NOMENCLATURE

DHWP - Deepwater Hyperbaric Welding Program

DHWT - Deepwater Hyperbaric Welding Test (Skånevikfjorden/Kyle)

DnV - Det norske Veritas

EPS - Electronic Power Supply

F.L. - Fusion Line

HAZ - Heat Affected Zone

MMA - Manual Metal Arc Welding

MSW - Meters of Seawater

MIG - Metal Inert Gas Welding

OD - Outer Diameter

RH - Relative Humidity

RPS - Rectifier Power Supply

SPAR - Submarine Pipeline Alignment Rig

STSP - Statfjord transportation System Project

TD&S - Taylor Diving & Salvage Co. Inc.

TIG - Tungsten Inert Gas Welding

W.T. - Wall Thickness

FAILURE MECHANISMS AND DEVELOPMENT OF TESTING PROCEDURES FOR COMPONENTS IN HIGH PRESSURE OXYGEN SYSTEMS

E. Borse*, S. Stavdal and O. H. Solumsmoen****

*NORGAS A/S, Oslo, Norway
**Det norske Veritas, Hovik, Norway

ABSTRACT

Reported or known accidents both onshore and offshore have shown the significance of the physical fact that high pressure oxygen drastically reduces the self-ignition temperature of most materials used in components for oxygen systems.

The results from several test series carried out, where flexible hoses were subjected to high pressure oxygen shocks, showed that several makes of hoses intended for diving systems exploded and caught fire at pressure shocks below service pressure. Further, the tests showed the importance of removing any contaminations and the importance of choosing the "correct" cleaning procedure. However, despite using the most efficient cleaning method, the critical pressure for some of the hoses was not raised above service operating pressure.

INTRODUCTION

Testing of components for high pressure oxygen systems is new in Norway. It is only after Veritas built a laboratory for testing such components that systematic testing has been possible. Countries like Germany and USA have testing facilities, but even in these countries systematic testing that has been reported is only about 10-15 years old.

Experience with high pressure oxygen systems offshore is still limited; the experience onshore is larger though still not systematically analysed. This paper will try to discuss <u>why</u> systematic testing of components used in high pressure oxygen systems is important, <u>what</u> we are looking for and <u>how</u> the results of such testing can bring us closer to safer systems. The paper has therefore been divided into:

- case studies
- failure mechanisms
- testing facilities and methods.

Based on results from present tests the paper further discusses:

- establishment and transmittance of shocks
- importance of choice of materials
- effect of contaminations

and suggestions for future studies are given.

CASE STUDIES

The most important sources of information about possible failure causes are studies of accidents due to component failure. The case studies selected below are from USA, Germany and Norway. They have been carefully chosen to show different failures and different components used and should therefore be representative of the different types of components which can be expected to fail.

The American space accident in 1967 (Alger, 1971) started an extensive research into the effect of high pressure oxygen on different materials. Whether the accident itself was due to use of unsuitable material, contaminations or other causes will not be discussed here, but the research into these areas revealed the complexity of the safety aspects in high pressure oxygen systems. The research referred to was carried out by NASA (1975) and US Navy (Dorr, 1969).

Burnouts of 31 pressure regulating valves from one manufacturer in the period 1957-71 have been reported (Celles, 1973). Faulty operation and gaps in maintenance were suspected to be the main cause of failure, showing the importance of avoiding contaminations of f.inst. organic substances.

In another report (NTBL) the explosion of a pressure gage with a bourdon tube was investigated. The report stated that "the dead-end configuration (in pressure gages and regulators used in compressed gas installations) lends itself to a natural depository for contaminations" and "the reduced cross-sectional area is a contributing factor which tends to accelerate the entering shock waves with a resultant increase in temperature upon impact at the dead-end".

Three failures occuring in 1971 have been investigated by Bundesanstalt für Materialprüfung (BAM) in Berlin (Roch, 1971). One was a fire in a two-step pressure regulator, caused by an oilbased lubricant applied in order to smoothen the operation of a valve. The second failure was explosion of safety valves. After three valves had exploded, it was proved that the material used in the valves caught fire when exposed to a high pressure shock. The third failure was due to a lubricant which was approved for low pressure oxygen only. Material unsuitable for use in high pressure oxygen had also been used. The component was an online pressure reduction device.

It has been observed by both BAM and Veritas that components have caught fire and/or produced high temperature steel fragments which again have caused damage to the laboratory.

An accident on board a diving ship in 1975 in Norway was probably caused by a sudden opening of a ball-valve which then caused the explosion of a flexible hose. There was substantial damage. Due to this accident "Arbeidstilsynet" requested Dr. B. Vedeler, Veritas, to initiate a litterature survey (Borse, 1975) and later the construction of the oxygen test laboratory at Veritas (Stavdal, Borse, 1979).

Recorded diving accidents in Norway caused by technical failures are, however, few. The accident reporting system onshore is considerably better and some failures have been recorded. In one case a bundle of oxygen cylinders was subjected to rough handling without the necessary protective plugs fitted to the valve outlets. On valve opening, one hose exploded, most likely due to contaminations of the valve seat. One man was injured. In filling stations, similar accidents have occured, probably due to contaminations caused by improper handling of the equipment by the customer or the operator.

If, f.inst. a hose rupture occurs in an oxygen system, the system will be drained for gas. In an industrial plant where a continuous gas supply is not essential this will in most cases usually have small consequences.

FAILURE MECHANISMS

The most common failures of components used in high pressure oxygen systems are similar to failures of components in all other systems; leakages, valve stem-failures, valve seat failures etc. These failures can be caused by:

- design error as choice of unsuitable components
- maloperation
- lack of maintenance

As the case studies showed, all components can fail given the right conditions. These conditions are related to:

- the ability of high pressure oxygen to lower substantially the self-ignition temperature of most substances. This ability increases with pressure.
- the ability of high pressure oxygen to particularily lower the self ignition temperature of certain contaminations which when ignited, can cause gaskets or even metal to catch fire.
- the ability of oxygen to support a fire.

Failures due to ignition of gaskets, steel parts and flexible hoses are shown in Figs. 1 and 2.

Valves and pressure regulators with moving parts are treated with non-combustible lubricant after production. Use of wrong type of lubricant or using components meant for low pressure in high pressure systems can thus mean that an ignition can take place (Roch, 1971).

One failure mechanism, standing waves in a dead-end pipe, have been much discussed since the theoretical temperature increase is high. However, no experimental evidence exist showing that this failure mechanism is of great importance. Another failure mechanism is ignition of contaminations.

The most common contaminations are:

- organic matters
- dust
- non-metallic substances carried from other parts of the system
- steel chips

The main failure mechanism seems to be the combination of unsuitable non-metallic material, unsuitable type of lubricant or contaminations and high temperature caused by a sudden pressure increase.

TESTING FACILITIES AND METHODS

The present testing of components in Norway have been carried out by, among others, Veritas and NORGAS. Both have been carrying out experiments investigating the effect of sudden pressure release by valve opening and NORGAS has in addition carried out experiments investigating the self-ignition temperature of certain materials in atmospheric oxygen.

The pressure testing facilities at Veritas are by far the most sophisticated. They are located in a portable container which consists of two rooms (Fig. 3), the operator compartment and the test compartment. Between the compartments is a steel wall with a small plexiglass window. The operator remotely controls the valves and monitors the test from the operators compartment.

The high pressure oxygen system (Fig. 4) is supplied from four bottles, each containing 50 liters oxygen at 200 bar. The oxygen bottles are located outside the operator compartment. Oxygen is led through stainless steel tubing to the test bench. The oxygen can then be preheated to simulate elevated temperatures before it is led through a ball valve to the test specimen.

The ball valve is operated pneumatically and can be opened as fast as 5-10 ms. For safety reasons, a non-return valve is mounted before the preheater and a safety valve is mounted before the test specimen. A solenoid valve is used to release the pressure after each test (Fig. 5).

With the facilities described the following methods can be used for component testing:

1. Self-ignition temperature testing.

   The self-ignition temperature of most materials can be established using specific test procedures in high and low pressure oxygen. There are presently no facilities for testing in high pressure oxygen in Norway. There have been discussions on the accuracy of these tests, but for most purposes the results are sufficiently accurate.

2. Testing based on pressure shocks.

   A criteria for type approval of components has been suggested based on the experience obtained in the Veritas laboratory. The criteria are based on the establishment of pressure shocks and on DIN 8546-5.7.8 (Stavdal, Borse, 1979).

This test is as follows:

"A minimum of three specimens shall be subjected to 20 shocks of oxygen with the prescribed approval pressure. None of the test specimens shall rupture or show signs of depreciation.

The full oxygen pressure, with refill if necessary, shall act on the test specimen for 10 sec. Time between each shock will be 30 sec.

The pressure in the test specimen will be reduced to atmospheric pressure between each shock.

The oxygen used will be industrial grade and preheated to 60°C.

For hoses the test length is 60 cm".

The test length was decided upon after series of tests on hoses of variable length. These test results can be summarized as follows:

Hose length	Critical pressure
60 cm	90 bar
80 cm	90 bar
120 cm	110 bar

Shorter hoses than 60 cm have given no visible effect. The longer hoses need more supply of oxygen and with a limited revervoir, the long hoses tested will have a lower oxygen pressure after pressure release and thus a higher critical pressure.

ESTABLISHMENT AND TRANSMITTANCE OF SHOCK WAVES.

A shock-wave is by definition a discontinuity. It travels with a velocity equal to or greater than the speed of sound in the untouched media. When the shock-wave travels with a velocity greater than the speed of sound, this means that a large pressure discontinuity exists at the shock front. This discontinuity is a function of the velocity increase above the speed of sound.

The shock wave moves as shown in Fig. 6 and is being reflected back and forward. When a valve is opened between the high and low pressure region, it will cause a discharge of high pressure gas. This discharge will create a series of shock-waves which try to equalize the pressure. The sum of these shock-waves is the pressure wave traced on a scope. The form and size of this pressure wave is a function of the valve opening time and the area change of the valve per unit time. This varies with the type of valve, the ball valve generally being the most effective since the initial area change is larger than for most other valves. It can also be opened extremely quickly. For testing purposes the ball valve is excellent since it gives close to the worst possible case. A system must be designed for possible malfunctions even though a ball valve is not used in most high pressure oxygen systems. A valve can fail or a valve seat can be blown away causing extreme pressure build-ups even from other types of valves (as spindle valves). Using a ball valve for testing purposes is therefore necessary in order to simulate the worst case.

The pressure wave formed when a valve is suddenly opened can steepen and thus cause a larger pressure discontinuity at the shock front. This is not important in short test specimens.

Most materials or even contaminations will not ignite in oxygen at ambient temperatures (say 20°C). Their self-ignition temperature is generally high (c.f. Table 1). If the self-ignition temperature has been lowered to e.g. 200°C, only a limited amount of heat is necessary to cause ignition. This heat can be generated by a pressure shock. A shock-wave will generally be of short duration, as if the valve was opened and immediately closed. Normally, a valve will stay open long enough to allow the pressure to be equalized and the resulting high temperature will last for sufficient time for ignition to occure.

A sudden discharge of oxygen gas can be undercritical, critical or overcritical, but is in most cases overcritical. This generates peak pressure at the shock front well above the reservoir pressure. This is shown in Fig. 7.

It has, however, not been proved whether the peak pressure created above reservoir pressure is of sufficiant duration to be important.

The final temperature can be calculated based on assumption of

- isentropic compression
- adiabatic compression
- polytropic compression
- applying blast wave theory with assumptions according to the accuracy required.
- including heat loss and absorbtion of energy in the specimen by increased tensile and radial strain energy.

The valve in the VERITAS test laboratory opens in 5 to 10 ms. For a 60 cm long hose or pipe this is equal to the time it takes for the shock to travel through the hose. This is shown in Fig. 6. Maximum pressure and temperature, occurs therefore after the first part of the pressure-wave has reached the tube end and before this part of the wave has been reflected from the other end. Too short test hose is thus not preferable. The half period of the pressure wave equals approximately valve opening time and therefore purely theoretical considerations based on isentropic or adiabatic compression are incorrect. The calculated curves shown in Fig. 8 indicate temperature increase of the order of 800°C. Temperature measurements suggest that the actual temperature increase is only of the order of 200-400°C. Even assuming that 10-20% of the energy is absorbed as strain energy, the theoretical calculations are too conservative.

IMPORTANCE OF CHOICE OF MATERIAL

Table 1 shows how the self-ignition temperature of the most common materials and metals decreases in atmospheric and high pressure oxygen. The values in Table 1 are taken from different sources discussed by Borse (1975). Of particular importance are the non-metallic materials. These materials are used in flexible hoses, gaskets and valve seats. The self-ignition temperature for most materials can in air be considered reasonably high; the self-ignition temperatures in oxygen at elevated pressures are generally low. It is therefore important to use materials tested and approved for used in high pressure oxygen.

CONTAMINATIONS

Table 2 shows examples of tests of flexible hoses with a static burts pressure of 1500 bar. When subjected to high pressure oxygen shocks the hoses failed at 70-100 bar. If high pressure nitrogen was used, no failure occured up to 200 bar. It is thus not the tensile or radial stress caused by the shock that is the main cause of failure, it is the ignition of the hose material itself or the contaminations present.

Further, Table 2 shows that all hoses failed after 30 testruns at 90 bar tested in the condition they were received. When they were rinsed, the results improved. Now 20% and 66% of the hoses survived 30 testruns at 90 and 80 bars respectively. When they were brushed the hoses either failed early or survived 30 testruns. The reason for the early failure is suspected to be contamination caused by the brushing liquid. Brushing and rinsing afterwards proved that the tube was good for a much higher pressure than originally expected.

The question raised was how long time will pass until a hose in normal service will be contaminated by other parts of the oxygen system if not <u>all</u> parts were thoroughly rinsed. This can be the case in an industrial or diving oxygen system.

The results show the importance of not only removing loose dust and gases by purging the oxygen system, but using solvents to remove fluid or dusts which stick to the tube walls. It can further be necessary to use some means e.g. brushing to remove large particles or particles which can not be dissolved by cleaning fluid.

The results are, however, part of a larger research program and must, until further evidence has been collected be treated as such.

FUTURE STUDIES

It is evident for everyone working with these problems that any research work are greatly appreciated by both the manufacturers of equipment and the consumer with whom the cooperation is needed in order to obtain sufficiently valid results.

Using the test facilities and methods described it is intended to carry out a series of tests on both hoses, valves and fittings, with a special emphasize on choice of materials, importance of various types of contaminations and the various cleaning procedures.

REFERENCES

Alger, R.S., and Nicolas, J.R. (1971). Survey of fires in hyprobaric and hyperbaric chambers. _Naval Ordance Laboratory report._

ASRDI (1975). Oxygen Technology Survey. _NASA SP-report vols 1 to 4._

Belles, F.E. (1973). High Pressure Oxygen utilization by NASA. _Paper presented at the 74th national meeting of the AICHE._

Borse, E. (1975. Safety of High Pressure Oxygen Systems. _Det norske Veritas report no. 75-10-M._

Dorr, V.A. and Schreiner, H.R. (1969). Safety in Diving Atmospheres. _U.S. Navy report_

Roch, K.H. (1971). Gleitmittel für Sauerstoffarmaturen. _Amts- und Mitteilungsblätt der BAM, 6 (1970/71)._

Roch, K.H. (1976). Unfallgefahren durch Ausbrennen von Sauerstoffarmaturen. _Moderne Unfallverhufung, 20._

NTBL Project A-570 (1974-75).

Stavdal, S. and Borse, E. (1979). Description of DnV Oxygen Test Laboratory. _Det norske Veritas report 79-199._

Stavdal, S. and Borse, E. (1979). Results from Preliminary Testing of Components for use in High Pressure Oxygen Systems. _Det norske Veritas, report 79-200._

Wegener, W. (1969). Erhøhung der Verbrennungsintensität von Organischen Stoffen durch Sauerstoff. _Moderne Unfallverhufung, 13._

Material	In air	In oxygen	In oxygen at elevated pressure
Hard PVC	515	325	250 (50 atm)
Vulcan fibre	455	305	200 (at 150 atm)
PTFE (polytetrafluor-ethylen or teflon)	>600°C	510	470 (at 250 atm)
Example of gasket material	450-500	260-320	180-200
Pure iron		930	
Steels		1200-1300	
Alloyed steels		800-1300	
18-8 stainless steel		1000	
Aluminium		2000 (assumed)	
Aluminium alloys		<1000	
Copper alloys		900-1100	
Other metals as low as		500-800	

Table 1. Example of self-ignition temperature in °C.
(In general, aluminium, magnesium and titanium are considered to have higher potential fire severity than steel and nickel, while copper has a low potential fire severity).

90 bar release pressure	Early rupture (0-10 testruns)	Rupture (11-29 testruns)	No rupture (30 testruns)
As received	66%	33%	0%
Rinsed	40%	40%	20%
Brushed	-	-	-
Brushed and rinsed	0%	0%	100%

80 bar release pressure	Early rupture (0-10 testruns)	Rupture (11-29 testruns)	No rupture (30 testruns)
As received	-	-	-
Rinsed	17%	17%	66%
Brushed	40%	0%	60%
Brushed and rinsed	-	-	-

Table 2. Oxygen shock tests on flexible hoses.
%-age rupture for various cleaning methods.
(%-age refer to %-age of specimens).

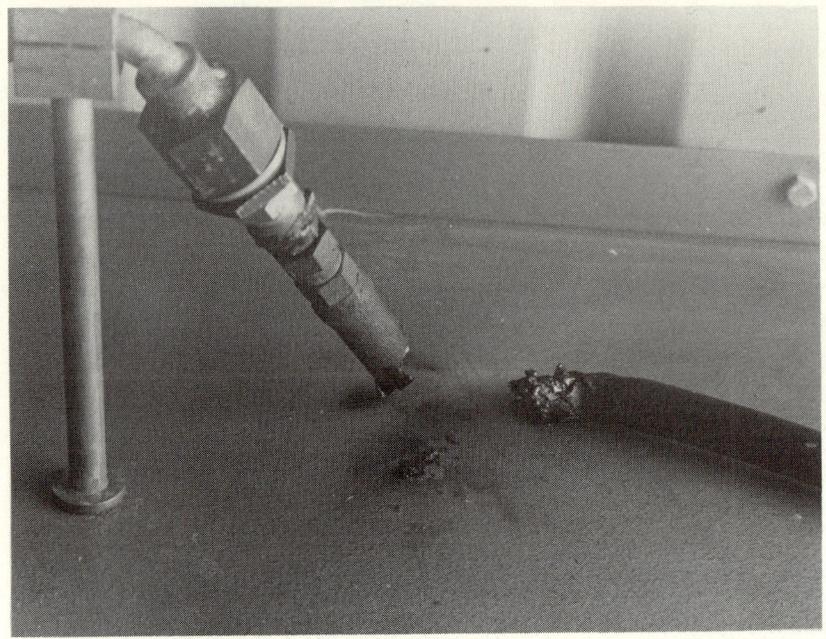

Fig. 1. Test specimen after 200 bar shock test with oxygen.

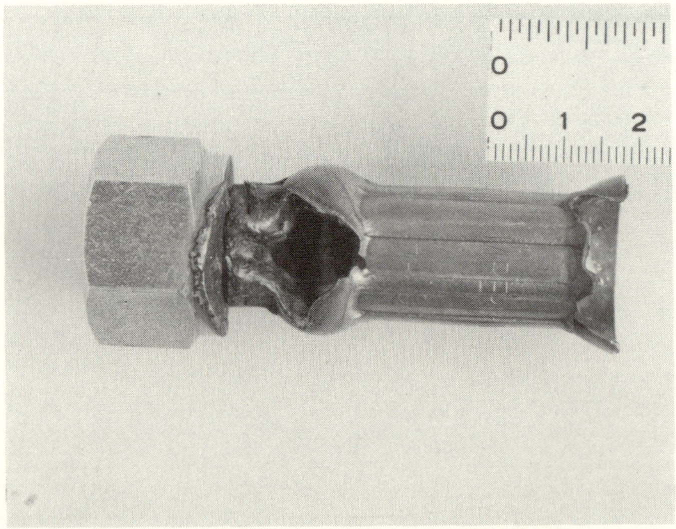

Fig. 2. Burnt-out steel coupling after oxygen shock test.

Container with oxygen shock test laboratorium.

Operator compartment

Fig. 3

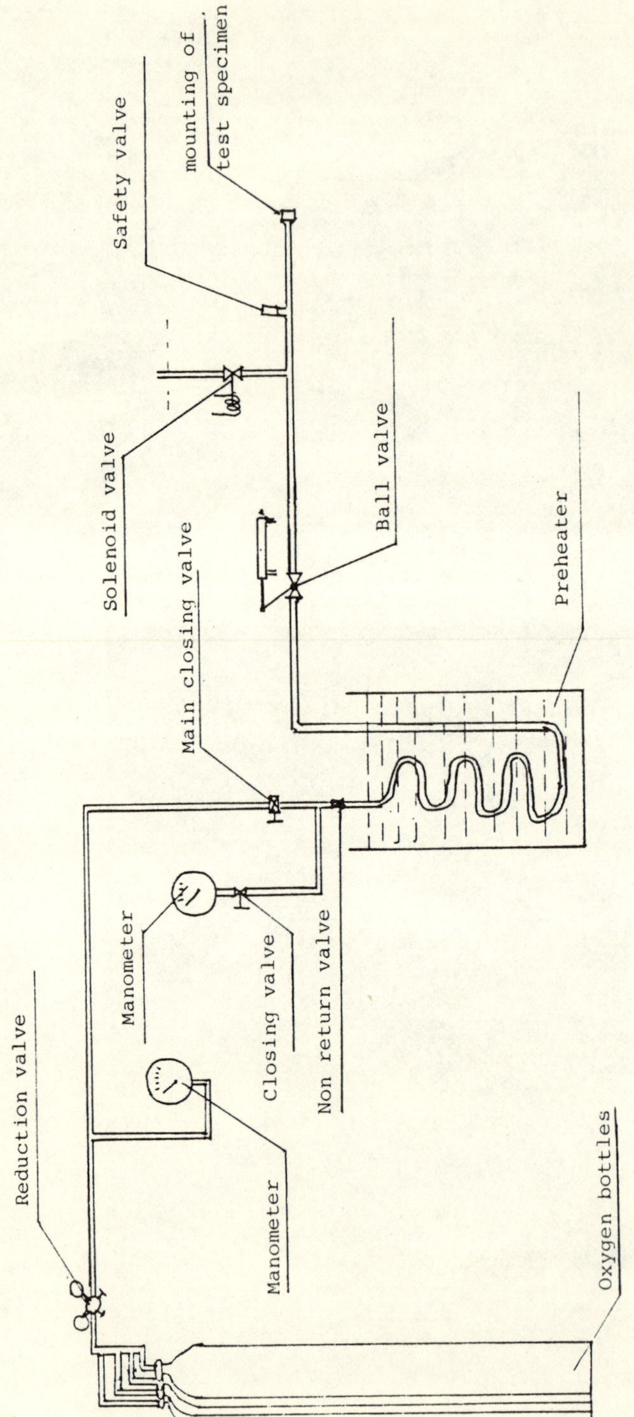

Fig. 4 Piping arrangement of oxygen shock test laboratorium.

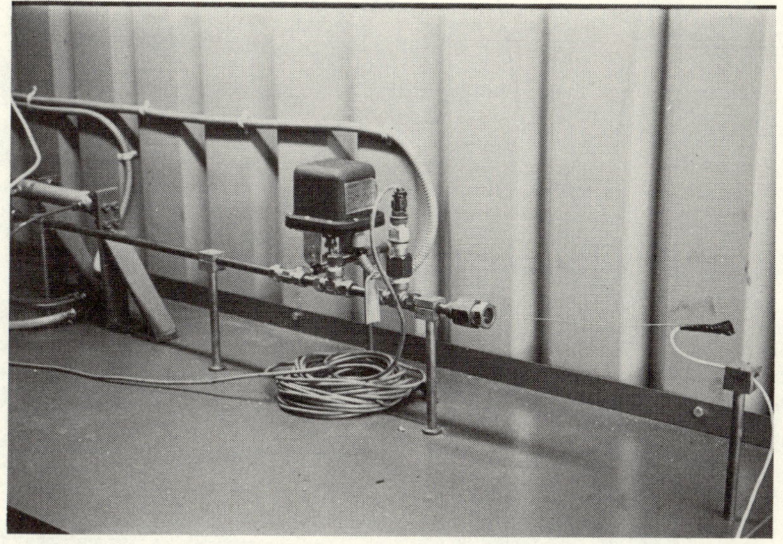

Fig. 5. Connection end of high pressure system with release valve.

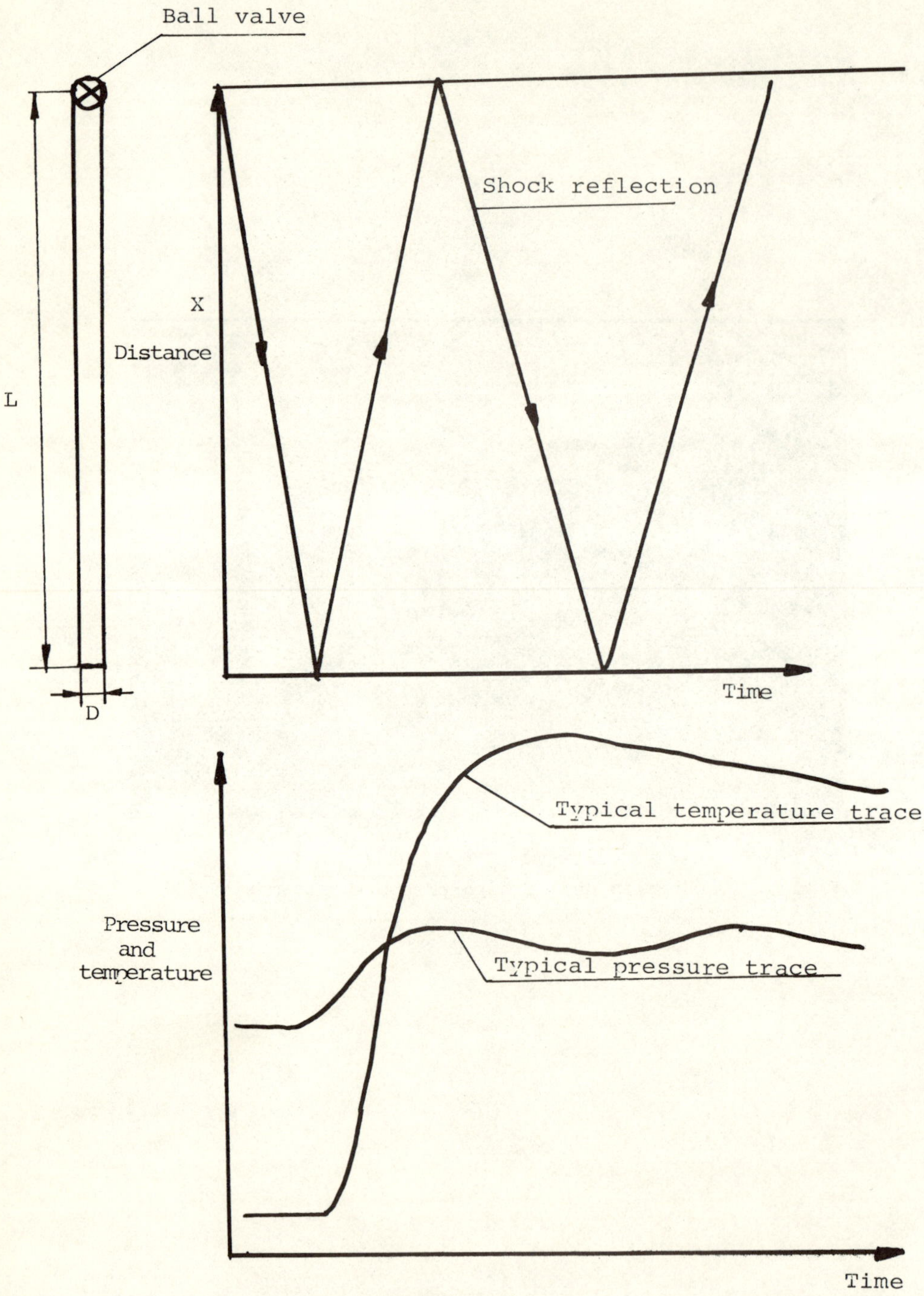

Fig. 6. Showing shock reflection (above) and typical pressure and temperature traces.

over-critical
critical
Reservoir
pressure
Pressure
under-critical
Time

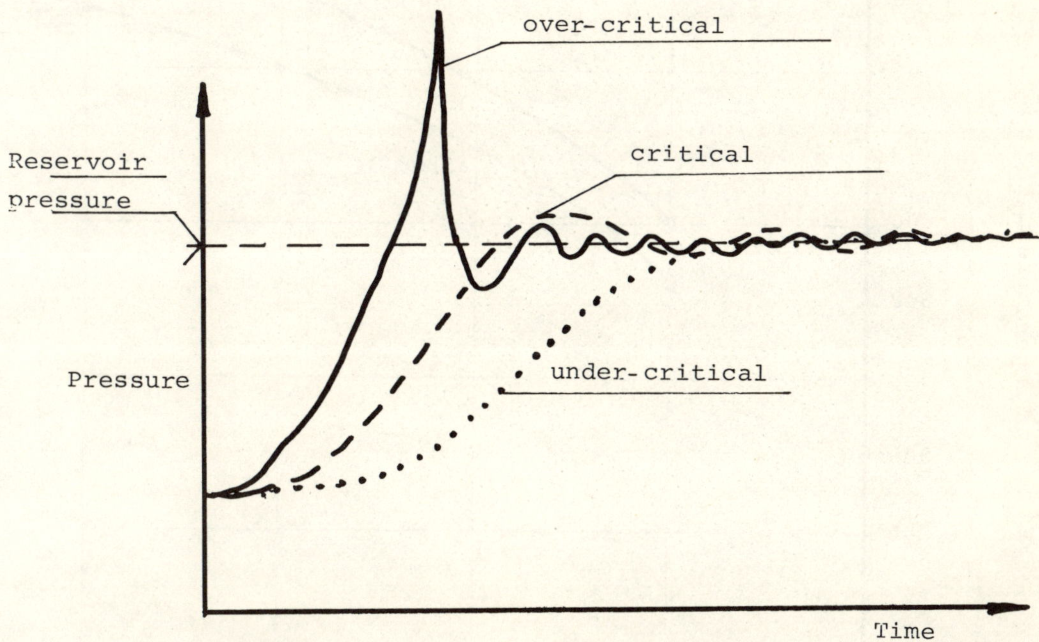

Fig. 7

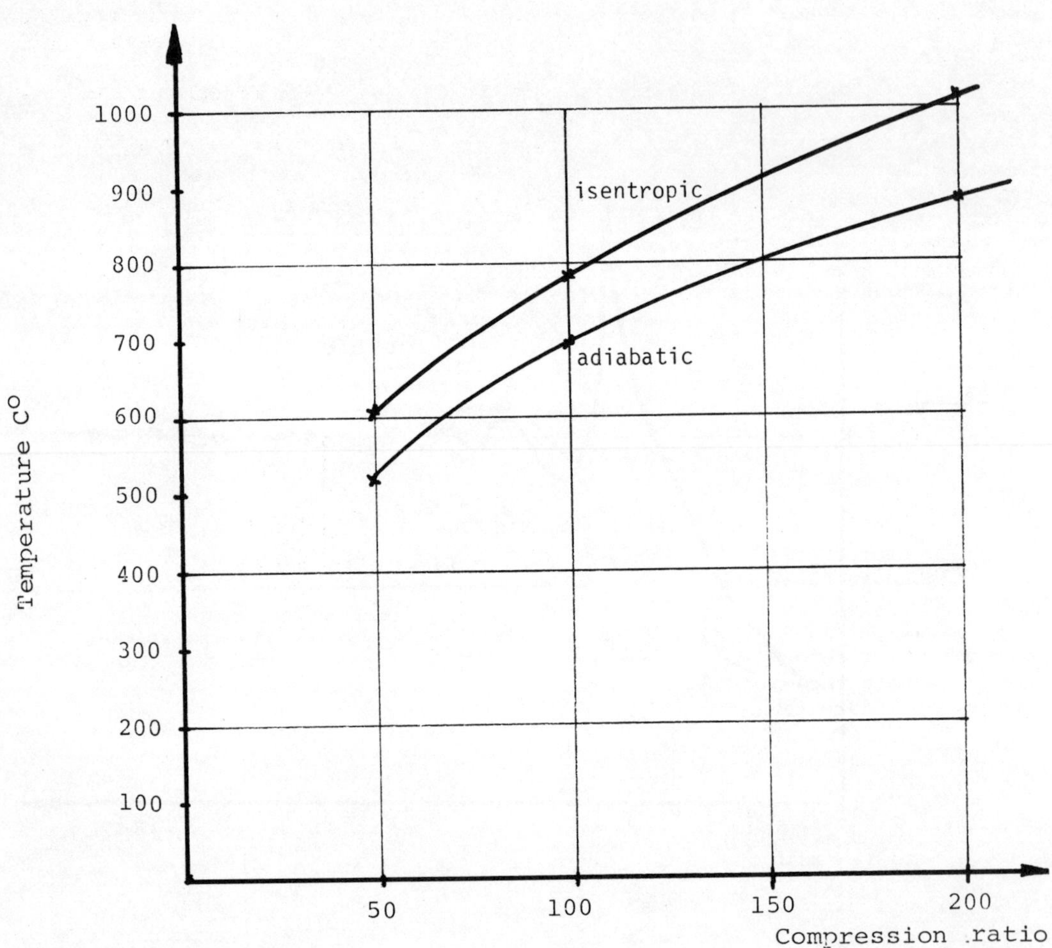

Fig. 8. Calculated temperatures after compression (corrected for loss of reservoir pressure).

MODEL TESTING OF OFFSHORE STRUCTURES

K. M. Gisvold

Norwegian Hydrodynamic Laboratories, Trondheim, Norway

ABSTRACT

Model testing of various types of offshore structures and systems has become a major concern of hydrodynamic laboratories all over the world. The field is rapidly expanding and developing into a speciality of its own with special facilities and instrumentation being constructed.

A number of basic hydrodynamic effects that are poorly understood and lacking adequate theoretical description have been disregarded in traditional ship model testing because their importance has been negligible. However, such effects may play an important role with regard to certain offshore structures. Serious scale effects are present in jacket-type structures based on tubular members. The elasticity and deformation of the structure itself becomes important as it interacts with the hydrodynamical loading. Instabilities like vortex shedding must be considered, and a number of nonlinearities affect the response caracteristics of the structures, invalidating traditional frequency domain spectral methods. The environment must also be modelled at a higher level of complexity and accuracy than previously necessary.

A new Norwegian testing facility is presently being constructed to meet these demands, and its main design philosophy and particulars are provided.

KEY WORDS

Model testing, scale effects, offshore structures, hydroelasticity, nonlinearities, testing facilities, directional waves.

1. INTRODUCTION

The increasing interest in the development of the resources of the hydrospace, be it in the water, on the seafloor or under the bottom of the sea, has lead to the birth of a whole new industry over the last 20 years. The considerable interest in offshore oil and gas has accelerated this development, but other resources such as sea bottom minerals, wave energy extraction, thermal energy conversion and aquaculture have added to the momentum of this development.

As a consequence, a number of the hydrodynamic laboratories all over the world have experienced a change in the scope and complexity of their tasks during the last decade. This is true both for the ship model testing laboratories and for the hydraulic laboratories. The two types of facilities have approached this development from their respective traditional positions. Today, these laboratories are working on much the same market. As a consequence, mergers between such institutions are being discussed in several countries.

From the point of view of ship model tanks the traditional clients used to be shipbuilders and shipowners, and the problems were related to the performance of ships in calm water, confined to the measurement of forces on ship models in towing tanks. Around 20 years ago seakeeping tests started to become more common, and such phenomena as added resistance and propulsion in waves, extreme ship motions, slamming, shipping of green water etc. were added to the répertoire. It became necessary to model the sea surface, and new special facilities and instrumentation were needed.

Manoeuvering started to gain increasing attention for civilian vessels in the same period and led to the development of facilities for this particular purpose. Following this development was the recognition of the influence of the human reaction mechanisms on the handling of ships and the impossibility of scaling these. It became necessary to develop techniques based on mathematical models implemented in manoeuvering simulators.

The ocean engineering activities marks another advance in diversification and complexity. The object is no longer a ship with a forward motion as the main purpose, but may be any floating or fixed structure or strange composition of structures and elements, at rest or in motion. The experiments that are performed in this field today may roughly be divided into three categories:

I Direct measurements of forces stresses, motions, accelerations etc.

II Testing and verification of basic phenomena and their theoretical descriptions. A result may be semi-empirical coefficients for use in numerical computation models. The traditional PMM test for manoeuvering is an early example of this. A relevant ocean engineering example is the drag and inertia coefficients used in Morison's equation.

III Studying the behaviour of vessels, platforms and complex systems without exact measurements, through a realistic simulation of the environment and the system's behaviour. Filming or video recordings at scale corrected speed is used, allowing study of operational feasibility, verification of procedures, training of operational crews or simply gaining a better understanding of the problem at hand.

The last class of experiments seems to be particular to offshore engineering and is posing new demands on the laboratories for their ability to simulate and record the behaviour of such systems.

2. TYPES OF STRUCTURES AND TESTS

A large majority of the ocean systems under study at the moment are related to the activities of the offshore oil and gas industry. However, a significant portion also deals with other problem areas such as ocean mining in very deep water, wave energy extraction, and ocean thermal energy conversion. To facilitate a brief survey of the types of structures that are involved a division may be made into fixed, floating and compliant structures. In addition there is a large group of miscellaneous systems and structures that defy simple categorization.

2.1 Fixed Structures

Such structures are used mainly for production at established oil and gas fields in shallow or moderate water depths up to 200 - 250 meters.

Space frame structures such as jackets are examined with respect to wave loads on the total structure and on the individual members both in the survival condition and in moderate sea states for fatigue analysis. Also studied is the transport from the construction yard to the site (Fig. 1), the launching of the platform from the barge and the upending process, determining the optimal sequence of ballast fulfilling stability and operational requirements.

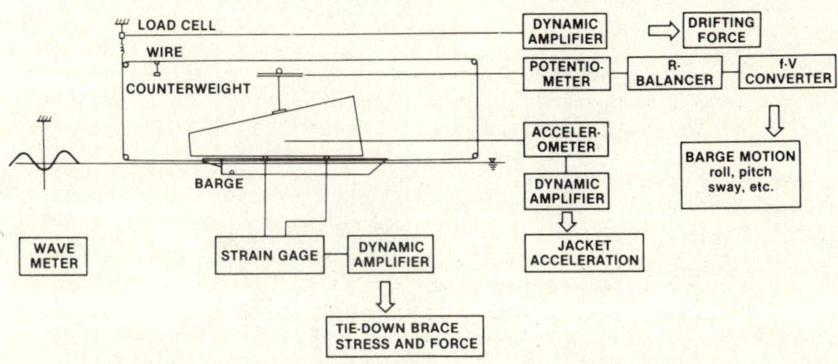

Fig. 1. Jacket transport.

The large concrete gravity platforms have gained a wide reputation, and the Condeep platform for the Statfjord B field, timed for installation in August 1981 (Fig. 2), may serve as an outstanding example. This class of structures consists of a large caisson, on which a number of columns are placed carrying the deck payload structure. Model test programs typically study the transport to the installation site, determining the necessary tugboat power (Nielsen, 1979). Problems are encountered due to the propeller race reflection from the large structure (Fig. 3). The actual immersion of the structure, including its stability during set-down, the flow induced erosion and pressure variations around its base and the total wave and current-induced forces and overturning moments on the stationary structure may also be studied.

2.2 Floating Structures

Floating structures are utilized for drilling, storage and construction activities, such as barges for crane operations and pipelaying, platforms and ships for exploratory drilling and transport. The Aker H-3 platform is a wellknown successful exploratory drilling rig whereas the Exxon Caisson Vessel (Fig. 4) represents an exiting new possibility for a floating deepwater production terminal in 1.000 meters of water. For such structures, the environmental loads and the induced motions must be determined both for the operating and the survival conditions. Systems for position keeping based on dynamic positioning or moorings must be incorporated into the model tests. Tests must include the equipment pertaining to the operations performed from the structure, such as the risers connected to a drillship or the stinger and pipe connected to a pipelaying barge.

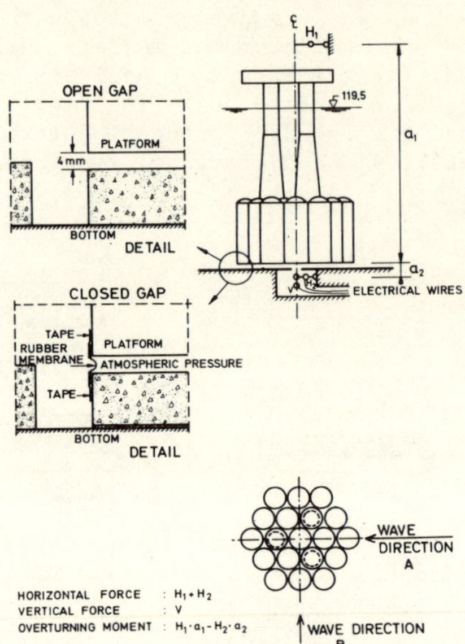

Fig. 2. Condeep Stat B.

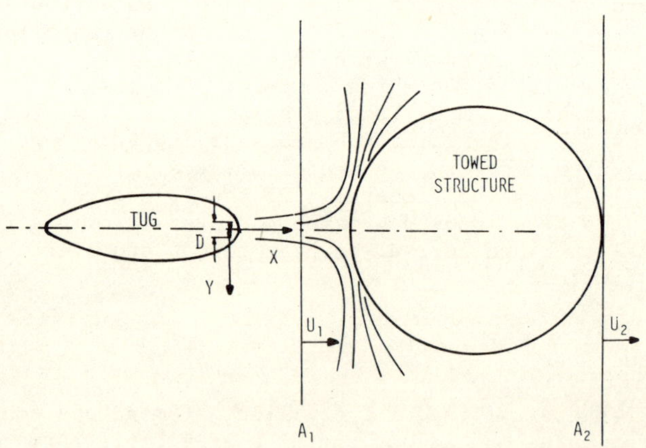

Fig. 3. Reflection of propeller race.

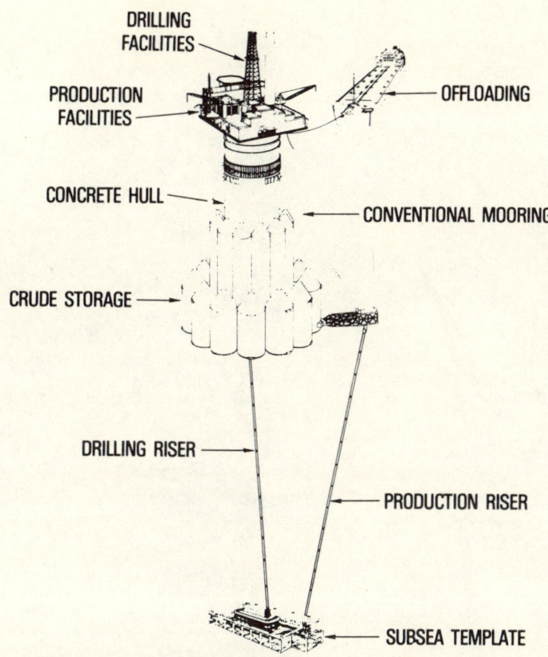

Fig. 4. Exxon Caisson Vessel.

2.3 Compliant (flexible) Structures

Compliant structures is a common denominator for structures that have less than six degrees of freedom without being completely fixed in space. Such structures have been developed to serve as loading platforms for offshore oil and gas or to serve as permanent production facilities for water depths exceeding 250 meters. The vertically moored platform (Fig. 5) and the Exxon Guyed Tower (Fig. 6) are examples of structures being seriously considered today for deepwater fields on the Norwegian Continental Shelf. For such structures, the motions and forces induced both in the survival and operational conditions are studied. The dynamic stability under wave and current action is the subject of thorough investigation.

Floating, moored structures will require testing with respect to the efficiency of the mooring geometry, necessary redundancy, forces and motions of the system components, anchor holding capacities etc. Such experiments demand large basins with advanced environmental simulation systems.

A number of single point mooring systems of various types have been developed (Fig. 7) to provide early, low-cost transport of oil. Such mooring systems must be tested alone under survival conditions and with the attached tanker under operating conditions, to determine the motions of the moored ship and the associated loads on the hoses, hawsers and their connections.

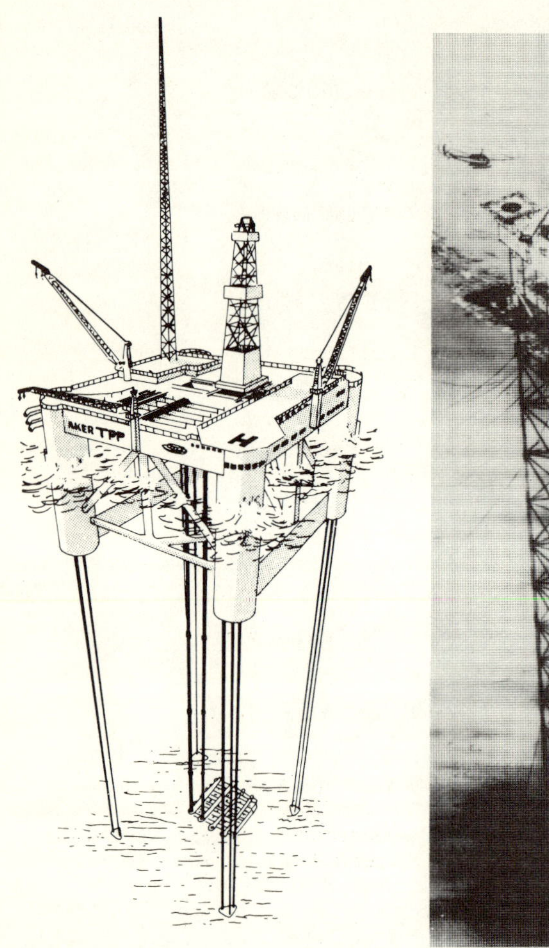

Fig. 5. Tension-leg platform.

Fig. 6. Exxon Guyed Tower.

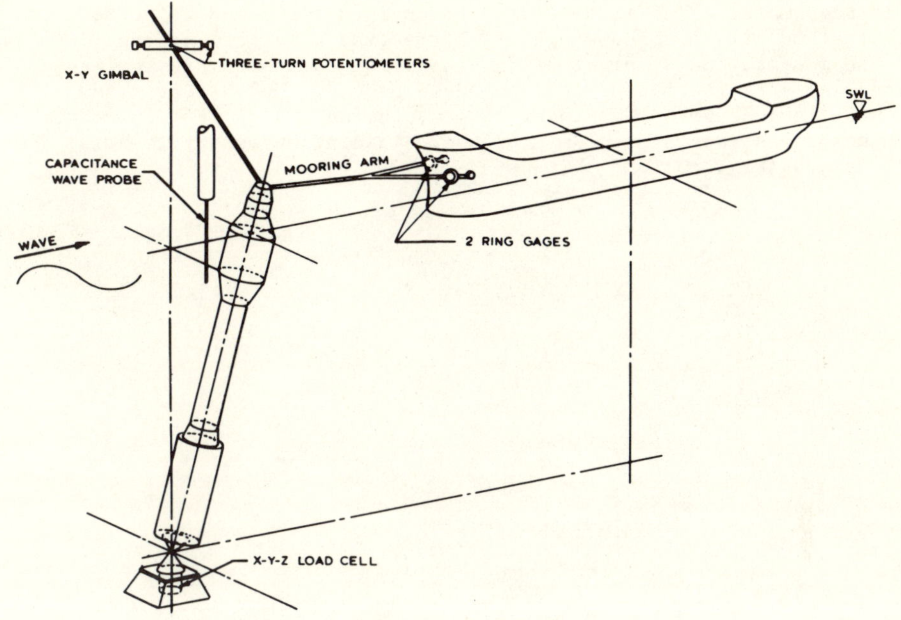

Fig. 7. Articulated loading platform.

2.4 Miscellaneous Systems

A number of other systems related to the oil and gas exploration will require model studies. Underwater storage tanks must be examined with respect to transport and installation procedures determining allowable weather conditions for the operation and the correct buoyancy procedures. Systems devoted to extracting the energy from the waves are extensively studied for their efficiency (Fig. 8).

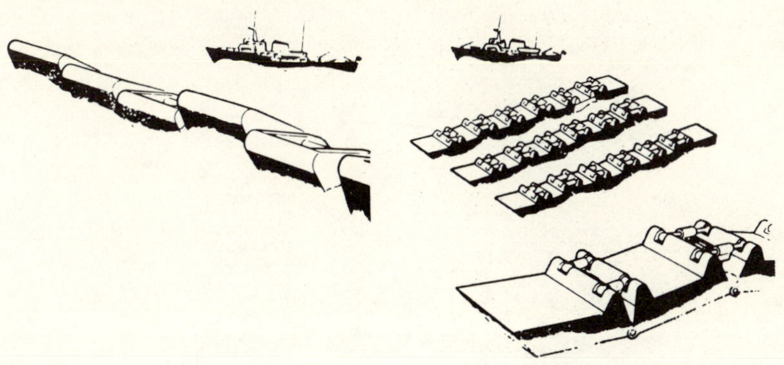

(a) Salter Duck Wave Energy Device (b) Cockerell Raft Wave Energy Device

Fig. 8. Wave energy devices.

Other systems for extracting energy from the sea such as the ocean thermal energy conversion plants (OTEC) must also be studied, although the requirements for extremely deep water makes it necessary to go into the ocean to perform the model tests (Fig. 9). Deepsea mining is attracting a lot of international attention at the moment, and a number of international companies are tooling up for the commercial exploitation of the manganese nodules. Model testing of such systems must include both the mining vessel, the subsea collection module with attached hoses and the ship to ship offshore loading necessary to bring the minerals ashore (Fig. 10).

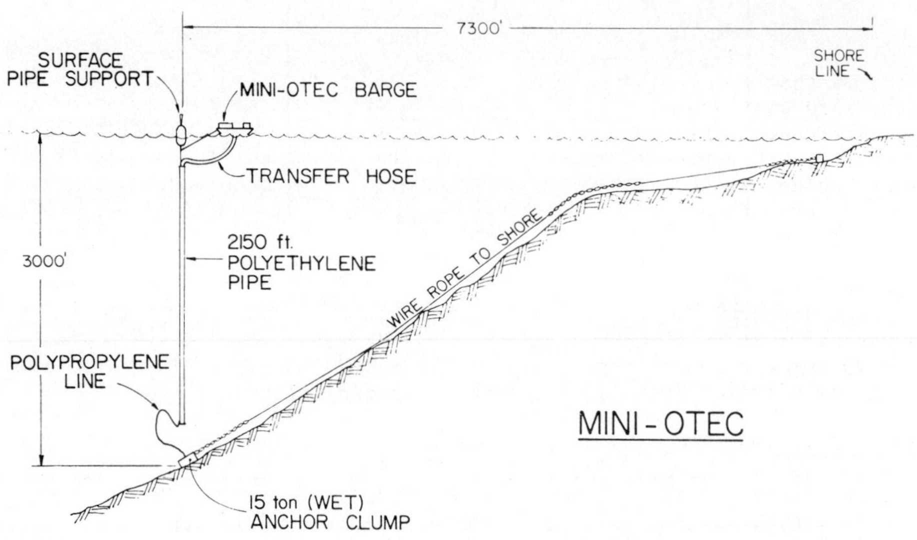

Fig. 9. OTEC.

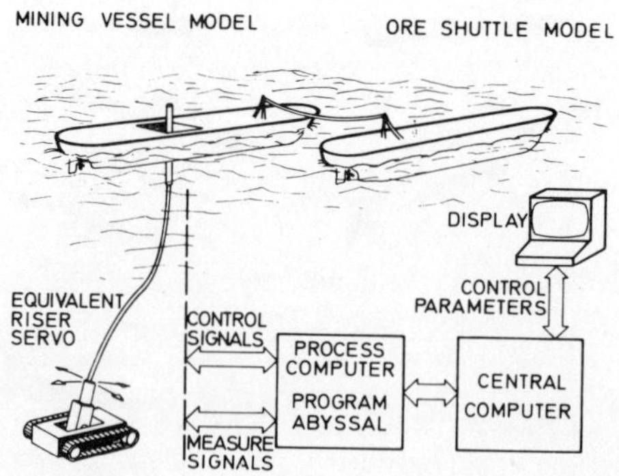

Fig. 10. Deep sea mining model test.

Of particular concern has been the environmental impact of the oil exploration activity and technological safeguards in case of offshore blowouts or major tanker accidents. Oil containment devices such as oil booms and oil skimmers have been the subject of extensive investigations on behalf of national authorities and companies supplying such equipment, to determine operational criteria and improve the designs.

The list of possible single- or multi-element systems that can be thought up for ocean exploration purposes is in principle infinite, but the above serves to illustrate the scope and complexity of the tasks that are being set for the laboratories.

3. PROBLEMS PARTICULAR TO OCEAN ENGINEERING TESTING

The problems that confront the laboratories when performing testing of such structures and systems may vary widely. There are problems that exist because they have not been observed before or because they are not adequately described theoretically. Other problems stem from the inability of the model testing techniques to cope with all scaling requirements simultaneously, or the difficulties in contructing the models themselves. Difficulties are also encountered because adequate facilities are not available and make-shift solutions are used.

Some problems are due to the fact that hydrodynamics is an integrated part of a much broader spectrum of professional insight that must be applied in an integrated manner to solve the problems. Most laboratories are not organized in such a way or do not possess the necessary scope of professional backgrounds that they can cope with this requirement.

3.1 Scale Effects

The scaling laws that are important for these types of model tests, be it of mechanical or hydrodynamic models, may be expressed in terms of a number of dimensionless parameters:

$$\frac{V}{\sqrt{gL}} = Fn, \text{ Froude number (inertial similarity)} \quad (1)$$

$$\frac{VL}{\nu} = Rn, \text{ Reynolds number (friction similarity)} \quad (2)$$

$$\frac{\sigma}{E} = \frac{F}{L^2 E} = Ho, \text{ Hooke number (elastic similarity)} \quad (3)$$

$$\frac{F}{\rho V^2 L^2} = Ne, \text{ Newton's number (static load similarity)} \quad (4)$$

$$\sqrt{\frac{Ho}{Ne}} = V\sqrt{\frac{\rho}{E}} = Ca, \text{ Cauchy's number (dynamic load similarity)} \quad (5)$$

where

V - velocity
L - characteristic model length
ν - kinematic viscosity
ρ - mass density
E - Young's modulus
g - gravity acceleration
F - force

As long as the model is tested in water it is impossible to get correct Froude and Reynolds numbers at the same time. In traditional ship model testing this is solved by scaling according to the Froude number. Various extrapolation schemes, friction lines, are applied to correct for the discrepancy, or so called scale effect, that appears in the frictional resistance. A history of model - full scale correlation going back a hundred years, has made this method very reliable with regards to conventional ship resistance. Unexpected problems do appear from time to time, however, when flow conditions are unstable and separation occurs, such as in the afterbody of full ships or in ships with hard chines.

Fluid loading on ocean structures is traditionally determined on the basis of Morisons's equation:

$$F = \tfrac{1}{2}\rho C_D u|u|A + C_M \cdot \rho V\dot{u} \tag{6}$$

where

- A - projected area of the structure
- V - structure volume
- C_D - drag coefficient
- C_M - inertia coefficient
- ρ - mass density
- u - water particle velosity

This formula contains two semi-empirical coefficients, the drag coefficient, C_D and the inertia coefficient, C_M. Depending on the type of structure and environmental conditions in question, the drag or inertial components of the force may dominate. This is illustrated in Fig. 11 for various types of offshore structures, when considering wave loading in the splash zone.

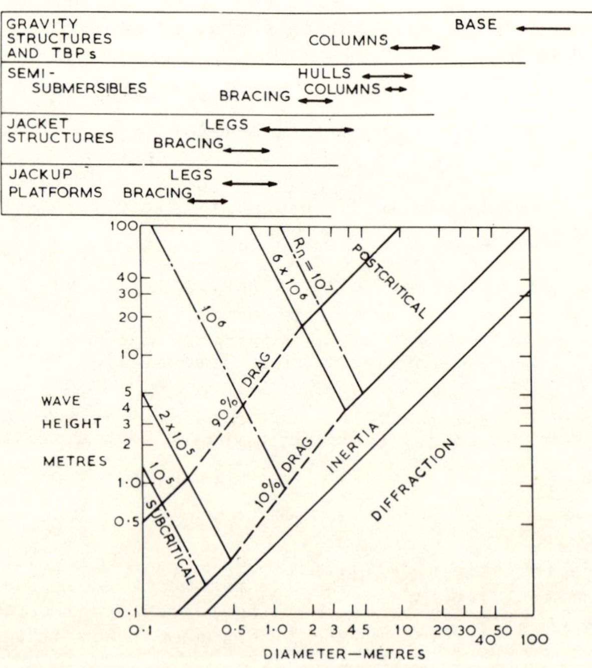

Fig. 11. Drag/Inertia loading on platforms.

The drag coefficient exhibits a distinct sub-division into regimes depending on Reynolds number, as shown in Fig. 12. The problem is, that whereas most full-size structures experience Reynolds numbers well above 10^6, model tests, even at exceedingly large scale, will be in the subcritical region and may accordingly overpredict the drag forces by a factor of 2 or more. This is particularly true for loading on members of tubular shape with a diameter that is relatively small compared to the wave height or motion amplitude.

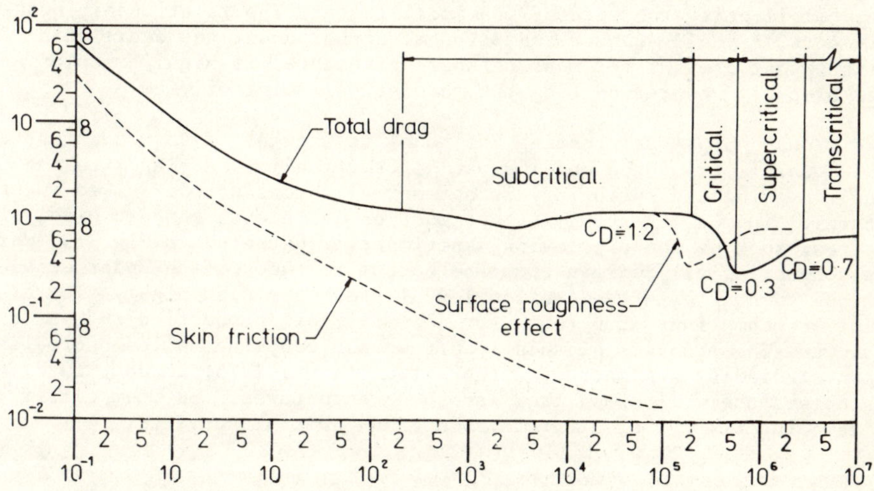

Fig. 12. C_D-R_n interaction.

There is considerable uncertainty in C_D when oscillating or permanent current is added, such as in a wave or in waves combined with current. Much research is going into this problem at the moment. In addition to the transient effects of oscillation, there will be flow interaction between adjacent tubular members due to shielding and wake effects.

All these unknowns add up to a very unclear picture of the effect of the Reynold's number on slender structures, which limits the use of model experiments for jacket type structures. The optimum technique for handling this problem must evidently be some combination of model experiments and computation that defines reliable predictors for the viscous forces.

A number of full scale measurement programs conducted on stationary ocean structures at great cost, have failed to give consistent values for the C_M and C_D coefficients of Morison's equation, and the effect of vortex shedding is thought to be one of the main reasons for this.

The phenomenon of vortex shedding is caused by flow instabilities around structural members, leading to eddy shedding and resulting in serious vibrations of the structure. The effects of vortex shedding and the prediction of the phenomenon is the subject of intense research. This phenomenon scales according to the Strouhal number:

$$S_n = f \cdot \frac{D}{u} \qquad (7)$$

where

D - member diameter
u - flow velocity
f - vortex shedding frequency

and may accordingly appear in model scale, but not in full scale or vice versa, further complicating the life of the model tester. The relationship between drag coefficients, Reynolds number and Strouhal number under the oscillating flow conditions that prevail for most offshore structures is highly complex and is not yet properly understood from the theoretical point of view.

In traditional ship model testing the model is thought of as a stiff structure and the possibility of the model's own deformation interacting with the fluid loading (whipping and springing) is not normally considered. Not so in offshore structures. A number of problems that must be dealt with concern highly flexible structures, such as the pipelaying experiment illustrated in Fig. 13. Marine risers conducting oil and gas from the bottom to the surface under exposed conditions such as with the tension leg platform are also highly flexible. Other examples are compliant structures for deepwater oil production such as the Exxon Guyed Tower. The problems of this load/response coupling called hydroelasticity has been studied by a number of authors (Wereldsma, 1975; Taudeu, 1978). Similar type problems are well known from aerospace structures, the most classical examples being flutter of aircraft, wings and control surfaces.

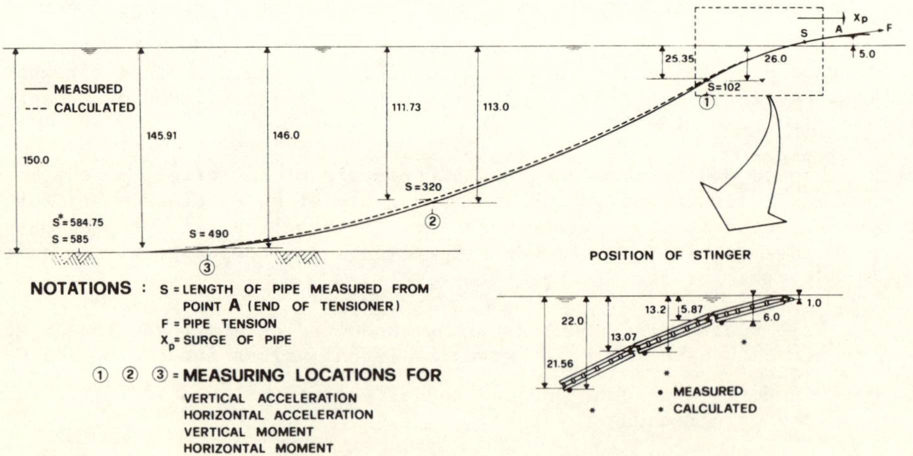

Fig. 13. Pipelaying experiment.

For the model testing of such systems it becomes necessary to bring into effect similarity laws of eq. 3, 4 and 5, relating to the elastic, static load and dynamic load similarity. Normally the Froude number is taken as a starting point for scaling, determining a length scale, λ. The bending stiffness EI then has to be reduced with a factor of λ^5 in order to obtain elastic similarity. If that is the case, the requirements for dynamic load similarity and static load similarity are also fulfilled.

When the structure is simple and the scale ratio does not lead to impractical wall thicknesses, the structure can be geometrically scaled, which leads to a reduction of I with factor of λ^4, in which case E also must be reduced by a factor λ. This implies finding a suitable model material. Frequently, it is not easy to make a perfect geometric similitude of the structure in model scale, in which case one has to resort to a composite model consisting of several different materials. In this case it is no longer possible to measure such items as stresses directly on the structure. However, the crossectional forces can be measured and stresses derived from these. Sometimes it is not even possible to construct a continuous elasticity in the model, in which case one has to resort to a segmented model with the elasticity imposed at points (hinges). The number of such hinges and their stiffness must be found through calculation of the natural frequencies of the segmented model compared to the natural frequencies of the full size structure, which implies finite element calculation or full scale measurements.

For large-volume structures, viscous effects play an insignificant role, but nonlinear potentional flow effects can be important. Under survival conditions, the wave height becomes so large that wave theory based on small wave amplitude theory can no longer be applied. This necessitates tests to be performed in waves with realistic shape and with correct representation of the extreme wave kinematics. Theoretical and experimental studies on breaking waves in deep water have been conducted (Kjeldsen and Myrhaug, 1979) in order to establish new criteria for the design of offshore structures against extreme forces caused by breaking waves. Figure 14 shows a deepwater plunging breaker as recreated in the laboratory.

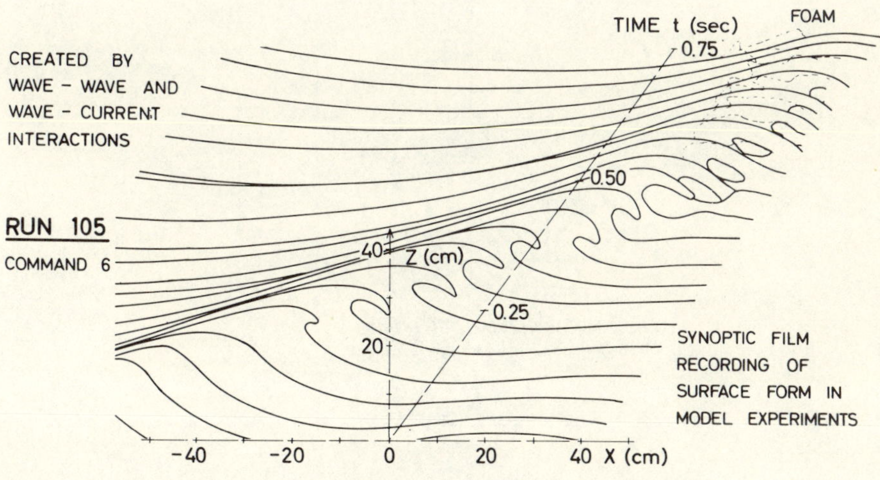

Fig. 14. Deep water plunging breaker.

The proper representation of the wave environment is very important to avoid problems in the modelling of nonlinear response phenomena. One such response phenomenon is the second order drift force (Faltinsen and Michelsen, 1974) which is of small magnitude and constant in regular waves, but varying at low frequencies in irregular waves. Such low natural frequencies occur in moored offshore systems, tanker loading terminals, vertically moored platforms and other similar systems, which therefore may be influenced significantly by drift forces in random irregular seas. Mathematically the phenomenon can be explained as wave modulation caused by frequency differences in the wave spectrum.

In order to display these effects in model tests it is important that there is a virtually continuous spread of energy over all frequencies in the spectrum. The model spectrum must not be composed of a too small number of components. Figure 15 shows some recent experiments performed to assess the effect of a simplified spectrum on the response of a floating caisson type platform (Næss, 1977). A good explanation of such forces on moored systems is given in (Løken and Olsen, 1979) where the effect of the nonlinear restoring force from the mooring lines also is taken into account. As long as the wave spectrum is continuous, model tests will predict such effects reliably.

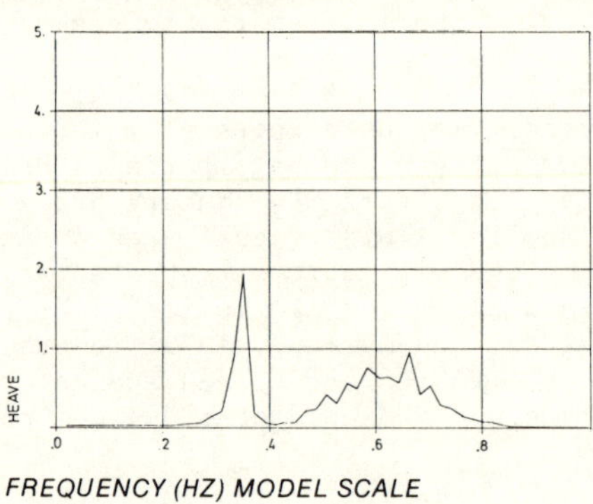

FREQUENCY (HZ) MODEL SCALE
a. «White noise» – wave spectrum.

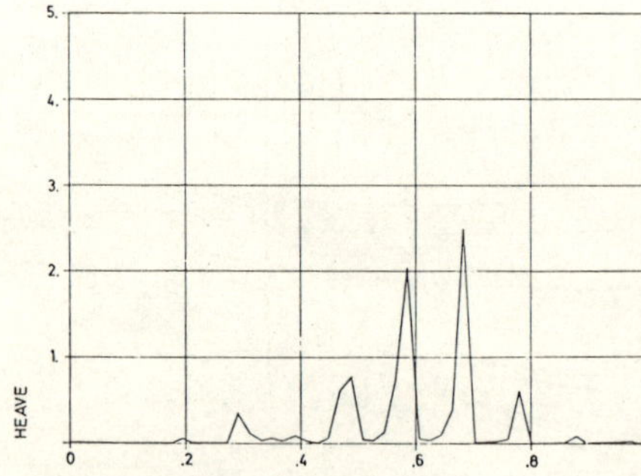

FREQUENCY (HZ) MODEL SCALE
b. 16-component wave spectrum.

Fig. 15. Effect of simplified wave spectrum.

The concept of the wave energy spectrum is today extensively used in ship and marine structure design. This spectrum provides the energy content of the sea distributed over the wave frequencies, produced by the Fourier transformation of the auto-correlation function. This concept has been found useful and valuable for predicting a range of responses in ships and marine structures, as long as these reponses could be reasonably assumed to be linearly dependant on the wave amplitude.

However, during the Fourier transformation a part of the information contained in the original time series record is lost to the user. In reality there is a second spectrum coming out of the Fourier transformation, namely the phase spectrum. Unfortunately, it does not have a suitable form for practical application and is normally disregarded. The information which is lost in this process relates to the sequence in which waves appear, since it is possible for an infinite number of different wave sequences to have the same wave energy spectrum. The effect of wave grouping, i.e. the sequence in which the waves appear, is important in certain offshore structure applications, for instance when dealing with the behaviour of moored structures and the determination of maximum amplitudes of motions. Figure 16 shows an example of a recorded wave group during a North Sea storm. Data on wave groups and the theory behind them are starting to appear (Goda, 1976; Rye, 1979) and their experimental recreation is being studied at the moment.

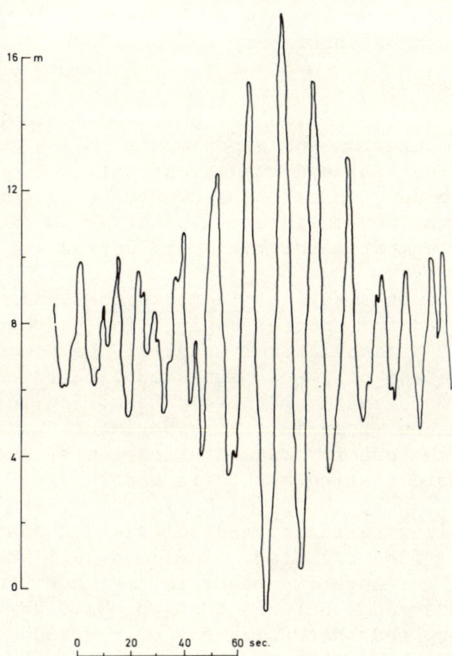

Fig. 16. An example of a recorded wave group during a storm in the North Sea.

4. COMPUTE OR TEST?

Rapid advances are being made in the area of computer modelling and simulation. New methods and techniques have increased our ability to compute today what was impossible 10 years ago. However, the advances in basic theory and physical understanding that makes it possible to formulate useable computer programs, are still lagging behind the purely numerical abilities. Thus, some of the presently available computer methods and techniques may be based on doubtful simplifications and empiricisms that can produce inaccurate results when used inadvertently on problems for which the underlying assumptions were not valid.

In spite of this drawback, theoretical methods may be viewed as reliable tools for certain types of problems, and their field is gradually increasing. They are, however, at the present limited to problems within the boundaries of existing model or full scale experience, frequently using coefficients or parameters derived from experiments or field measurement, and using models of computation based on simplifications and limitations established through such experiments. For new developments outside the area of such experience, the use of computer tools alone is doubtful.

This is today a problem area for some testing laboratories, as new competitors that do not possess experimental facilities, but a well developed program library instead, are coming in on the market. Reactions are diverse, ranging from denouncing all types of computations as inferior to model tests, to developing and offering these services as an extension of the experimental services. There is little doubt in my mind that a policy along the latter course must be pursued by the successful offshore testing laboratory of the future - but with an extra twist to it.

This extra something must be to utilize the unique opportunity that the laboratories possess to fuse computation and experiments into a coherent service package that is both cost-optimal and result-optimal to the customer. One result of such a policy may be to develop performance prediction programs that rely on model test data for their proper calibration, another to develop model testing philosophies that use the computer interactively during the test itself.

An example of this are parameter estimation techniques, based on theories and methods originally developed for control engineering purposes. These methods, which basically consist of a mathematical model of the process containing unknown parameters, make iterative use of the on-line coupling to the experimental process to determine these parameters, thereby resulting in a calibrated prediction model for the phenomenon. Some initial experiments have been performed utilizing this type of technique for maneouvering tests, and further efforts to develop this into a standard testing method is underway.

It is thus in the symbiosis with theory and numerical computation that the model testing of the future will find its place. The less well known the physical phenomenon to be studied, the more important is the model testing part of the task. It is from this cross-roads between that which at any time can be adequately described by theory and that which can only be observed, that future developments will come.

5. NEW DEMANDS FOR TESTING FACILITIES

It is clear that testing of ocean structures requires both a highly advanced state of technology when it comes to facilities, equipment and instrumentation as well as a high level of experience and theoretical competence on the part of the responsible scientists.

It is evidently of great importance to be able to simulate the environment as realistically as possible. Ideally, such simulation should include all major influences, such as wind, waves and current. So far, this has been impossible to achieve under controlled laboratory conditions, and one has been forced to limit the experimental modelling to the major influences. Wave action has normally been considered the most significant factor and has been modelled the best way possible.

There is now a demand that the sea be more fully described with respect to the directionality of its energy. The experimental recreation of multi-directional seas (short-crested seas), in which the complex combination of components arriving from different directions is taken into account, is now considered necessary in order to assess the proper behaviour of a number of offshore systems. These include loading terminals with attached tankers, moored platforms, the behaviour and efficiency of wave energy devices. Directional spectra (Fig. 17) are presently being recorded at some selected locations (Black, 1979), and their effect on the fluid loading on structures has been shown to be very important. Depending on the spreading function of the directional spectrum the loading on a single pile can decrease by up to 39% as the directional spread increase from unidirectional to omnidirectional (Sharma and Dean, 1979).

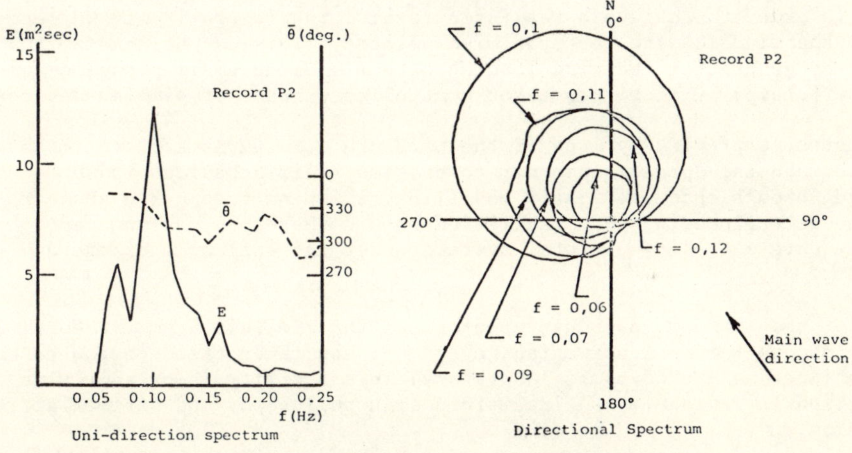

Fig. 17. Directional spectrum.

Facilities recreating directional spectra in laboratories are extremely limited in number and size at the moment. On a small scale, two different approaches have been tried in the United Kingdom. One, based on creating a focus of limited size, using wavemakers arranged in semicircle, has been employed by the Hydraulic Research Station at Wallingford (Huntington and Gilberg, 1979). Another system has been developed at the University of Edinburgh, extending the snake wavemaker concept to mixed frequency operation. This is simple in principle, but involves separate control of a high number of individual flaps. The principle of the system is more fully described by (Næser, 1979). In order to generate a complete directional spectrum, the control signal S_r for each flap must be the sum of the components corresponding to each frequency f_n and direction ϕ_m and may be written as:

$$S_r = \sum_n \sum_m T_n a_{nm} \mathrm{Cos}(k_n \mathrm{Sin}\theta_m X_r - 2\pi f_n t) \tag{8}$$

where T_n is the wave flap transfer function, relating the input signal amplitude to the wave amplitude. The system as installed and employed at Edinburgh for work on wave energy devices has proved so successful that it has been adopted in an enlarged format for the new Ocean Laboratory in Trondheim.

In addition to a more complete modelling of the wave environment it is necessary to take into account at the same time, the influences of current and wind. The effect of current on many offshore structures is quite critical and may be more important than the wave forces. The interaction between current and waves is also important, modifying the wave shape and creating a combined effect on the structure which cannot be modelled by superimposing separate current and wave experiments. There is consequently a demand for facilities that can model any current direction and intensity in combination with directional waves.

Wind is also a dominating loading factor for a number of offshore structures, notably semisubmersible platforms and other floating structures with a large exposed profile. It is therefore necessary to model wind loads simultaneously with the waves and the current. Wind loads scale mainly according to Reynolds number. In order to achieve a realistic total load picture it may be necessary to model the wind loading, not the wind velocity. This can be done by a battery of fans; by generating the wind loading directly using small propellers mounted on the structure; by micro-processor controlled winches, or similar devices.

As demonstrated previously many of the offshore structures and systems have a large geographical spread. In order to examine their behaviour it is necessary to provide enough space for their modelling, allowing reasonable scale ratios. Future ocean engineering test facilities should therefore be large enough to allow adequate modelling of total systems under the influence of the environmental parameters.

In view of the tremendous industry potential that follows from the North Sea development program, the Norwegian government has taken the decision to build an ocean engineering hydrodynamic laboratory. This facility is presently under construction in Trondheim and its main design philosophy and particulars are given below.

6. THE NEW OCEAN LABORATORY AT TRONDHEIM

When building a new laboratory for this kind of model testing, an all-important consideration is that the facility must cover needs not only for the immediate future, where the tasks may be more or less defined, but also for the more diffuse long term challenges.

Based on the considerations given previously in this paper and many other evaluations, the following design aims were laid down for the new facility:

- It must provide a total environmental simulation including wind, waves and current, with particularly strict demands placed on the wave generating capability for the production of short crested waves.

- It must provide space for the modelling of large complex systems and allow testing of a number of typical structures at significantly larger scale than available at existing laboratories.

- It must provide water depths variable from zero to the largest reasonable depths considered for exploitation in the next 20 years, with reasonable scale limitations.

- It must retain flexibility for future changes in measuring systems, instrumentation, computing equipment and basic facility equipment such as wave makers, current generators and wind batteries.

In the course of a long and exciting design process, in which a large number of people have taken part, certain central problems were identified as requiring particular solution:

- The long range monitoring of position and motions of floating models with a high accuracy, necessitated by the decision to forfeit a measuring carriage in favour of using the money to increase basin size and the environmental reproduction systems.

- The miniaturization of transducers, amplifiers, recorders and transmitters to be placed on free floating models.

- The design of a data telemetry package to send data back and forth between model and shore, chiefly without physical connection.

- A current generating system that can operate at various water depths in any given direction, creating low levels of turbulence.

- A wave generating system capable of producing short-crested waves with a controlled directional spread of the energy according to recorded and theoretical spectra.

- An adjustable bottom to provide precise depth conditions for any experiment.

- A suitable wind generation system.

The laboratory as it is now being constructed consists of two main units:

A. An extension of the present Ship Model Tank, at the existing width of 10.5 meters for approx. 85 meters at a water depth of 10 meters.

B. An Ocean Basin with water area dimensions 50 x 80 meters and a maximum of 9.1 meters water depth, provided with adjustable bottom, short-crested wave generating capability, current generation system and pollution filtering devices.

Figure 18 shows a model view of the whole laboratory complex at Tyholt as it will appear in February 1981, and Fig. 19 gives a general plan of the facility.

The building cost is presently estimated at NOK 110 mill. Of this cost, approximately NOK 80 mill. will be a grant from the government and NOK 30 mill. will be capital borrowed and depreciated at commercial rates. The first step of construction was started on December 7th, 1977, following Parliament approval of the budget in October 1977. The actual construction is divided into two phases, the first comprising the extension of the towing tank and the second the new Ocean Basin, new workshops, etc. The extended towing tank has already gone into operation in April 1979 and the basin is scheduled for completion in February 1981.

Fig. 18. Model of the Tyholt Laboratory Complex in Trondheim.

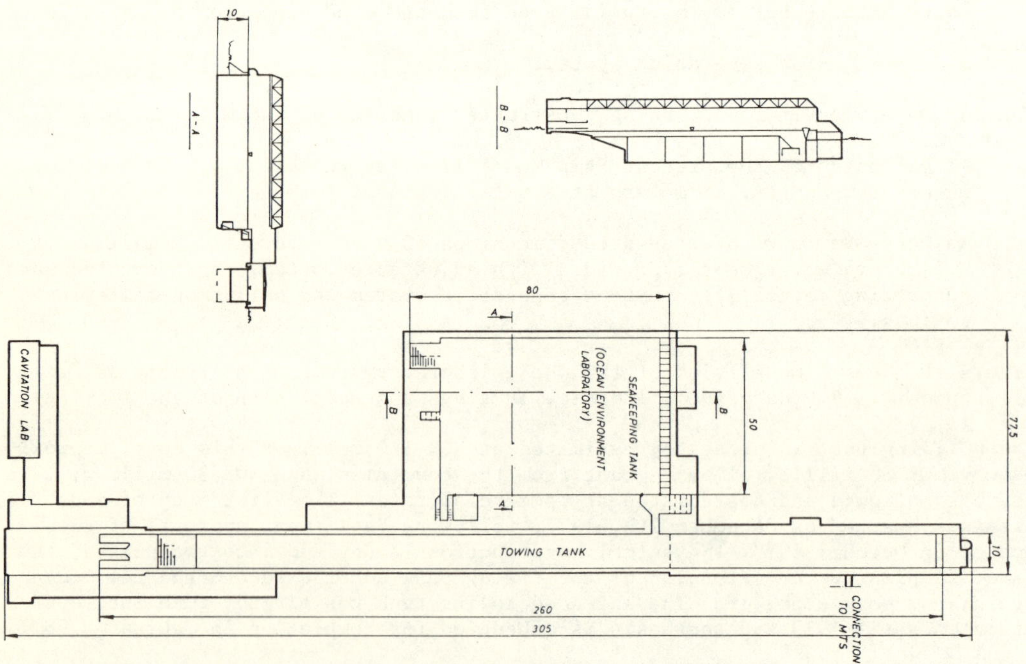

Fig. 19. General plan of the facility.

REFERENCES

Black, J. L. (1979). Hurricane ELOISE Directional Wave Energy Spectra, Offshore Technology Conference, Paper 3645, Vol. III, 2067-2074, Houston.

Faltinsen, O., and F.C. Michelsen (1974). Motions of Large Structures in Waves at Zero Froude Number. Proc. Int. Symposium on Dynamics of Marine Vehicles and Structures in Waves, Inst. Mech. Eng., London.

Goda, Y. (1976). On Wave Groups. Proc. Boss -76 Conf., Vol. I, 115-128, Trondheim.

Huntington, S.W., and G. Gilbert (1979). Extreme Forces in Short Crested Seas, Offshore Technology Conference, Paper 3595, Vol. III, 2075-2084, Houston.

Kjeldsen, S.P., and D. Myrhaug (1979). Breaking Waves in Deep Water and Resulting Wave Forces, Offshore Technology Conference, Paper 3646, Vol. IV, 2515-2522, Houston.

Løken, A.E., and O.A. Olsen (1979). The Influence of Slowly Varying Wave Forces on Mooring Systems, Offshore Technology Conference, Paper 3626, Vol. IV, 2325-2336, Houston.

Nielsen, F.G. (1979). On the Influence of the Propeller Race on Large, Towed Structures, Offshore Technology Conference, Paper 3560, Vol. III, 1773-1780, Houston.

Næser, H. (1979). Generation of Uniform Directional Spectra in a Wave Basin using the Natural Diffraction of Waves, POAC 79, Proc. Vol. I, 621-632, Trondheim.

Næss, Å., and R. Børresen (1977). On Experimental Prediction of Low-Frequency Oscillations of Moored Offshore Structures, Offshore Technology Conference, Paper 2879, Vol. III, 7-18, Houston.

Rye, H. (1979). Wave Parameter Studies and Wave Groups, Proc. Int. Conf. on Sea Climatology, Paris.

Sharma, J., and R.G. Dean (1979). SeconOrder Directional Seas and Associated Wave Forces, Offshore Technology Conference, Paper 3645, Vol. IV, 2505-2514, Houston.

Taudeu, P. (1978). Dynamic Response of Flexible Offshore Structures to Regular Waves, Offshore Technology Conference, Paper 3160, Vol. II, 975-987, Houston.

Wereldsma, R. (1975). Fundamentals of Experiments on Models of Elastic Seaborne Structures, a Tankery Problem, International Towing Tank Conference, Ottawa.

AUTHOR INDEX